W9-BZN-889

COMPLEX
VARIABLES
AND
APPLICATIONS

McGRAW-HILL
BOOK COMPANY
New York
St. Louis
San Francisco
Düsseldorf
Johannesburg
Kuala Lumpur
London
Mexico
Montreal
New Delhi
Panama
Paris
São Paulo
Singapore
Sydney
Tokyo
Toronto

RUEL V. CHURCHILL
Professor Emeritus of Mathematics
University of Michigan

JAMES W. BROWN
Professor of Mathematics
University of Michigan—Dearborn

ROGER F. VERHEY
Professor of Mathematics
University of Michigan—Dearborn

Complex Variables and Applications

THIRD EDITION

This book was set in Times New Roman.
The editors were Jack L. Farnsworth and Madelaine Eichberg;
the cover was designed by Joseph Gillians;
the production supervisor was Sam Ratkewitch.
The drawings were done by Reproduction Drawings Ltd.
Kingsport Press, Inc. was printer and binder.

Library of Congress Cataloging in Publication Data

Churchill, Ruel Vance, date
Complex variables and applications.

First ed. published in 1948 under title: Introduction to complex variables and applications.
Bibliography: p.
1. Functions of complex variables. I. Brown, James W., joint author. II. Verhey, Roger F., joint author. III. Title.
QA331.C524 1974 515′.9 73–22380
ISBN 0–07–010855–2

COMPLEX
VARIABLES
AND
APPLICATIONS

2 3 4 5 6 7 8 9 0 KPKP 7 9 8 7 6 5 4

CONTENTS

PREFACE

This book is a revision of the second edition published in 1960 by the first author, R. V. Churchill. That edition has served, just as the first one did, as a textbook for a one-term introductory course in the theory and applications of functions of a complex variable. This revision preserves the basic content and structure of the earlier editions.

The authors' efforts in the revision have been directed primarily toward improvement of the exposition and clarification of definitions and statements of results. In some instances the order of topics has been changed to achieve better continuity of the subject matter, and a number of footnotes referring to results from the calculus of real variables have been added. Various new exercises have been included in order to further develop certain of the topics in the text. Other exercises have been changed to achieve greater clarity.

Among the more obvious specific changes are the earlier introduction of Euler's formula, the inclusion of a section on the Riemann sphere, motivation of the complex exponential function as an entire function which is equal to its derivative, a more careful use of the point at infinity in the development of linear fractional transformations, the inclusion of a new section on sequences of complex numbers, and a detailed treatment of the argument principle and Rouché's theorem.

As was the case in the earlier editions, the first objective of this revision is to develop in a rigorous and self-contained manner those parts of the theory which are prominent in the applications of the subject. The second objective is to furnish an introduction to applications of residues and conformal mapping. Special emphasis is given to the use of conformal mapping in solving boundary value problems which arise in studies of heat conduction, electrostatic potential, and fluid flow. Hence the book may be considered as a companion volume to R. V. Churchill's "Fourier Series and Boundary Value Problems" and "Operational Mathematics," where other classical methods of solving boundary value problems are treated. The second book just mentioned also contains applications of residues in connection with Laplace transforms.

The first nine chapters of this book, with various substitutions from the remaining chapters, have for many years formed the content of a three-hour course given each term at the University of Michigan. The classes have consisted mainly of seniors and graduate students majoring in mathematics, engineering, or one of the physical sciences. The students have usually completed one term of advanced calculus. Some of the material is not covered in the lectures and is left for the students to read on their own. If it is desired to cover applications of conformal mapping to boundary value problems early in the course, Chapters 8 and 9 can be introduced immediately after Chapter 4 on elementary mappings.

Most of the basic results are stated as theorems followed by examples and exercises which illustrate those results. A bibliography of other and, in many cases, more advanced books is provided in Appendix 1. A table of conformal transformations useful in applications appears in Appendix 2.

In preparing this revision, the authors have taken advantage of improvements suggested by various students and colleagues and hereby express their appreciation. They are also indebted to Catherine A. Rader for her skillful typing of the manuscript.

RUEL V. CHURCHILL
JAMES W. BROWN
ROGER F. VERHEY

COMPLEX
VARIABLES
AND
APPLICATIONS

1
COMPLEX NUMBERS

In this chapter we survey the basic algebraic and geometric structure of the complex number system. We assume various corresponding properties of real numbers to be known.

1. Definition

Complex numbers z can be defined as ordered pairs

$$z = (x,y) \tag{1}$$

of real numbers x and y, with operations of addition and multiplication to be specified below. Complex numbers of the form $(0,y)$ are called *pure imaginary numbers*. The real numbers x and y in expression (1) are called the *real and imaginary parts* of z, respectively; and we write

$$\operatorname{Re} z = x, \qquad \operatorname{Im} z = y. \tag{2}$$

We say that two complex numbers (x_1,y_1) and (x_2,y_2) are equal whenever they have the same real parts and the same imaginary parts. That is,

$$(x_1,y_1) = (x_2,y_2) \qquad \text{if and only if} \qquad x_1 = x_2 \text{ and } y_1 = y_2. \tag{3}$$

The operations of addition $(z_1 + z_2)$ and multiplication $(z_1 z_2)$ are defined for complex numbers $z_1 = (x_1,y_1)$ and $z_2 = (x_2,y_2)$ by the equations

$$(x_1,y_1) + (x_2,y_2) = (x_1 + x_2, y_1 + y_2), \tag{4}$$

$$(x_1,y_1)(x_2,y_2) = (x_1x_2 - y_1y_2,\ y_1x_2 + x_1y_2). \tag{5}$$

In particular, $(x,0) + (0,y) = (x,y)$ and $(0,1)(y,0) = (0,y)$. Hence

$$(x,y) = (x,0) + (0,1)(y,0). \tag{6}$$

Any ordered pair $(x,0)$ is to be identified as the real number x, and so the set of complex numbers includes the real numbers as a subset. Furthermore, the operations defined by equations (4) and (5) become the usual operations of addition and multiplication when restricted to the real numbers:

$$(x_1,0) + (x_2,0) = (x_1 + x_2,0),$$

$$(x_1,0)(x_2,0) = (x_1x_2,0).$$

The complex number system is thus a natural extension of the real number system.

Thinking of a real number as either x or $(x,0)$ and *letting i denote the pure imaginary number* $(0,1)$, we can rewrite equation (6) as

$$(x,y) = x + iy. \tag{7}$$

With the convention $z^2 = zz$, $z^3 = zz^2$, etc., we note that $i^2 = (0,1)(0,1) = (-1,0)$; that is,

$$i^2 = -1.$$

In view of identity (7), equations (4) and (5) become

$$(x_1 + iy_1) + (x_2 + iy_2) = (x_1 + x_2) + i(y_1 + y_2), \tag{8}$$

$$(x_1 + iy_1)(x_2 + iy_2) = (x_1x_2 - y_1y_2) + i(y_1x_2 + x_1y_2). \tag{9}$$

Observe that the right-hand sides of these equations can be obtained by formally manipulating the terms on the left as if they involved only real numbers and by replacing i^2 by -1 when it occurs.

2. Algebraic Properties

Various properties of addition and multiplication of complex numbers are the same as for real numbers. We list here the more basic of these algebraic properties and verify a few of them.

The commutative laws

$$z_1 + z_2 = z_2 + z_1, \qquad z_1z_2 = z_2z_1 \tag{1}$$

and the associative laws

$$(z_1 + z_2) + z_3 = z_1 + (z_2 + z_3), \qquad (z_1z_2)z_3 = z_1(z_2z_3) \tag{2}$$

follow easily from the definitions of addition and multiplication of complex numbers and the fact that real numbers obey these laws. For example,

$$\begin{aligned} z_1 + z_2 = (x_1,y_1) + (x_2,y_2) &= (x_1 + x_2, y_1 + y_2) = (x_2 + x_1, y_2 + y_1) \\ &= (x_2,y_2) + (x_1,y_1) = z_2 + z_1. \end{aligned}$$

The proofs of the rest of the above laws, as well as the distributive law

$$z_1(z_2 + z_3) = z_1z_2 + z_1z_3, \tag{3}$$

are left as exercises.

According to the commutative law for multiplication, $iy = yi$; hence we can write either

$$z = x + iy \qquad \text{or} \qquad z = x + yi.$$

The additive identity $0 = (0,0)$ and the multiplicative identity $1 = (1,0)$ for real numbers carry over to the entire complex number system. That is,

$$z + 0 = z \qquad \text{and} \qquad z \cdot 1 = z \tag{4}$$

for every complex number z. Furthermore, 0 and 1 are the only complex numbers with these properties; verification of this fact is left to the exercises.

There is associated with each complex number $z = (x,y)$ an additive inverse

$$-z = (-x, -y); \tag{5}$$

that is, $-z$ is a complex number such that $z + (-z) = 0$. There is, moreover, only one such additive inverse for any given z. Additive inverses are used to define subtraction:

$$z_1 - z_2 = z_1 + (-z_2). \tag{6}$$

So if $z_1 = (x_1,y_1)$ and $z_2 = (x_2,y_2)$, then

$$z_1 - z_2 = (x_1 - x_2, y_1 - y_2) = (x_1 - x_2) + i(y_1 - y_2). \tag{7}$$

Likewise, for any *nonzero* complex number $z = (x,y)$, there is a number z^{-1} such that $zz^{-1} = 1$. This multiplicative inverse is less obvious than the additive one. To find it, we write $z^{-1} = (u,v)$ and seek real numbers u and v in terms of x and y such that $(x,y)(u,v) = (1,0)$. We see that u and v must be solutions of the pair

$$xu - yv = 1, \qquad yu + xv = 0$$

of simultaneous equations; and simple computation yields the unique solutions

$$u = \frac{x}{x^2 + y^2}, \qquad v = \frac{-y}{x^2 + y^2}.$$

The multiplicative inverse of $z = (x,y)$ is then

$$z^{-1} = \left(\frac{x}{x^2 + y^2}, \frac{-y}{x^2 + y^2}\right) \qquad (z \neq 0). \tag{8}$$

Division by a nonzero complex number is now defined:

$$\frac{z_1}{z_2} = z_1 z_2^{-1} \qquad (z_2 \neq 0). \tag{9}$$

Observe that, if $z_1 = (x_1,y_1)$ and $z_2 = (x_2,y_2)$, equations (8) and (9) yield

$$\begin{aligned} \frac{z_1}{z_2} &= \left(\frac{x_1x_2 + y_1y_2}{x_2^2 + y_2^2}, \frac{y_1x_2 - x_1y_2}{x_2^2 + y_2^2}\right) \\ &= \frac{x_1x_2 + y_1y_2}{x_2^2 + y_2^2} + i\frac{y_1x_2 - x_1y_2}{x_2^2 + y_2^2} \qquad (z_2 \neq 0). \end{aligned} \tag{10}$$

Division is not defined when $z_2 = 0$; note that $z_2 = 0$ means that $x_2^2 + y_2^2 = 0$, and this is not permitted in expressions (10).

Other properties of multiplication and division of complex numbers are now easy to show. In particular, if $z_1 = 1$ in equation (9), we see that

$$\frac{1}{z_2} = z_2^{-1}.$$

Thus, equation (9) may be rewritten as

$$\frac{z_1}{z_2} = z_1\left(\frac{1}{z_2}\right) \qquad (z_2 \neq 0). \tag{11}$$

Using the formulas for the product and quotient or the fact that a multiplicative inverse of a given complex number is unique, we can derive

$$\frac{1}{z_1z_2} = \left(\frac{1}{z_1}\right)\left(\frac{1}{z_2}\right) \qquad (z_1 \neq 0, z_2 \neq 0, z_1z_2 \neq 0). \tag{12}$$

With the aid of equations (11) and (12), the formulas

$$\frac{z_1 + z_2}{z_3} = \frac{z_1}{z_3} + \frac{z_2}{z_3}, \qquad \frac{z_1z_2}{z_3z_4} = \left(\frac{z_1}{z_3}\right)\left(\frac{z_2}{z_4}\right) \tag{13}$$

$$(z_3 \neq 0, z_4 \neq 0, z_3z_4 \neq 0)$$

can also be obtained.

Computations like the following are now justified:

$$\left(\frac{1}{2-3i}\right)\left(\frac{1}{1+i}\right) = \frac{1}{5-i}\frac{5+i}{5+i} = \frac{5+i}{26} = \frac{5}{26} + \frac{1}{26}i.$$

Another important property is that *if the product z_1z_2 is zero, then so is at least one of the factors z_1 and z_2.* To prove this, suppose $z_1z_2 = 0$ and $z_1 \neq 0$. Writing $z_1 = (x_1,y_1)$ and $z_2 = (x_2,y_2)$, we have

$$x_1x_2 - y_1y_2 = 0 \qquad \text{and} \qquad y_1x_2 + x_1y_2 = 0 \tag{14}$$

where at least one of the numbers x_1 and y_1 is not zero. Let us now solve the homogeneous simultaneous equations (14) for x_2 and y_2.

The determinant of the coefficients is $x_1{}^2 + y_1{}^2$. Since this is nonzero, it follows that $x_2 = y_2 = 0$; that is, $z_2 = 0$. Hence if $z_1z_2 = 0$, either $z_1 = 0$ or $z_2 = 0$, or possibly both are equal to 0.

Another way to state the above property is that if both z_1 and z_2 are nonzero, then so is z_1z_2. The condition $z_1z_2 \neq 0$ in equation (12) and the one $z_3z_4 \neq 0$ in the second of equations (13) are therefore superfluous.

Finally, we remark that the usual ordering of real numbers cannot be extended to the entire complex number system. Thus, *a statement such as $z_1 < z_2$ has meaning only if both z_1 and z_2 are real.*

EXERCISES

1. Verify: (*a*) $(\sqrt{2} - i) - i(1 - \sqrt{2}i) = -2i$; (*b*) $(2, -3)(-2, 1) = (-1, 8)$; (*c*) $(3, 1)(3, -1)(\frac{1}{5}, \frac{1}{10}) = (2,1)$; (*d*) $\dfrac{1+2i}{3-4i} + \dfrac{2-i}{5i} = -\dfrac{2}{5}$; (*e*) $\dfrac{5}{(1-i)(2-i)(3-i)} = \dfrac{1}{2}i$; (*f*) $(1-i)^4 = -4$.
2. Show that each of the two numbers $z = 1 \pm i$ satisfies the equation $z^2 - 2z + 2 = 0$.
3. Solve the equation $z^2 + z + 1 = 0$ by writing $z = (x,y)$ and solving $(x,y)^2 + (x,y) + (1,0) = (0,0)$ for x and y. Check your solution.

   Suggestion: Note that $y \neq 0$ since no real number x satisfies the equation $x^2 + x + 1 = 0$.
4. Prove that multiplication is commutative, as stated in the second of equations (1), Sec. 2.
5. Prove the associative laws (2), Sec. 2.
6. Prove the distributive law (3), Sec. 2.
7. Prove that $z(z_1 + z_2 + z_3) = zz_1 + zz_2 + zz_3$.
8. Show that the complex numbers 0 and 1 are the only additive and multiplicative identities, respectively.

Suggestion: Write $z = (x,y)$ and look for all complex numbers (u,v) such that $(x,y) + (u,v) = (x,y)$. Similarly, look for all complex numbers (u,v) such that $(x,y)(u,v) = (x,y)$ when $z \neq 0$.

9. Use the idea in the suggestion of Exercise 8 to show that $-z$ is the only additive inverse of a given complex number z.

10. Prove that (*a*) $\text{Im}\,(iz) = \text{Re}\,z$; (*b*) $\text{Re}\,(iz) = -\text{Im}\,z$; (*c*) $1/(1/z) = z\ (z \neq 0)$.

11. Prove formula (12), Sec. 2.

12. Establish the first of formulas (13), Sec. 2.

13. Establish the second of formulas (13), Sec. 2, and use it to prove that

$$\frac{zz_1}{zz_2} = \frac{z_1}{z_2} \qquad (z \neq 0,\ z_2 \neq 0).$$

14. Prove that $(z_1z_2)(z_3z_4) = (z_1z_3)(z_2z_4)$.

15. Prove that if $z_1z_2z_3 = 0$, at least one of the three factors is zero.

16. Prove that $(1 + z)^2 = 1 + 2z + z^2$.

17. Use induction to prove the binomial formula

$$(1 + z)^n = 1 + \frac{n}{1!}z + \frac{n(n-1)}{2!}z^2 + \cdots + \frac{n(n-1)(n-2)\cdots(n-k+1)}{k!}z^k + \cdots + z^n$$

where n is a positive integer.

3. Cartesian Coordinates

It is natural to associate the complex number $z = x + iy$ with a point in the plane whose cartesian coordinates are x and y. Each complex number corresponds to just one point, and conversely. The number $-2 + i$, for instance, is represented by the point $(-2,1)$ (Fig. 1). The number z can also be thought of as the directed line segment, or vector, from the origin to the point (x,y). When used for the purpose of displaying the numbers $z = x + iy$ geometrically, the xy plane is called the *complex plane* or the *z plane*. The x axis is called the *real axis*, the y axis being called the *imaginary axis*.

According to the definition of the sum of two complex numbers $z_1 = x_1 + iy_1$ and $z_2 = x_2 + iy_2$, the number $z_1 + z_2$ corresponds to the point

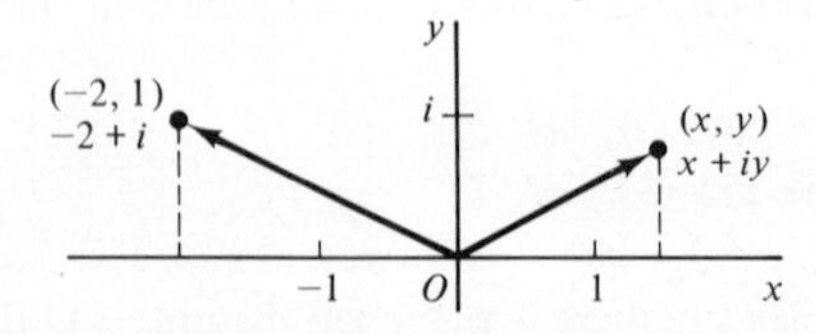

FIGURE 1

$(x_1 + x_2, y_1 + y_2)$. It also corresponds to a vector with those coordinates as its components. Hence $z_1 + z_2$ may be obtained vectorially as shown in Fig. 2. The difference $z_1 - z_2$ is represented by a directed line segment from the point (x_2,y_2) to the point (x_1,y_1) (Fig. 3).

The *modulus*, or *absolute value*, of a complex number $z = x + iy$ is defined as the nonnegative real number $\sqrt{x^2 + y^2}$ and is denoted by $|z|$; that is,

$$|z| = \sqrt{x^2 + y^2}. \tag{1}$$

The number $|z|$ is the distance between the point (x,y) and the origin, and it reduces to the usual absolute value in the real number system when $y = 0$. Note that, while the statement $z_1 < z_2$ is in general meaningless, $|z_1| < |z_2|$ means that the point corresponding to z_1 is closer to the origin than is the point corresponding to z_2.

The distance between the points representing the complex numbers z_1 and z_2 is given by $|z_1 - z_2|$. This is seen directly from expression (7), Sec. 2, and definition (1) which gives us

$$|z_1 - z_2| = \sqrt{(x_1 - x_2)^2 + (y_1 - y_2)^2}.$$

The complex numbers corresponding to the points lying on the circle with center (0,1) and radius 3, for example, satisfy the equation $|z - i| = 3$, and conversely. We refer to this set of points simply as the circle $|z - i| = 3$.

The real numbers $|z|$, Re z, and Im z are related by the equation

$$|z|^2 = (\text{Re } z)^2 + (\text{Im } z)^2 \tag{2}$$

as well as the inequalities

$$|z| \geqq |\text{Re } z| \geqq \text{Re } z, \qquad |z| \geqq |\text{Im } z| \geqq \text{Im } z. \tag{3}$$

The *complex conjugate*, or simply the conjugate, of a complex number $z = x + iy$ is defined as the complex number $x - iy$, denoted by $\bar{z}$; that is,

$$\bar{z} = x - iy. \tag{4}$$

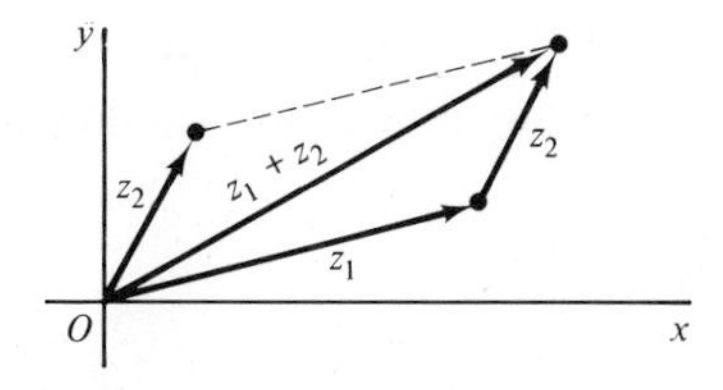

FIGURE 2

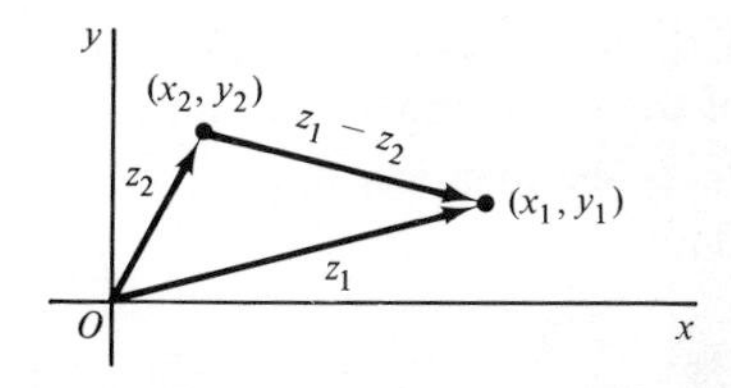

FIGURE 3

The number $\bar{z}$ is represented by the point $(x,-y)$ which is the reflection in the real axis of the point (x,y) representing z. Note that $\bar{\bar{z}} = z$ and $|\bar{z}| = |z|$ for all z.

If $z_1 = x_1 + iy_1$ and $z_2 = x_2 + iy_2$, then

$$\overline{z_1 + z_2} = (x_1 + x_2) - i(y_1 + y_2) = (x_1 - iy_1) + (x_2 - iy_2).$$

So the conjugate of the sum is the sum of the conjugates:

$$\overline{z_1 + z_2} = \bar{z}_1 + \bar{z}_2. \tag{5}$$

In like manner, it is easy to show that

$$\overline{z_1 - z_2} = \bar{z}_1 - \bar{z}_2, \tag{6}$$

$$\overline{z_1 z_2} = \bar{z}_1 \bar{z}_2, \tag{7}$$

$$\overline{\left(\frac{z_1}{z_2}\right)} = \frac{\bar{z}_1}{\bar{z}_2} \qquad (z_2 \neq 0). \tag{8}$$

The sum $z + \bar{z}$ of a complex number and its conjugate is the real number $2 \operatorname{Re} z$, and the difference $z - \bar{z}$ is the pure imaginary number $i2 \operatorname{Im} z$. Hence we have the identities

$$\operatorname{Re} z = \frac{z + \bar{z}}{2}, \qquad \operatorname{Im} z = \frac{z - \bar{z}}{2i}. \tag{9}$$

An important identity relating conjugates to moduli is

$$z\bar{z} = |z|^2, \tag{10}$$

each side being $x^2 + y^2$. It provides, for example, another way of determining the quotient z_1/z_2 in equation (10), Sec. 2. The procedure is to multiply both numerator and denominator by $\bar{z}_2$ so that the denominator becomes the real number $|z_2|^2$. An illustration is

$$\frac{-1 + 3i}{2 - i} = \frac{-1 + 3i}{2 - i}\,\frac{2 + i}{2 + i} = \frac{-5 + 5i}{5} = -1 + i.$$

4. The Triangle Inequality

Various properties of moduli are easily obtained by means of equation (10) of the preceding section and known relations involving moduli and conjugates.

We mention that

$$|z_1 z_2| = |z_1|\,|z_2|, \tag{1}$$

$$\left|\frac{z_1}{z_2}\right| = \frac{|z_1|}{|z_2|} \qquad (z_2 \neq 0). \tag{2}$$

To prove property (1), we simply write

$$|z_1 z_2|^2 = (z_1 z_2)\overline{(z_1 z_2)} = (z_1 \bar{z}_1)(z_2 \bar{z}_2) = |z_1|^2 |z_2|^2 = (|z_1||z_2|)^2$$

and recall that a modulus is never negative. The proof of property (2) is similar.

By using this general technique, we furnish here an algebraic derivation of the *triangle inequality*,

$$|z_1 + z_2| \leqq |z_1| + |z_2|, \tag{3}$$

which is geometrically evident in Fig. 2. Indeed, it is merely a statement that the length of one side of a triangle is less than or equal to the sum of the lengths of the other two sides.

We start the proof by writing

$$|z_1 + z_2|^2 = (z_1 + z_2)\overline{(z_1 + z_2)} = (z_1 + z_2)(\bar{z}_1 + \bar{z}_2).$$

Multiplying out the right-hand side, we have

$$|z_1 + z_2|^2 = z_1\bar{z}_1 + (z_1\bar{z}_2 + \overline{z_1\bar{z}_2}) + z_2\bar{z}_2.$$

But

$$z_1\bar{z}_2 + \overline{z_1\bar{z}_2} = 2\,\mathrm{Re}\,(z_1\bar{z}_2) \leqq 2|z_1\bar{z}_2| = 2|z_1||z_2|;$$

and so

$$|z_1 + z_2|^2 \leqq |z_1|^2 + 2|z_1||z_2| + |z_2|^2,$$

or

$$|z_1 + z_2|^2 \leqq (|z_1| + |z_2|)^2.$$

Inasmuch as moduli are nonnegative, inequality (3) follows.

The triangle inequality is easily extended to include any number of summands. That is, we can write

$$|z_1 + z_2 + z_3| \leqq |z_1 + z_2| + |z_3| \leqq |z_1| + |z_2| + |z_3|$$

which is generalized by means of induction to

$$\left|\sum_{k=1}^{n} z_k\right| \leq \sum_{k=1}^{n} |z_k| \qquad (n = 1, 2, \ldots). \tag{4}$$

A lower bound for $|z_1 + z_2|$ is given by the inequality

$$||z_1| - |z_2|| \leqq |z_1 + z_2|. \tag{5}$$

To see this, we write

$$|z_1| = |(z_1 + z_2) + (-z_2)| \leqq |z_1 + z_2| + |-z_2|,$$

which means that

$$|z_1| - |z_2| \leqq |z_1 + z_2|. \tag{6}$$

We thus obtain inequality (5) when $|z_1| \geqq |z_2|$. If $|z_1| < |z_2|$, we need only interchange z_1 and z_2 in inequality (6) to get

$$-(|z_1| - |z_2|) \leqq |z_1 + z_2|,$$

which is the desired result.

Inequality (5) and the triangle inequality combine to become

$$||z_1| - |z_2|| \leqq |z_1 + z_2| \leqq |z_1| + |z_2|. \tag{7}$$

EXERCISES

1. Locate the numbers $z_1 + z_2$ and $z_1 - z_2$ vectorially when
(*a*) $z_1 = 2i$, $z_2 = \frac{2}{3} - i$; (*b*) $z_1 = (-\sqrt{3}, 1)$, $z_2 = (\sqrt{3}, 0)$;
(*c*) $z_1 = (-3, 1)$, $z_2 = (1,4)$; (*d*) $z_1 = x_1 + iy_1$, $z_2 = x_1 - iy_1$.

2. Show that the point representing $(z_1 + z_2)/2$ is the midpoint of the line segment joining the points for z_1 and z_2.

3. Show that (*a*) $\overline{\bar{z} + 3i} = z - 3i$; (*b*) $\overline{iz} = -i\bar{z}$; (*c*) $\overline{\dfrac{(2+i)^2}{3-4i}} = 1$;
(*d*) $|(2\bar{z} + 5)(\sqrt{2} - i)| = \sqrt{3}|2z + 5|$.

4. Prove relations (2) and (3), Sec. 3, algebraically and interpret them geometrically.

5. Prove that $\sqrt{2}|z| \geqq |\text{Re } z| + |\text{Im } z|$.

6. Derive identities (6), (7), and (8) in Sec. 3.

7. Prove that (*a*) z is real if and only if $\bar{z} = z$; (*b*) z is either real or pure imaginary if and only if $(\bar{z})^2 = z^2$.

8. Prove that (*a*) $\overline{z_1 z_2 z_3} = \bar{z}_1 \bar{z}_2 \bar{z}_3$; (*b*) $\overline{(z^4)} = (\bar{z})^4$.

9. Prove property (2), Sec. 4.

10. Show that when $z_2 z_3 \neq 0$,

$$(a) \quad \overline{\left(\frac{z_1}{z_2 z_3}\right)} = \frac{\bar{z}_1}{\bar{z}_2 \bar{z}_3}; \quad (b) \quad \left|\frac{z_1}{z_2 z_3}\right| = \frac{|z_1|}{|z_2||z_3|}.$$

11. Prove (*a*) inequality (4), Sec. 4; (*b*) the inequalities

$$||z_1| - |z_2|| \leqq |z_1 - z_2| \leqq |z_1| + |z_2|.$$

12. In each case sketch the set of points determined by the given condition:
(*a*) $|z - 1 + i| = 1$; (*b*) $|z + i| \leqq 3$; (*c*) $\text{Re}(\bar{z} - i) = 2$;
(*d*) $|z - i| = |z + i|$.

13. Prove that if $|z_2| \neq |z_3|$, then

$$\left|\frac{z_1}{z_2 + z_3}\right| \leqq \frac{|z_1|}{||z_2| - |z_3||}.$$

14. Let R be a positive constant and let z_0 be a fixed complex number. Show that the equation for the circle of radius R centered at $-z_0$ may be written

$$|z|^2 + 2\,\mathrm{Re}\,(\bar{z}_0 z) + |z_0|^2 = R^2.$$

15. Using equations (9), Sec. 3, show that the hyperbola $x^2 - y^2 = 1$ can be written as $z^2 + \bar{z}^2 = 2$.

16. Argue geometrically that $|z - 4i| + |z + 4i| = 10$ is an ellipse, and then show this algebraically.

5. Polar Coordinates

Let r and θ be polar coordinates of the point (x,y) corresponding to a nonzero complex number $z = x + iy$. Since

$$x = r\cos\theta \qquad \text{and} \qquad y = r\sin\theta, \tag{1}$$

z can be written in *polar form* as

$$z = r(\cos\theta + i\sin\theta). \tag{2}$$

For example,

$$1 + i = \sqrt{2}\left[\cos\left(\frac{\pi}{4}\right) + i\sin\left(\frac{\pi}{4}\right)\right] = \sqrt{2}\left[\cos\left(-\frac{7\pi}{4}\right) + i\sin\left(-\frac{7\pi}{4}\right)\right].$$

The number r is the length of the vector representing z; that is, $r = |z|$. The number θ is called an *argument* of z, and we write $\theta = \arg z$. Geometrically, $\arg z$ is any angle, measured in radians, that z makes with the positive real axis when z is interpreted as a directed line segment from the origin (Fig. 4). Hence it has any one of an infinite number of real values differing by integral multiples of 2π. These values can be determined from the equation

$$\tan\theta = \frac{y}{x} \tag{3}$$

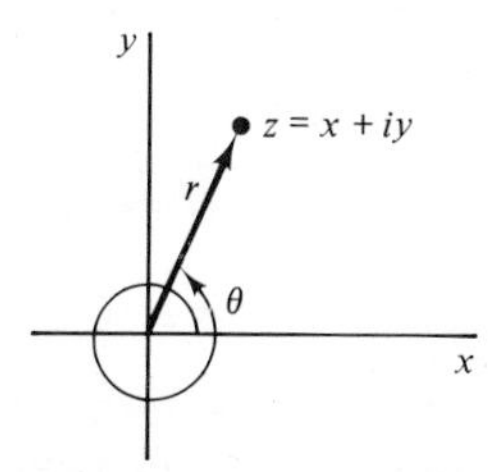

FIGURE 4

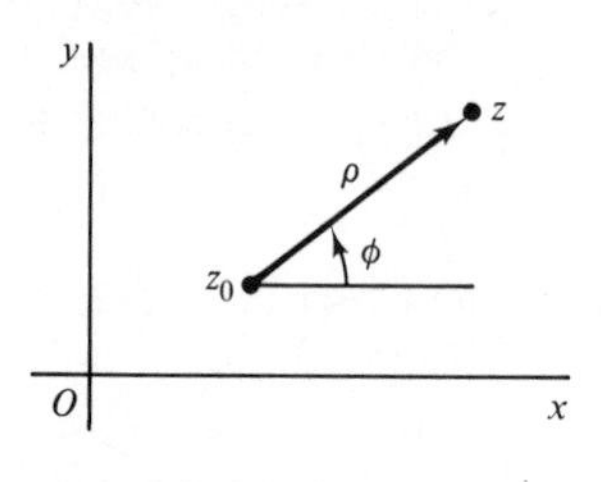

FIGURE 5

where the quadrant containing the point corresponding to z must be specified. For any given nonzero z the *principal value* of $\arg z$, denoted by $\operatorname{Arg} z$, is defined as that unique value of $\arg z$ such that $-\pi < \arg z \leqq \pi$.

If $z = 0$, equation (3) does not apply and θ is undefined. In the remainder of this section, the complex numbers whose polar forms are used are understood to be nonzero even when that condition is not explicitly stated.

When $z \neq z_0$, the representation

$$z - z_0 = \rho(\cos\phi + i\sin\phi)$$

of $z - z_0$ in polar form can be interpreted geometrically as indicated in Fig. 5. That is, $\rho = |z - z_0|$ and is the distance between the points representing z and z_0, while $\phi = \arg(z - z_0)$ and is the angle of inclination of the vector representing $z - z_0$.

We now turn to an important identity involving arguments:

$$\arg(z_1 z_2) = \arg z_1 + \arg z_2. \tag{4}$$

It is to be interpreted as saying that any argument of $z_1 z_2$ is the sum of an argument of z_1 and an argument of z_2; and, conversely, any argument of z_1 plus any argument of z_2 is an argument of $z_1 z_2$. Identity (4) is not always valid when *arg* is replaced by *Arg*; we need only write $z_1 = -1$ and $z_2 = i$ to see this.

To obtain identity (4), we start with z_1 and z_2 in polar form:

$$z_1 = r_1(\cos\theta_1 + i\sin\theta_1), \qquad z_2 = r_2(\cos\theta_2 + i\sin\theta_2).$$

Now

$$z_1 z_2 = r_1 r_2[(\cos\theta_1\cos\theta_2 - \sin\theta_1\sin\theta_2) + i(\sin\theta_1\cos\theta_2 + \cos\theta_1\sin\theta_2)],$$

and this reduces to the polar form of the product:

$$z_1 z_2 = r_1 r_2[\cos(\theta_1 + \theta_2) + i\sin(\theta_1 + \theta_2)]. \tag{5}$$

So any argument of z_1 plus any argument of z_2 is an argument of $z_1 z_2$ (Fig. 6).

On the other hand, consider any argument of the product $z_1 z_2$. In view of expression (5), this argument must be of the form $\theta_1 + \theta_2 + 2n\pi$ for some

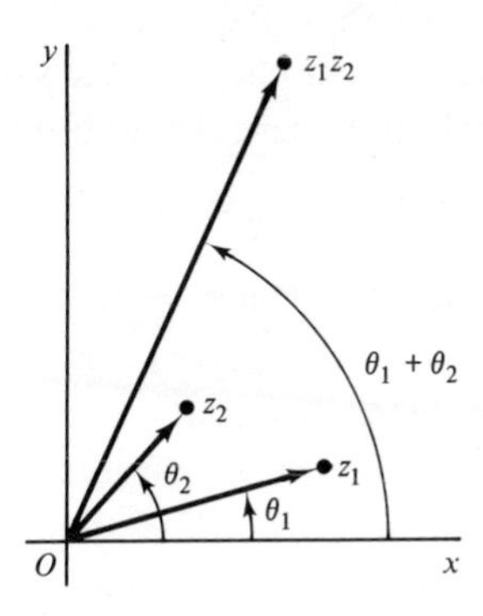

FIGURE 6

integer n. In identity (4) we can then take $\arg z_1 = \theta_1$ and $\arg z_2 = \theta_2 + 2n\pi$, for example. This completes the proof of that identity.

Note that when a nonzero complex number $z = r(\cos\theta + i\sin\theta)$ is multiplied by i, the directed line segment from the origin representing iz is obtained by rotating the one for z through a right angle in the positive, or counterclockwise, direction without changing its length. This is because, according to equation (5),

$$iz = \left(\cos\frac{\pi}{2} + i\sin\frac{\pi}{2}\right)r(\cos\theta + i\sin\theta)$$
$$= r\left[\cos\left(\theta + \frac{\pi}{2}\right) + i\sin\left(\theta + \frac{\pi}{2}\right)\right].$$

From equation (5) it is also evident that the polar form for the multiplicative inverse of a nonzero complex number

$$z = r(\cos\theta + i\sin\theta)$$

is

$$z^{-1} = \frac{1}{r}\left[\cos(-\theta) + i\sin(-\theta)\right], \tag{6}$$

the product of these polar forms being unity. Since $z_1/z_2 = z_1 z_2^{-1}$, we thus have the following expression for the quotient of two nonzero complex numbers:

$$\frac{z_1}{z_2} = \frac{r_1}{r_2}\left[\cos(\theta_1 - \theta_2) + i\sin(\theta_1 - \theta_2)\right]. \tag{7}$$

It is often convenient to let $e^{i\theta}$ denote the expression $\cos\theta + i\sin\theta$:

$$e^{i\theta} = \cos\theta + i\sin\theta. \tag{8}$$

This is known as *Euler's formula*. The choice of the symbol $e^{i\theta}$ will be motivated later on in Sec. 21. We simply note here the additivity property

$$e^{i\theta_1}e^{i\theta_2} = e^{i(\theta_1+\theta_2)} \tag{9}$$

which is obtained from formula (5) when $r_1 = r_2 = 1$. That is, when $z_1 = e^{i\theta_1}$ and $z_2 = e^{i\theta_2}$, then $z_1 z_2 = e^{i(\theta_1+\theta_2)}$. Property (9) is, of course, analogous to the corresponding property for e^x where x is real.

Observe that, by the additivity property, $e^{i\theta}e^{-i\theta} = 1$. The number $e^{-i\theta}$ is therefore the multiplicative inverse of $e^{i\theta}$, and so $1/e^{i\theta} = e^{-i\theta}$.

It follows from expressions (2) and (8) that any nonzero complex number z can be written

$$z = re^{i\theta}; \tag{10}$$

and, according to formula (6), its multiplicative inverse is

$$z^{-1} = \frac{1}{r}\, e^{-i\theta}. \tag{11}$$

Also, when $z_1 = r_1 e^{i\theta_1}$ and $z_2 = r_2 e^{i\theta_2}$, formulas (5) and (7) take the forms

$$z_1 z_2 = r_1 r_2\, e^{i(\theta_1+\theta_2)} \tag{12}$$

and

$$\frac{z_1}{z_2} = \frac{r_1}{r_2}\, e^{i(\theta_1-\theta_2)}, \tag{13}$$

respectively.

6. Powers and Roots

Integral powers of a nonzero complex number $z = re^{i\theta}$ are given by the formula

$$z^n = r^n e^{in\theta} \qquad (n = 0, \pm 1, \pm 2, \ldots). \tag{1}$$

This is easily proved when $n = 1, 2, \ldots$ by means of induction and the use of property (9), Sec. 5; the formula also holds when $n = 0$ with the convention that $z^0 = 1$. If $n = -1, -2, \ldots$, on the other hand, we define z^n by the equation $z^n = (z^{-1})^{-n}$. It then follows from formula (11), Sec. 5, and the fact that formula (1) is valid for positive powers that

$$z^n = \left(\frac{1}{r}\right)^{-n} e^{i(-n)(-\theta)} = r^n e^{in\theta}.$$

Hence formula (1) is valid for all integral powers.

Observe that when $r = 1$, formula (1) becomes

$$(e^{i\theta})^n = e^{in\theta} \qquad (n = 0, \pm 1, \pm 2, \ldots), \tag{2}$$

or

$$(\cos\theta + i\sin\theta)^n = \cos n\theta + i\sin n\theta \qquad (n = 0, \pm 1, \pm 2, \ldots), \tag{3}$$

which is known as *de Moivre's theorem*.

Formula (1) is useful, for example, in computing roots of nonzero complex numbers. To illustrate, let us solve the equation

$$z^n = 1 \qquad (n = 1, 2, \ldots) \tag{4}$$

and thus find the nth roots of unity. Inasmuch as $z \neq 0$, we may write $z = re^{i\theta}$ and look for values of r and θ such that

$$(re^{i\theta})^n = 1,$$

or

$$r^n e^{in\theta} = 1e^{i0}.$$

Now if two complex numbers are equal, their moduli are equal. If, moreover, they are expressed in polar form, the arguments used in those expressions can differ at most by an integral multiple of 2π. Hence

$$r^n = 1 \qquad \text{and} \qquad n\theta = 0 + 2k\pi$$

where k is any integer ($k = 0, \pm 1, \pm 2, \ldots$). Consequently, $r = 1$ and $\theta = 2k\pi/n$, and we have the *distinct* solutions

$$z = e^{i(2k\pi/n)} \qquad (k = 0, 1, 2, \ldots, n - 1)$$

to equation (4). That is, the complex numbers

$$\cos\frac{2k\pi}{n} + i\sin\frac{2k\pi}{n} \qquad (k = 0, 1, 2, \ldots, n - 1) \tag{5}$$

are nth roots of unity. Because of the periodicity of the cosine and sine, no further roots are obtained with other values of k.

So the number of nth roots of unity is n. Geometrically, these roots correspond to points lying at the vertices of a regular polygon of n sides. This polygon is inscribed in the unit circle centered at the origin, and it has one vertex at the point corresponding to the root $z = 1$. If we write

$$\omega_n = \cos\frac{2\pi}{n} + i\sin\frac{2\pi}{n}, \tag{6}$$

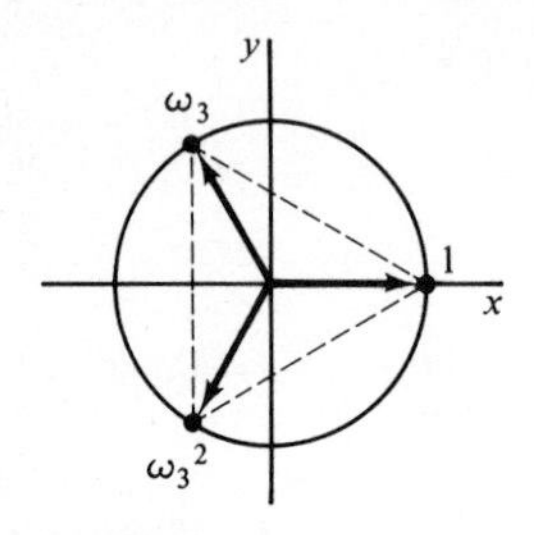

FIGURE 7

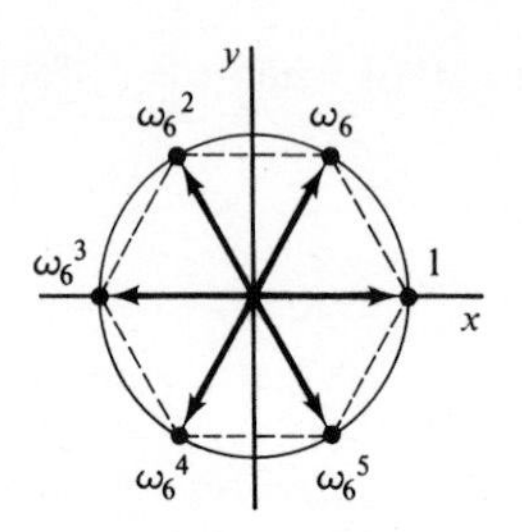

FIGURE 8

then, by de Moivre's theorem, the nth roots of unity are simply $1, \omega_n, \omega_n^2, \ldots, \omega_n^{n-1}$. Note that $\omega_n^n = 1$. See Fig. 7 for the interpretation of the three cube roots of 1 as vertices of an equilateral triangle. Fig. 8 illustrates the case $n = 6$.

The above discussion is readily extended to finding the nth roots of any nonzero complex number $w = \rho(\cos\phi + i\sin\phi)$. Those roots are

$$\sqrt[n]{\rho}\left(\cos\frac{\phi + 2k\pi}{n} + i\sin\frac{\phi + 2k\pi}{n}\right) \qquad (k = 0, 1, 2, \ldots, n-1) \tag{7}$$

where $\sqrt[n]{\rho}$ denotes the positive nth root of the real number ρ. Geometrically, $\sqrt[n]{\rho}$ is the length of each of the vectors representing the n roots. An argument of one of those nth roots is ϕ/n and arguments of the other roots are obtained by adding integral multiples of $2\pi/n$. Note that if z_0 is any particular nth root of w, the set of all nth roots may be written

$$z_0, z_0\omega_n, z_0\omega_n^2, \ldots, z_0\omega_n^{n-1}$$

where ω_n is defined in equation (6). This is because multiplication of any nonzero complex number by ω_n corresponds to increasing the argument of that number by $2\pi/n$.

We shall let $w^{1/n}$ denote any of the nth roots of a nonzero complex number w. If, in particular, w is a positive real number ρ, the symbol $\rho^{1/n}$ denotes any of the roots and the symbol $\sqrt[n]{\rho}$ used in expression (7) is reserved for the one positive root.

EXERCISES

1. Find one value of arg z when

(*a*) $z = \dfrac{-2}{1 + i\sqrt{3}}$; (*b*) $z = \dfrac{i}{-2 - 2i}$; (*c*) $z = (\sqrt{3} - i)^6$.

Ans. (*a*) $2\pi/3$; (*c*) π.

2. Use the polar form to show that
(*a*) $i(1 - i\sqrt{3})(\sqrt{3} + i) = 2 + i2\sqrt{3}$; (*b*) $5i/(2 + i) = 1 + 2i$;
(*c*) $(-1 + i)^7 = -8(1 + i)$; (*d*) $(1 + i\sqrt{3})^{-10} = 2^{-11}(-1 + i\sqrt{3})$.

3. In each case find all the roots and exhibit them geometrically:
(*a*) $(2i)^{1/2}$; (*b*) $(-i)^{1/3}$; (*c*) $(-1)^{1/3}$; (*d*) $8^{1/6}$.

Ans. (*a*) $\pm(1 + i)$; (*b*) $i, (\pm\sqrt{3} - i)/2$;
(*d*) $\pm\sqrt{2}, (1 \pm i\sqrt{3})/\sqrt{2}, (-1 \pm i\sqrt{3})/\sqrt{2}$.

4. Prove formula (1), Sec. 6, when $n = 1, 2, \ldots$.

5. Find a value of arg z when $z_1 \neq 0$ and $z_2 \neq 0$: (*a*) $z = z_1/z_2$;
(*b*) $z = z_1{}^n$ $(n = 1, 2, \ldots)$; (*c*) $z = z_1{}^{-1}$.

Ans. (*a*) $\arg z_1 - \arg z_2$; (*b*) $n \arg z_1$; (*c*) $-\arg z_1$.

6. Show that $\arg \bar{z} = -\arg z$, where $z \neq 0$. Determine further restrictions needed on z such that $\operatorname{Arg} \bar{z} = -\operatorname{Arg} z$.

Ans. z is not a negative real number.

7. Let z be a nonzero complex number and n a negative integer $(n = -1, -2, \ldots)$. Using the definition $z^n = (z^{-1})^{-n}$ given in Sec. 6, show that we can also write $z^n = (z^{-n})^{-1}$.

8. Find the four roots of the equation $z^4 + 4 = 0$ and use them to factor $z^4 + 4$ into quadratic factors with real coefficients.

Ans. $(z^2 + 2z + 2)(z^2 - 2z + 2)$.

9. Use de Moivre's theorem to derive the following trigonometric identities:
(*a*) $\cos 3\theta = \cos^3 \theta - 3 \cos \theta \sin^2 \theta$; (*b*) $\sin 3\theta = 3 \cos^2 \theta \sin \theta - \sin^3 \theta$.

10. Obtain expression (7), Sec. 6.

11. Given that $z_1 z_2 \neq 0$, use the polar form to prove that $\operatorname{Re}(z_1 \bar{z}_2) = |z_1||z_2|$ if and only if $\theta_1 - \theta_2 = 2n\pi$ $(n = 0, \pm 1, \pm 2, \ldots)$, where $\theta_1 = \arg z_1$ and $\theta_2 = \arg z_2$.

12. Given that $z_1 z_2 \neq 0$, use the result in Exercise 11 to prove that $|z_1 + z_2| = |z_1| + |z_2|$ if and only if $\theta_1 - \theta_2 = 2n\pi$ $(n = 0, \pm 1, \pm 2, \ldots)$, where $\theta_1 = \arg z_1$ and $\theta_2 = \arg z_2$. Verify this statement geometrically.

13. Given that $z_1 z_2 \neq 0$, use the result in Exercise 11 to prove that $|z_1 - z_2| = ||z_1| - |z_2||$ if and only if $\theta_1 - \theta_2 = 2n\pi$ $(n = 0, \pm 1, \pm 2, \ldots)$, where $\theta_1 = \arg z_1$ and $\theta_2 = \arg z_2$. Verify this statement geometrically.

14. Establish the identity

$$1 + z + z^2 + \cdots + z^n = \frac{1 - z^{n+1}}{1 - z} \qquad (z \neq 1);$$

then derive *Lagrange's trigonometric identity*,

$$1 + \cos \theta + \cos 2\theta + \cdots + \cos n\theta = \frac{1}{2} + \frac{\sin [(n + \frac{1}{2})\theta]}{2 \sin (\theta/2)} \qquad (0 < \theta < 2\pi).$$

Suggestion: To establish the first identity, write $S = 1 + z + z^2 + \cdots + z^n$ and consider the difference $S - zS$.

15. Show that if z is any nth root of unity other than unity itself, then $1 + z + z^2 + \cdots + z^{n-1} = 0$.

Suggestion: Use the first identity in Exercise 14.

16. Prove that the usual quadratic formula solves the quadratic equation $az^2 + bz + c = 0$ when the coefficients a, b, and c are complex numbers.

7. Regions in the Complex Plane

In this section we are concerned with sets of complex numbers, or points, and their closeness to one another. Our basic tool is the concept of an ε *neighborhood*, or simply neighborhood,

$$|z - z_0| < \varepsilon \tag{1}$$

of a given point z_0 consisting of all points z lying inside but not on a circle centered at z_0 and with a specified positive radius ε (Fig. 9).

A point z_0 is said to be an *interior point* of a set S whenever there is some neighborhood of z_0 which contains only points of S; it is called an *exterior point* of S when there exists a neighborhood of it containing no points of S. If z_0 is neither of these, it is a *boundary point* of S. A boundary point is therefore a point all of whose neighborhoods contain points both in S and not in S. The totality of all boundary points is called the *boundary* of S. The circle $|z| = 1$, for instance, is the boundary of each of the sets

$$|z| < 1 \qquad \text{and} \qquad |z| \leqq 1. \tag{2}$$

A set is *open* if it contains none of its boundary points. It is left as an exercise to show that a set is open if and only if each of its points is an interior point. A set is *closed* if it contains all of its boundary points; and the *closure* $\bar{S}$ of a set S is the closed set consisting of all points in S together with the boundary

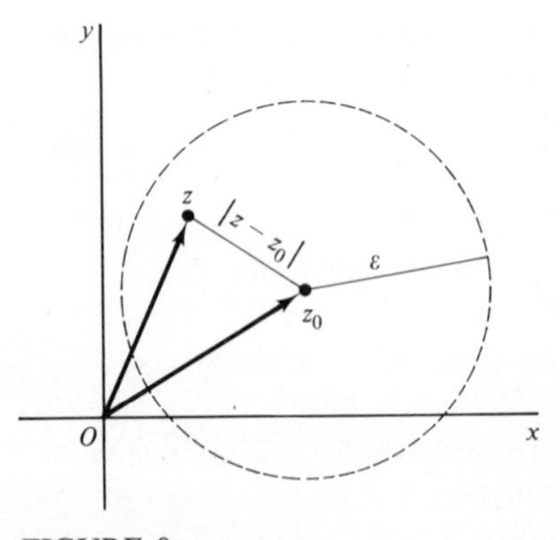

FIGURE 9

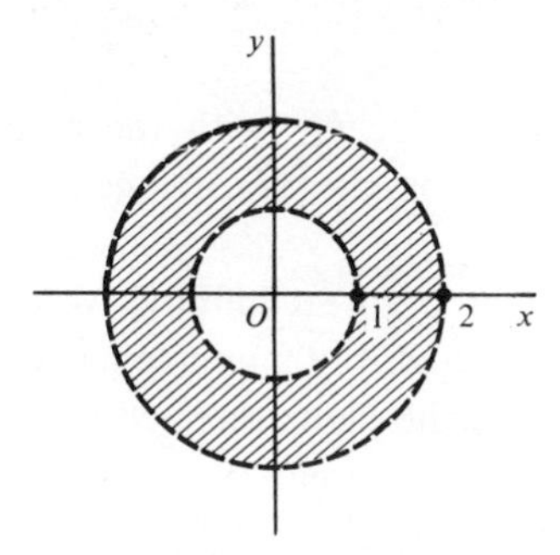

FIGURE 10

of S. Note that the first of sets (2) is open, and the second is the closure of both those sets.

Some sets are, of course, neither open nor closed. In order for a set to be not open, there must be a boundary point which is contained in the set; and if a set is not closed, there exists a boundary point not contained in the set. Observe that the set $0 < |z| \leqq 1$ is neither open nor closed. The set of all complex numbers is, on the other hand, both open and closed; for there are no boundary points.

An open set S is *connected* if each pair of points in it can be joined by a polygonal path, consisting of a finite number of line segments joined end to end, which lies entirely in S. The open set $|z| < 1$ is connected. The annulus $1 < |z| < 2$ is, of course, open and it is also connected (Fig. 10). An open set that is connected is called a *domain*. Note that any neighborhood is a domain. A domain together with some, none, or all of its boundary points is referred to as a *region*.

A set S is *bounded* if every point of S lies inside some circle $|z| = R$; otherwise, it is *unbounded*. In the next section we shall pursue the concept of an unbounded set.

Finally, a point z_0 is said to be an *accumulation point* of a set S if each neighborhood of z_0 contains at least one point of S distinct from z_0. It follows that if a set S is closed, then it contains each of its accumulation points. For if an accumulation point z_0 were not in S, it would be a boundary point of S; but this contradicts the fact that a closed set contains all of its boundary points. It is left as an exercise to show that the converse is in fact true. Thus, a set is closed if and only if it contains each of its accumulation points.

Evidently a point z_0 is *not* an accumulation point of a set S whenever there exists some neighborhood of z_0 which does not contain points of S distinct from z_0. Note that the origin is the only accumulation point of the set $z_n = i/n$ $(n = 1, 2, \ldots)$.

8. The Point at Infinity

It is sometimes convenient to include with the complex plane *the point at infinity*, denoted by ∞. The complex plane together with this point is called the *extended complex plane*.

In order to visualize the point at infinity, we can think of the complex plane as passing through the equator of a unit sphere centered at the point $z = 0$ (Fig. 11). To each point z in the plane there corresponds exactly one point P on the surface of the sphere. The point P is determined by the intersection of the line through the point z and the north pole N of the sphere with that surface. In like manner, to each point P on the surface of the sphere, other than the north pole N, there corresponds exactly one point z in the plane. By letting the point N of the sphere correspond to the point at infinity, we obtain a one to one correspondence between the points of the sphere and the points of the extended complex plane. The sphere is known as the *Riemann sphere*, and the correspondence is called a *stereographic projection.*

Observe that the exterior of the unit circle centered at the origin in the complex plane corresponds to the upper hemisphere with the equator and the point N deleted. Moreover, for each small positive number ε, those points in the complex plane exterior to the circle $|z| = 1/\varepsilon$ correspond to points on the sphere close to N. We thus call the set $|z| > 1/\varepsilon$ an ε *neighborhood*, or neighborhood, of ∞.

Whenever every neighborhood of ∞ contains at least one point of a given set S in the complex plane, we say that ∞ is an accumulation point of S. As an illustration, ∞ is an accumulation point of the set $z_n = ni$ $(n = 0, 1, 2, \ldots)$ as well as of the domain $\operatorname{Im} z > 0$. Note that a set S is unbounded (Sec. 7) if and only if ∞ is one of its accumulation points.

Let us agree that in referring to a point z we mean a point in the *finite* plane. If the point at infinity is intended, it will be specifically mentioned.

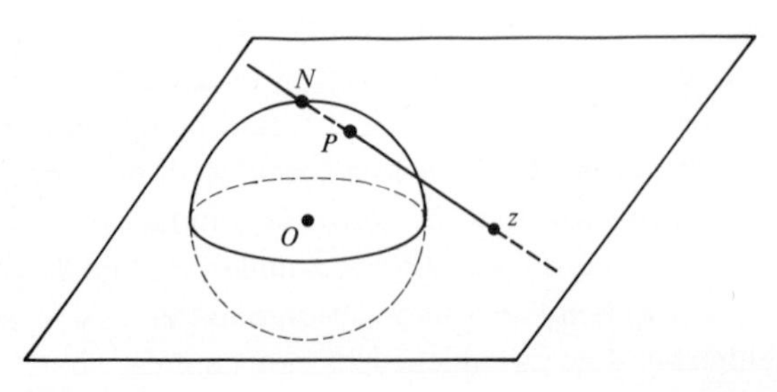

FIGURE 11

EXERCISES

1. In each case sketch the set and determine whether it is a domain:
(*a*) $|z-2+i| \leqq 1$; (*b*) $|2z+3|>4$; (*c*) $\operatorname{Im} z>1$;
(*d*) $|\operatorname{Im} z|>1$; (*e*) $|z|>0, 0 \leqq \arg z \leqq \pi/4$; (*f*) $|z-4| \geqq |z|$;
(*g*) $0<|z-z_0|<\delta$ where z_0 is a fixed point and δ is a positive number.
Ans. (*b*), (*c*), (*g*) domains.

2. Which sets in Exercise 1 are neither open nor closed?
Ans. (*e*).

3. Which sets in Exercise 1 are bounded?
Ans. (*a*); (*g*).

4. In each case sketch the closure of the set: (*a*) $|z|>0, -\pi<\arg z<\pi$; (*b*) $|\operatorname{Re} z|<|z|$; (*c*) $\operatorname{Re}(1/z) \leqq 1/2$; (*d*) $\operatorname{Re}(z^2)>0$.

5. Let S be the open set consisting of all points z such that $|z|<1$ or $|z-2|<1$. Show why S is not connected.

6. Show that a set S is open if and only if each point in S is an interior point.

7. Determine the accumulation points of each of the following sets:
(*a*) $z_n = i^n$ $(n=1, 2, \ldots)$; (*b*) $z_n=(1/n)i^n$ $(n=1, 2, \ldots)$;
(*c*) $|z|>1, 0 \leqq \arg z<\pi/2$; (*d*) $z_n=(-1)^n(1+i)(n-1)/n$ $(n=1, 2, \ldots)$.
Ans. (*a*) none; (*b*) 0; (*d*) $\pm(1+i)$.

8. Prove that if a set contains each of its accumulation points, it must be a closed set.

9. Show that any point z_0 of a domain is an accumulation point of that domain.

10. Show that a finite set of points $z_1, z_2, \ldots, z_n$ cannot have any accumulation points.

11. Show that in the extended plane the point at infinity is an accumulation point of the set $\operatorname{Re} z>0$ as well as the set $\operatorname{Re} z<0$.

2

ANALYTIC FUNCTIONS

We now consider functions of a complex variable and develop a theory of differentiation for them. The main goal of the chapter is to introduce analytic functions; they play a central role in complex analysis.

9. Functions of a Complex Variable

Let S be a set of complex numbers. A *function* f defined on S is a rule which assigns to each z in S a complex number w. The number w is called a *value* of f at z and is denoted by $f(z)$; that is,

$$w = f(z).$$

The set S is called the *domain of definition* of f. Although the domain of definition is often a domain (Sec. 7), it need not be.

It is not always convenient to use different notation to distinguish between a given function and its values. For example, the function f defined on the set $\operatorname{Im} z > 1$ by means of the equation $w = 1/z$ is also referred to as the function $1/z$ defined on $\operatorname{Im} z > 1$. The function is described by stating its value at each point in the domain of definition.

When the domain of definition is not mentioned, we agree that the largest possible set is to be taken. Thus, if we speak of the function $1/z$, the domain of definition is understood to be the set of all nonzero points in the plane.

In the theory of complex variables there arise *multiple-valued functions*, or functions which can assume more than one value at a specific point. An example is $z^{1/2}$ which has two values at each nonzero point in the complex plane. Our study of multiple-valued functions will usually involve certain single-valued functions where just one of the possible values assigned to each point is taken. Let us agree that *the term function signifies a single-valued function* unless the contrary is clearly indicated.

Suppose that $w = u + iv$ is the value of a function f at $z = x + iy$; that is,

$$u + iv = f(x + iy).$$

Each of the real numbers u and v depends on the real variables x and y. If, for instance, $f(z) = z^2$, we have

$$f(z) = (x + iy)^2 = x^2 - y^2 + i2xy;$$

hence

$$u = x^2 - y^2 \qquad \text{and} \qquad v = 2xy.$$

This illustrates how a function of the complex variable z can be expressed in terms of a pair of real-valued functions of the real variables x and y:

$$f(z) = u(x,y) + iv(x,y). \tag{1}$$

If, on the other hand, $u(x,y)$ and $v(x,y)$ are two given real-valued functions of the real variables x and y, equation (1) can be used to define a function of the complex variable $z = x + iy$. For example, given the real-valued functions indicated below, we can write

$$f(z) = y \int_0^\infty e^{-xt}\, dt + i \sum_{n=0}^{\infty} y^n. \tag{2}$$

The domain of definition of f here is understood to be the semi-infinite strip $x > 0$, $-1 < y < 1$; for the improper integral and the infinite series both converge only when x and y are so restricted.

If in equation (1) the number $v(x,y)$ is always zero, then the number $f(z)$ is always real. An example of such a *real-valued function of a complex variable* is $f(z) = |z|^2$.

If n is zero or a positive integer and if $a_0, a_1, a_2, \ldots, a_n$ are complex constants, the function

$$P(z) = a_0 + a_1 z + a_2 z^2 + \cdots + a_n z^n \qquad (a_n \neq 0)$$

is a *polynomial* of degree n. Note that the sum here has a finite number of terms and the domain of definition is the entire z plane. Quotients of polynomials, $P(z)/Q(z)$, are called *rational functions* and are defined at each z except where $Q(z) = 0$. Polynomials and rational functions constitute elementary, but important, classes of functions of a complex variable.

10. Mappings

Properties of a real-valued function of a real variable are often exhibited by the graph of the function. But when $w = f(z)$ where z and w are complex, no such convenient graphical representation of the function f is available because each of the numbers z and w is located in a plane rather than on a line. We can, however, display some information about the function by indicating pairs of corresponding points $z = (x,y)$ and $w = (u,v)$. To do this, it is generally simpler to draw the z and w planes separately.

When a function f is thought of in this way, it is often referred to as a *mapping* or *transformation*. The *image* of a point z in the domain of definition S is the point $w = f(z)$, and the set of images of all points in a set T which is contained in S is called the image of T. The image of the entire domain of definition S is called the *range* of f. The *inverse image* of a point w is the set of all points z in the domain of definition of f which have w as their image. The inverse image of a point may contain just one point, many points, or none at all. The last case occurs, of course, when w is not in the range of f.

Terms such as *translation*, *rotation*, and *reflection* are used to convey dominant geometric characteristics of certain mappings. In such cases it is sometimes convenient to consider the z and w planes to be the same. The mapping $w = z + 1$, for instance, can be thought of as a translation of each point z to a position one unit to the right. The mapping $w = iz$ rotates each nonzero z counterclockwise through an angle $\pi/2$, and the mapping $w = \bar{z}$ transforms each point z into its reflection in the real axis.

More information is usually exhibited by sketching images of curves and regions than by simply indicating images of individual points. As an illustration, the function

$$f(z) = \sqrt{x^2 + y^2} - iy$$

maps points on the circle $x^2 + y^2 = c^2$, where $c \geqq 0$, into the line $u = c$ because $u = \sqrt{x^2 + y^2}$. Since $v = -y$ and y assumes all values from $-c$ to c, the image of the circle is in fact the line segment $u = c$, $-c \leqq v \leqq c$ (Fig. 12). Since the points $z = (x,y)$ and $-\bar{z} = (-x,y)$ have the same image w, each point on that line segment, except the end points, is the image of two points on the circle.

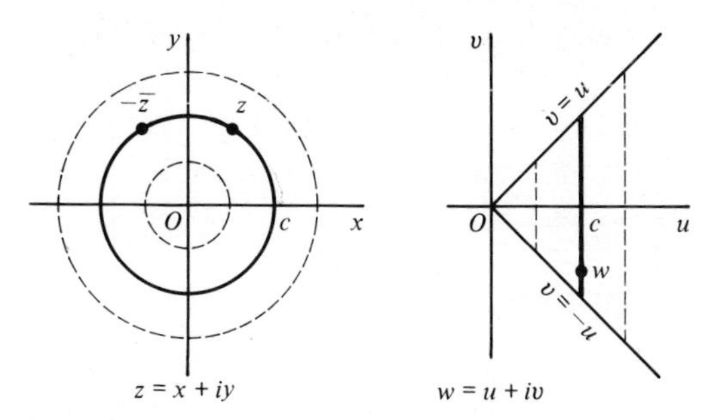

FIGURE 12

The domain of definition of the function is the entire z plane, and each point z lies on such a circle since c is any nonnegative constant. The image of each circle is a line segment as described above, and every such line segment is the image of one of the circles. Hence the range of the function is the quadrant

$$u \geqq 0, \qquad -u \leqq v \leqq u.$$

11. Limits

Let a function f be defined at all points in some neighborhood of z_0, except possibly for the point z_0 itself. The statement that the *limit* of f as z appraoches z_0 is a number w_0,

$$\lim_{z \to z_0} f(z) = w_0, \tag{1}$$

means that the point $w = f(z)$ can be made arbitrarily close to w_0 if we choose a point z which is close enough to z_0 but distinct from it. We now express the definition of limit in a precise and usable form.

Statement (1) means that for each positive number ε there is a positive number δ such that

$$|f(z) - w_0| < \varepsilon \qquad \text{whenever} \qquad 0 < |z - z_0| < \delta. \tag{2}$$

Geometrically, this definition says that for each ε neighborhood $|w - w_0| < \varepsilon$ of w_0 there is a δ neighborhood $|z - z_0| < \delta$ of z_0 such that the images of all points in the δ neighborhood, with the possible exception of z_0, lie in the ε neighborhood (Fig. 13). Note that these image points need not constitute the entire ε neighborhood. The points z, however, constitute the entire domain $0 < |z - z_0| < \delta$. Note too that z is allowed to approach z_0 in an arbitrary manner, not just from some particular direction.

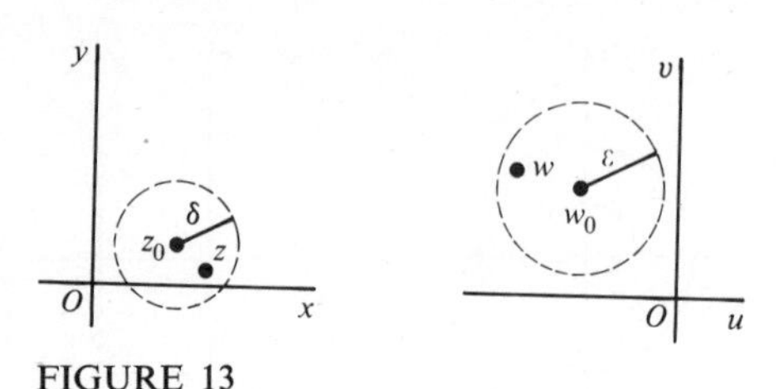

FIGURE 13

Definition (2) requires that f be defined at all points in some neighborhood of z_0, with the possible exception of the point z_0 itself. Such a neighborhood, of course, always exists when z_0 is an interior point of a region on which f is defined. We can extend the definition of limit to the case when z_0 is a boundary point of that region by agreeing that the first of inequalities (2) need be satisfied by only those points z which lie in *both* the domain $0 < |z - z_0| < \delta$ and the region.

While definition (2) provides a means of testing whether a given point is a limit, it does not directly provide a method of determining that limit. Theorems on limits, presented in the following section, will enable us to actually find many limits.

Let us now consider the function $f(z) = iz/2$ defined on the open disk $|z| < 1$ and show that

$$\lim_{z \to 1} \frac{iz}{2} = \frac{i}{2},$$

the point $z = 1$ being on the boundary of the domain of definition. Observe that when z is in the region $|z| < 1$,

$$\left| f(z) - \frac{i}{2} \right| = \left| \frac{iz}{2} - \frac{i}{2} \right| = \frac{|z - 1|}{2}.$$

Hence, for any such z and any positive number ε,

$$\left| f(z) - \frac{i}{2} \right| < \varepsilon \qquad \text{whenever} \qquad 0 < |z - 1| < 2\varepsilon.$$

Condition (2) is thus satisfied when δ is equal to 2ε (Fig. 14) or any smaller positive number.

To further illustrate definition (2), we show next that

$$\lim_{z \to 2i} (2x + iy^2) = 4i \qquad (z = x + iy). \tag{3}$$

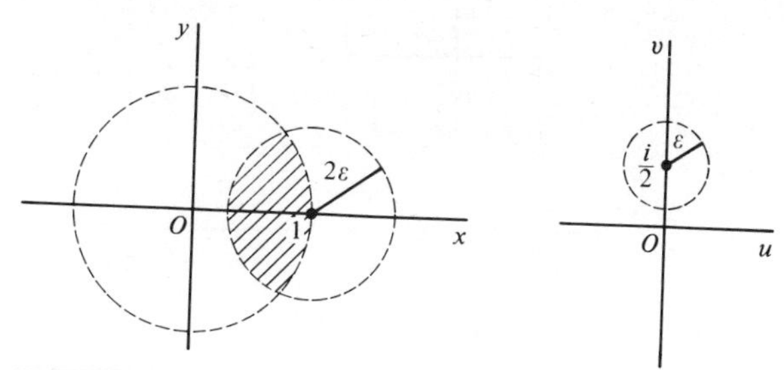

FIGURE 14

For each positive number ε we must find a positive number δ such that

$$|2x + iy^2 - 4i| < \varepsilon \qquad \text{whenever} \qquad 0 < |z - 2i| < \delta. \tag{4}$$

To do this, we write

$$|2x + iy^2 - 4i| \leqq 2|x| + |y^2 - 4| = 2|x| + |y-2|\,|y+2|$$

and note that the first of inequalities (4) will then be satisfied when

$$2|x| < \frac{\varepsilon}{2} \qquad \text{and} \qquad |y-2|\,|y+2| < \frac{\varepsilon}{2}.$$

The first of these inequalities is, of course, satisfied when $|x| < \varepsilon/4$. To establish conditions on y such that the second one holds, we observe that

$$|y+2| = |(y-2)+4| \leqq |y-2| + 4 < 5$$

if we restrict y so that $|y - 2| < 1$. Hence

$$|y-2|\,|y+2| < \left(\frac{\varepsilon}{10}\right)5 = \frac{\varepsilon}{2}$$

whenever $|y-2| < \min\{\varepsilon/10, 1\}$, where $\min\{\varepsilon/10, 1\}$ denotes the minimum, or lesser, of the two numbers $\varepsilon/10$ and 1. An appropriate value of δ is now easily seen (Fig. 15) from the conditions that $|x|$ must be less than $\varepsilon/4$ and that $|y-2|$ must be less than $\min\{\varepsilon/10, 1\}$:

$$\delta = \min\left\{\frac{\varepsilon}{10}, 1\right\}.$$

We have tacitly assumed that when a limit of a function f exists at a point z_0, it is unique. This is indeed the case. For suppose that

$$\lim_{z \to z_0} f(z) = w_0 \qquad \text{and} \qquad \lim_{z \to z_0} f(z) = w_1$$

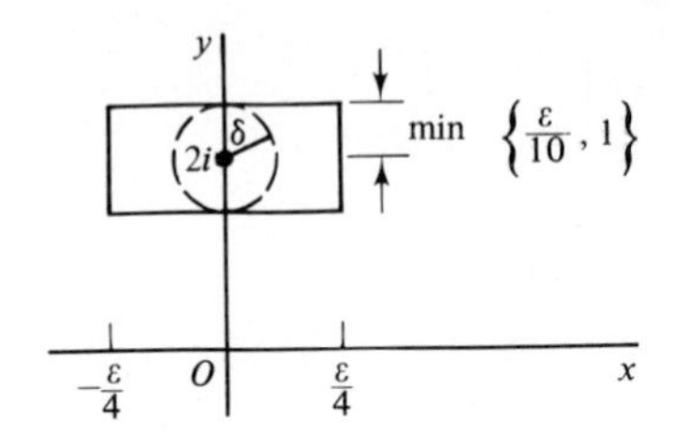

FIGURE 15

where $w_0 \neq w_1$. It follows that for an arbitrary positive number ε there are positive numbers δ_0 and δ_1 such that

$$|f(z) - w_0| < \varepsilon \qquad \text{whenever} \qquad 0 < |z - z_0| < \delta_0$$

and

$$|f(z) - w_1| < \varepsilon \qquad \text{whenever} \qquad 0 < |z - z_0| < \delta_1.$$

Hence if we write $\varepsilon = |w_0 - w_1|/2$ and let δ denote the smaller of the two numbers δ_0 and δ_1, we find that when $0 < |z - z_0| < \delta$,

$$\begin{aligned} |w_0 - w_1| &= |[f(z) - w_1] - [f(z) - w_0]| \\ &\leqq |f(z) - w_1| + |f(z) - w_0| < 2\varepsilon = |w_0 - w_1|. \end{aligned}$$

But the inequality $|w_0 - w_1| < |w_0 - w_1|$ is impossible, and we conclude that a limit is unique.

Finally, we remark that a meaning to the statement

$$\lim_{z \to z_0} f(z) = w_0 \tag{5}$$

is readily given when either z_0 or w_0, or possibly each of these numbers, is the point at infinity. In definition (2) we simply replace the appropriate neighborhoods of z_0 and w_0 by neighborhoods of ∞. For example, the statement

$$\lim_{z \to \infty} f(z) = w_0 \tag{6}$$

means that for each positive number ε there is a positive number δ such that

$$|f(z) - w_0| < \varepsilon \qquad \text{whenever} \qquad |z| > \frac{1}{\delta}.$$

That is, the point $f(z)$ lies in the ε neighborhood $|w - w_0| > \varepsilon$ of w_0 whenever z lies in the δ neighborhood $|z| > 1/\delta$ of the point at infinity.

For an illustration, observe that

$$\lim_{z \to \infty} \frac{1}{z^2} = 0$$

since

$$\left|\frac{1}{z^2} - 0\right| < \varepsilon \qquad \text{whenever} \qquad |z| > \frac{1}{\sqrt{\varepsilon}}.$$

That is, we can write $\delta = \sqrt{\varepsilon}$.

The definition of limit for the case in which w_0 is ∞ in statement (5), as well as the case in which z_0 and w_0 are both ∞, is left to the exercises where specific examples are included.

12. Theorems on Limits

We can expedite our treatment of limits by establishing a connection between the limit of a function of a complex variable and limits of real-valued functions of two real variables. Limits of the latter type are treated in the calculus of real variables. We shall use their definition and properties freely.

Theorem 1. *Suppose that*

$$f(z) = u(x,y) + iv(x,y), \qquad z_0 = x_0 + iy_0, \qquad \textit{and} \qquad w_0 = u_0 + iv_0.$$

Then

$$\lim_{z \to z_0} f(z) = w_0 \tag{1}$$

if and only if

$$\lim_{(x,y)\to(x_0,y_0)} u(x,y) = u_0 \qquad \textit{and} \qquad \lim_{(x,y)\to(x_0,y_0)} v(x,y) = v_0. \tag{2}$$

To prove the theorem, let us first assume that statement (1) is true and show that conditions (2) follow. According to statement (1), there is for each positive number ε a positive number δ such that

$$|u(x,y) - u_0 + i[v(x,y) - v_0]| < \varepsilon \quad \text{whenever} \quad 0 < |x - x_0 + i(y - y_0)| < \delta.$$

Since

$$|u(x,y) - u_0| \leqq |u(x,y) - u_0 + i[v(x,y) - v_0]|$$

and

$$|v(x,y) - v_0| \leqq |u(x,y) - u_0 + i[v(x,y) - v_0]|,$$

it follows that

$$|u(x,y) - u_0| < \varepsilon \qquad \text{and} \qquad |v(x,y) - v_0| < \varepsilon$$
$$\text{whenever} \qquad 0 < (x - x_0)^2 + (y - y_0)^2 < \delta^2.$$

So conditions (2) hold.

Let us now start with conditions (2). For each positive number ε there exist positive numbers δ_1 and δ_2 such that

$$|u(x,y) - u_0| < \frac{\varepsilon}{2} \quad \text{whenever} \quad 0 < (x - x_0)^2 + (y - y_0)^2 < \delta_1{}^2$$

and

$$|v(x,y) - v_0| < \frac{\varepsilon}{2} \quad \text{whenever} \quad 0 < (x - x_0)^2 + (y - y_0)^2 < \delta_2{}^2.$$

Let δ denote the smaller of the two numbers δ_1 and δ_2. Since

$$|u(x,y) - u_0 + i[v(x,y) - v_0]| \leqq u\,|(x,y) - u_0| + |v(x,y) - v_0|,$$

we have

$$|u(x,y) + iv(x,y) - (u_0 + iv_0)| < \varepsilon \quad \text{whenever} \quad 0 < |x + iy - (x_0 + iy_0)| < \delta.$$

This is statement (1), and the proof of the theorem is complete.

Theorem 2. *Suppose that*

$$\lim_{z \to z_0} f(z) = w_0 \quad \textit{and} \quad \lim_{z \to z_0} F(z) = W_0. \tag{3}$$

Then

$$\lim_{z \to z_0} [f(z) + F(z)] = w_0 + W_0, \tag{4}$$

$$\lim_{z \to z_0} [f(z)F(z)] = w_0 W_0, \tag{5}$$

and if $W_0 \neq 0$,

$$\lim_{z \to z_0} \frac{f(z)}{F(z)} = \frac{w_0}{W_0}. \tag{6}$$

This fundamental theorem can be established directly from the definition (Sec. 11) of the limit of a function of a complex variable. But with the aid of Theorem 1 it follows almost immediately from theorems on limits of real-valued functions of two real variables.

Consider the proof of property (5), for example, and write

$$f(z) = u(x,y) + iv(x,y), \qquad F(z) = U(x,y) + iV(x,y),$$

$$z_0 = x_0 + iy_0, \qquad w_0 = u_0 + iv_0, \qquad W_0 = U_0 + iV_0.$$

Then, according to hypotheses (3) and Theorem 1, the limits as (x,y) approaches (x_0,y_0) of the functions u, v, U, and V exist and have the values u_0, v_0, U_0, and V_0, respectively. The real and imaginary parts of the function

$$f(z)F(z) = u(x,y)U(x,y) - v(x,y)V(x,y) + i[v(x,y)U(x,y) + u(x,y)V(x,y)]$$

therefore have the limits $u_0 U_0 - v_0 V_0$ and $v_0 U_0 + u_0 V_0$, respectively, as (x,y) approaches (x_0,y_0). Hence $f(z)F(z)$ has the limit

$$u_0 U_0 - v_0 V_0 + i(v_0 U_0 + u_0 V_0)$$

as z approaches z_0, and this is equal to $w_0 W_0$. Property (5) is thus proved.

From the definition of limit we see that for any z_0

$$\lim_{z \to z_0} z = z_0$$

since we can write $\delta = \varepsilon$ when $f(z) = z$. Then, by property (5) and induction,

$$\lim_{z \to z_0} z^n = z_0{}^n \qquad (n = 1, 2, \ldots).$$

Also,

$$\lim_{z \to z_0} c = c$$

where c is a complex constant. So, in view of properties (4) and (5), it follows that the limit of a polynomial

$$P(z) = a_0 + a_1 z + a_2 z^2 + \cdots + a_n z^n \qquad (a_n \neq 0)$$

as z approaches a point z_0 is the value of the polynomial at that point:

$$\lim_{z \to z_0} P(z) = P(z_0). \tag{7}$$

Another useful property of limits is that

$$\text{if} \quad \lim_{z \to z_0} f(z) = w_0, \qquad \text{then} \qquad \lim_{z \to z_0} |f(z)| = |w_0|. \tag{8}$$

This is easily proved by using the definition of limit and the fact that

$$||f(z)| - |w_0|| \leqq |f(z) - w_0|.$$

Finally, we remark that the results of this section involve points in the *finite* plane only. As indicated in Sec. 8, the point at infinity is not to be considered unless specifically mentioned.

EXERCISES

1. For each of the functions defined below describe the domain of definition that is understood:

(a) $f(z)=\dfrac{1}{z^2+1}$; (b) $f(z)=\text{Arg}\left(\dfrac{1}{z}\right)$; (c) $f(z)=\dfrac{z}{z+\bar{z}}$;

(d) $f(z)=\dfrac{1}{1-|z|^2}$.

Ans. (a) $z\neq\pm i$; (c) $\text{Re}\, z\neq 0$.

2. Sketch the domain of definition of the function

$$g(z)=\frac{y}{x}+\frac{i}{1-y}.$$

For all z in the domain of definition of the function f described in equation (2), Sec. 9, show that $g(z)=f(z)$.

3. Write the function $f(z)=z^3+z+1$ in the form $f(z)=u(x,y)+iv(x,y)$.

4. Suppose that $f(z)=x^2-y^2-2y+i(2x-2xy)$. Using identities (9), Sec. 3, express the right-hand side here in terms of z and simplify.

Ans. $\bar{z}^2+2iz$.

5. In Fig. 12 consider the point z as it traverses the circle $x^2+y^2=c^2$ counterclockwise, starting at the point $(c,0)$. Describe the corresponding path taken by the point $w=\sqrt{x^2+y^2}-iy$.

6. Let a, c, and z_0 denote complex constants. Use definition (2), Sec. 11, of limit to prove that (a) $\lim_{z\to z_0} c=c$; (b) $\lim_{z\to z_0}(az+c)=az_0+c$;

(c) $\lim_{z\to z_0}(z^2+c)=z_0{}^2+c$; (d) $\lim_{z\to z_0}\text{Re}\, z=\text{Re}\, z_0$; (e) $\lim_{z\to z_0}\bar{z}=\bar{z}_0$;

(f) $\lim_{z\to 1-i}[x+i(2x+y)]=1+i$; (g) $\lim_{z\to 0}(\bar{z}^2/z)=0$.

7. Prove statement (4) in Theorem 2, Sec. 12 (a) by using Theorem 1, Sec. 12, and properties of limits of real-valued functions; (b) directly from definition (2), Sec. 11, of limit.

8. Let n be a positive integer and let $P(z)$ and $Q(z)$ be polynomials where $Q(z_0)\neq 0$. Use Theorem 2, Sec. 12, and established limits to find

(a) $\lim_{z\to z_0}\dfrac{1}{z^n}\ (z_0\neq 0)$; (b) $\lim_{z\to i}\dfrac{iz^3-1}{z+i}$; (c) $\lim_{z\to z_0}\dfrac{P(z)}{Q(z)}$.

Ans. (a) $1/z_0{}^n$; (b) 0; (c) $P(z_0)/Q(z_0)$.

9. Write $\Delta z=z-z_0$ and show that $\lim_{z\to z_0}f(z)=w_0$ if and only if $\lim_{\Delta z\to 0}f(z_0+\Delta z)=w_0$.

10. Show that $\lim_{z\to z_0}f(z)g(z)=0$ if $\lim_{z\to z_0}f(z)=0$ and if there exists a positive number M such that $|g(z)|\leqq M$ for all z in some neighborhood containing z_0.

11. Interpret statement (5), Sec. 11, for the case in which (a) w_0 is the point at infinity; (b) both z_0 and w_0 are the point at infinity.

12. Using a definition of limit involving the point at infinity, prove that

(*a*) $\lim\limits_{z\to\infty} \dfrac{1}{z^2+1} = 0$; (*b*) $\lim\limits_{z\to 1} \dfrac{1}{(z-1)^3} = \infty$; (*c*) $\lim\limits_{z\to\infty} 3z^2 = \infty$.

13. Show that a limit of the type (6), Sec. 11, is unique.

14. Prove property (8), Sec. 12.

15. Let the function $f(z) = e^x e^{iy}$ $(z = x + iy)$ be defined on the entire complex plane. (*a*) Show that if $w_0 \neq 0$, there is an infinite number of points z in any neighborhood of ∞ such that $f(z) = w_0$. (*b*) Show that $\lim\limits_{z\to\infty} f(z)$ has no value, including ∞.

16. Another interpretation of the equation $w = f(z)$ is that of a *vector field* in the domain of definition of f. The equation assigns a vector w with components $u(x,y)$ and $v(x,y)$ to each point (x,y) in the domain of definition. Indicate graphically the vector fields represented by the equations

(*a*) $w = iz$; (*b*) $w = \dfrac{z}{|z|}$.

13. Continuity

A function f is *continuous* at a point z_0 if all three of the following conditions are satisfied:

$$\lim_{z\to z_0} f(z) \text{ exists,} \tag{1}$$

$$f(z_0) \text{ exists,} \tag{2}$$

$$\lim_{z\to z_0} f(z) = f(z_0). \tag{3}$$

Observe that statement (3) actually contains statements (1) and (2); for the existence of the quantity on each side of the equation there is implicit. Statement (3) says that for each positive number ε there is a positive number δ such that

$$|f(z) - f(z_0)| < \varepsilon \qquad \text{whenever} \qquad |z - z_0| < \delta. \tag{4}$$

A function of a complex variable is said to be continuous in a region R if it is continuous at each point in R.

If two functions are continuous at a point, their sum and product are also continuous at that point; their quotient is continuous at any such point where the denominator is not zero. These observations are immediate consequences of Theorem 2, Sec. 12. Note, too, that a polynomial is continuous in the entire plane because of equation (7), Sec. 12.

Let us now show directly from definition (4) that *a composition of continuous functions is continuous.* To be precise, let a function f be defined on a neighborhood of a point z_0 and let the image of that neighborhood be contained in a region on which a function g is defined. Then the composition $g[f(z)]$ is defined for all z in the neighborhood of z_0. Now if f is continuous at z_0 and g is continuous at the point $f(z_0)$, it follows that the composition $g[f(z)]$ is continuous at z_0. For, in view of the continuity of g, we know that for each positive number ε there is a positive number γ such that

$$|g[f(z)] - g[f(z_0)]| < \varepsilon \qquad \text{whenever} \qquad |f(z) - f(z_0)| < \gamma.$$

But corresponding to γ there exists a positive number δ such that the latter inequality is satisfied whenever $|z - z_0| < \delta$. The continuity of the composition is thus established.

From Theorem 1, Sec. 12, it follows that a function f of a complex variable is continuous at a point $z_0 = (x_0,y_0)$ if and only if its component functions u and v are continuous there.

With this result, we know, for example, that the function $f(z) = xy^2 + i(2x - y)$ is continuous everywhere in the complex plane because the component functions are polynomials in x and y and are therefore continuous at each point (x,y). Likewise, the function $f(z) = e^{xy} + i \sin (x^2 - 2xy^3)$ is continuous for all z because of the continuity of polynomials in x and y as well as the continuity of the exponential and sine functions.

Various properties of continuous functions of a complex variable can be deduced from corresponding properties of continuous real-valued functions of two real variables.[1]

Suppose, for example, that the function $f(z) = u(x,y) + iv(x,y)$ is continuous in a region R which is both closed and bounded. The function $\sqrt{[u(x,y)]^2 + [v(x,y)]^2}$ is then continuous in R and thus reaches a maximum value somewhere in that region. That is, f is *bounded* in R and $|f(z)|$ reaches a maximum value somewhere in R. To be precise, there exists a positive number M such that

$$|f(z)| \leqq M \qquad \text{for all } z \text{ in } R,$$

where equality holds for at least one such z.

Another result which follows from the corresponding one for real-valued functions of two real variables is that a function f which is continuous in a closed bounded region R is *uniformly continuous* there. That is, a single value of δ, independent of z_0, may be chosen such that condition (4) is satisfied at each point z_0 in R.

[1] For such properties quoted here, see, for example, A. E. Taylor and W. R. Mann, "Advanced Calculus," 2d ed., pp. 135–136 and 558–561, 1972.

14. Derivatives

Let f be a function whose domain of definition contains a neighborhood of a point z_0. We define the *derivative* of f at z_0, written $f'(z_0)$, by the equation

$$f'(z_0) = \lim_{z \to z_0} \frac{f(z) - f(z_0)}{z - z_0}, \tag{1}$$

provided the limit here exists. The function f is said to be *differentiable* at z_0 when its derivative at z_0 exists.

By expressing z in definition (1) in terms of the new complex variable

$$\Delta z = z - z_0,$$

we can write that definition as

$$f'(z_0) = \lim_{\Delta z \to 0} \frac{f(z_0 + \Delta z) - f(z_0)}{\Delta z}. \tag{2}$$

Note that, because f is defined on a neighborhood of z_0, the number $f(z_0 + \Delta z)$ is always defined for $|\Delta z|$ sufficiently small.

When taking form (2) of the definition of derivative, we often drop the subscript on z_0 and introduce the number

$$\Delta w = f(z + \Delta z) - f(z)$$

which denotes the change in the value $w = f(z)$ corresponding to a change Δz in the variable z. Then, if we write dw/dz for $f'(z)$, equation (2) becomes

$$\frac{dw}{dz} = \lim_{\Delta z \to 0} \frac{\Delta w}{\Delta z}. \tag{3}$$

Suppose, for example, that $f(z) = z^2$. For any point z,

$$\lim_{\Delta z \to 0} \frac{\Delta w}{\Delta z} = \lim_{\Delta z \to 0} \frac{(z + \Delta z)^2 - z^2}{\Delta z} = \lim_{\Delta z \to 0} (2z + \Delta z) = 2z$$

since $2z + \Delta z$ is a polynomial in Δz. Hence $dw/dz = 2z$, or $f'(z) = 2z$.

Let us now examine the function $f(z) = |z|^2$. Here

$$\frac{\Delta w}{\Delta z} = \frac{|z + \Delta z|^2 - |z|^2}{\Delta z} = \frac{(z + \Delta z)(\bar{z} + \overline{\Delta z}) - z\bar{z}}{\Delta z} = \bar{z} + \overline{\Delta z} + z\frac{\overline{\Delta z}}{\Delta z}.$$

When $z = 0$, $\Delta w/\Delta z = \overline{\Delta z}$; hence $dw/dz = 0$ at the origin. If the limit of $\Delta w/\Delta z$ exists when $z \neq 0$, that limit may be found by letting Δz approach 0 in any manner. If, in particular, Δz approaches 0 through real values, then $\overline{\Delta z} = \Delta z$ and the limit of $\Delta w/\Delta z$ is evidently $\bar{z} + z$. But if Δz approaches 0 through pure imaginary values, $\overline{\Delta z} = -\Delta z$ and the limit is found to be $\bar{z} - z$. Since any limit is unique, it follows that dw/dz does not exist when $z \neq 0$. Consequently, dw/dz exists only at the origin.

This example shows that a function can be differentiable at a certain point but nowhere else in any neighborhood of that point. It also shows that the real and imaginary parts of a function of a complex variable can have continuous partial derivatives of all orders at a point and yet the function may not be differentiable there. For the real and imaginary parts of $f(z) = |z|^2$ are

$$u(x,y) = x^2 + y^2 \qquad \text{and} \qquad v(x,y) = 0,$$

respectively.

Note that the function $f(z) = |z|^2$ is continuous at each point in the plane. So the continuity of a function at a point does not imply the existence of a derivative there. It is, however, true that *the existence of the derivative of a function at a point implies the continuity of the function there.* To see this, we assume that $f'(z_0)$ exists and write

$$\lim_{z \to z_0} [f(z) - f(z_0)] = \lim_{z \to z_0} \frac{f(z) - f(z_0)}{z - z_0} \lim_{z \to z_0} (z - z_0) = 0$$

from which it follows that

$$\lim_{z \to z_0} f(z) = f(z_0).$$

This is the statement of continuity of f at z_0.

15. Differentiation Formulas

Our definition of derivative is identical in form to that of the derivative of a real-valued function of a real variable. In fact, the basic differentiation formulas given below can be derived from our definition together with various theorems on limits by essentially the same steps used in the calculus of real variables. In these formulas the derivative of a function f at a point z will be denoted by either $f'(z)$ or $d[f(z)]/dz$, depending on which notation is more convenient.

Let c be a complex constant and let f be a function whose derivative exists at a point z. It is easy to show that

$$\frac{d}{dz} c = 0, \qquad \frac{d}{dz} z = 1, \qquad \frac{d}{dz} [cf(z)] = cf'(z). \tag{1}$$

Also, if the derivatives of two functions f and F exist at a point z, then

$$\frac{d}{dz}[f(z) + F(z)] = f'(z) + F'(z), \tag{2}$$

$$\frac{d}{dz}[f(z)F(z)] = f(z)F'(z) + F(z)f'(z), \tag{3}$$

and when $F(z) \neq 0$,

$$\frac{d}{dz}\left[\frac{f(z)}{F(z)}\right] = \frac{F(z)f'(z) - f(z)F'(z)}{[F(z)]^2}. \tag{4}$$

If n is a positive integer, then at each point z

$$\frac{d}{dz} z^n = nz^{n-1}. \tag{5}$$

This formula is also true when n is a negative integer if $z \neq 0$.

Let us derive formula (3). To do this, we write

$$\Delta w = f(z + \Delta z) - f(z), \qquad \Delta W = F(z + \Delta z) - F(z)$$

and find that

$$\frac{f(z + \Delta z)F(z + \Delta z) - f(z)F(z)}{\Delta z} = f(z)\frac{\Delta W}{\Delta z} + F(z)\frac{\Delta w}{\Delta z} + \Delta w \frac{\Delta W}{\Delta z}.$$

Note that f is continuous at z because it is differentiable there; so Δw tends to 0 as Δz approaches 0. In view of the limit theorems for sums and products, formula (3) then follows.

There is also a chain rule for differentiating composite functions. Suppose that f has a derivative at z_0 and g has a derivative at the point $f(z_0)$. Then the function $F(z) = g[f(z)]$ has a derivative at z_0, and

$$F'(z_0) = g'[f(z_0)]f'(z_0). \tag{6}$$

Note that if we write $w = f(z)$ and $W = g(w)$, then $W = F(z)$ and the chain rule becomes

$$\frac{dW}{dz} = \frac{dW}{dw}\frac{dw}{dz}.$$

To illustrate this, we compute the derivative of $(2z^2 + i)^5$. We write $w = 2z^2 + i$, $W = w^5$ and find that

$$\frac{d}{dz}(2z^2 + i)^5 = 5w^4 4z = 20z(2z^2 + i)^4.$$

To start the proof of formula (6), choose a specific point z_0 such that $f'(z_0)$ exists. Write $w_0 = f(z_0)$ and also assume that $g'(w_0)$ exists. Now there is some ε neighborhood $|w - w_0| < \varepsilon$ such that for all points w in that neighborhood we can define a function

$$\Phi(w) = \begin{cases} \dfrac{g(w) - g(w_0)}{w - w_0} - g'(w_0) & (w \neq w_0), \\ 0 & (w = w_0). \end{cases} \tag{7}$$

Note that, in view of definition (1), Sec. 14, of derivative,

$$\lim_{w \to w_0} \Phi(w) = 0. \tag{8}$$

Since $f'(z_0)$ exists and f is therefore continuous at z_0, we can choose a positive number δ such that the point $f(z)$ lies in the ε neighborhood $|w - w_0| < \varepsilon$ of w_0 if z lies in the δ neighborhood $|z - z_0| < \delta$ of z_0. Thus it is legitimate to replace w in expression (7) by $f(z)$ when z is any point in the neighborhood $|z - z_0| < \delta$. So with the substitution $w = f(z)$, as well as $w_0 = f(z_0)$, expression (7) yields

$$\frac{g[f(z)] - g[f(z_0)]}{z - z_0} = \{g'[f(z_0)] + \Phi[f(z)]\} \frac{f(z) - f(z_0)}{z - z_0} \qquad (0 < |z - z_0| < \delta), \tag{9}$$

where we must stipulate that $z \neq z_0$ so that we are not dividing by zero. Now f is continuous at z_0, and Φ is continuous at the point $w_0 = f(z_0)$ because of equation (8) and the fact that $\Phi(w_0) = 0$. The composition $\Phi[f(z)]$ is thus continuous at z_0, and

$$\lim_{z \to z_0} \Phi[f(z)] = 0.$$

So equation (9) becomes

$$F'(z_0) = g'[f(z_0)]f'(z_0)$$

in the limit as z approaches z_0.

EXERCISES

1. Show that the function $f(z) = |z|^2$ is continuous in the entire complex plane.

2. Using results in Sec. 15, show that the derivative of a polynomial

$$P(z) = a_0 + a_1 z + a_2 z^2 + \cdots + a_n z^n \qquad (n \geqq 1, a_n \neq 0)$$

exists everywhere and that $P'(z) = a_1 + 2a_2 z + \cdots + na_n z^{n-1}$.

3. Use results in Sec. 15 to find $f'(z)$ when (*a*) $f(z) = 3z^2 - 2z + 4$;
(*b*) $f(z) = (1 - 4z^2)^3$; (*c*) $f(z) = (z - 1)/(2z + 1)$ $(z \neq -1/2)$;
(*d*) $f(z) = [(1 + z^2)^4]/z^2$ $(z \neq 0)$.

4. Apply the definition of derivative directly to prove that $f'(z) = -1/z^2$ when $f(z) = 1/z$, provided $z \neq 0$.

5. Derive formula (2), Sec. 15.

6. Use either induction or the binomial formula (Exercise 17, Sec. 2) to derive formula (5), Sec. 15, for the derivative of z^n when n is a positive integer.

7. Extend the result of Exercise 6 to include the case in which n is a negative integer and $z \neq 0$.

8. Apply the definition of derivative to show that if $f(z) = \text{Re } z$, then $f'(z)$ does not exist anywhere.

9. Show that the function $f(z) = \bar{z}$ is nowhere differentiable.

10. Determine whether the function $f(z) = \text{Im } z$ has a derivative at any point.

11. Show that if a function f is continuous at a point z_0 in some domain and $f(z_0) \neq 0$, then there exists a neighborhood of z_0 throughout which $f(z) \neq 0$.

Suggestion: Write the first inequality in definition (4), Sec. 13, of continuity as $|f(z_0) - f(z)| < |f(z_0)|/2$, where $\varepsilon = |f(z_0)|/2$. Then note the contradiction if $f(z) = 0$ at some point z in every neighborhood of z_0.

12. Modify definition (4), Sec. 13, of continuity for the case in which $z_0 = \infty$. Do the same when $f(z_0) = \infty$ and also when z_0 and $f(z_0)$ are both the point at infinity. Then show that each of the following functions is *continuous everywhere in the extended plane*:

$$(a)\quad f(z) = \begin{cases} \dfrac{1}{z} & (z \neq 0, \infty), \\ \infty & (z = 0), \\ 0 & (z = \infty); \end{cases}$$

$$(b)\quad f(z) = \begin{cases} 2z + 1 & (z \neq \infty), \\ \infty & (z = \infty). \end{cases}$$

16. The Cauchy-Riemann Equations

Suppose that a function f is defined on a neighborhood of a point z_0 by the equation

$$f(z) = u(x,y) + iv(x,y). \tag{1}$$

In this section we obtain some necessary conditions on the component functions u and v in order that the derivative of f exist at z_0.

Assume that the derivative

$$f'(z_0) = \lim_{\Delta z \to 0} \frac{f(z_0 + \Delta z) - f(z_0)}{\Delta z} \tag{2}$$

exists. Writing $z_0 = x_0 + iy_0$ and $\Delta z = \Delta x + i\,\Delta y$, we then have, by Theorem 1 of Sec. 12, the expressions

$$\text{(3)} \qquad \operatorname{Re}[f'(z_0)] = \lim_{(\Delta x, \Delta y)\to(0,0)} \operatorname{Re}\left[\frac{f(z_0+\Delta z)-f(z_0)}{\Delta z}\right],$$

$$\text{(4)} \qquad \operatorname{Im}[f'(z_0)] = \lim_{(\Delta x, \Delta y)\to(0,0)} \operatorname{Im}\left[\frac{f(z_0+\Delta z)-f(z_0)}{\Delta z}\right]$$

where

$$\frac{f(z_0+\Delta z)-f(z_0)}{\Delta z} = \frac{u(x_0+\Delta x, y_0+\Delta y) - u(x_0,y_0) + i[v(x_0+\Delta x, y_0+\Delta y) - v(x_0,y_0)]}{\Delta x + i\Delta y}.$$

If, in particular, $\Delta z = \Delta x + i0$, the point $z_0 + \Delta z$ is $(x_0 + \Delta x, y_0)$ (Fig. 16) and

$$\operatorname{Re}[f'(z_0)] = \lim_{\Delta x\to 0} \frac{u(x_0+\Delta x,y_0) - u(x_0,y_0)}{\Delta x},$$

$$\operatorname{Im}[f'(z_0)] = \lim_{\Delta x\to 0} \frac{v(x_0+\Delta x,y_0) - v(x_0,y_0)}{\Delta x}.$$

That is,

$$\text{(5)} \qquad f'(z_0) = u_x(x_0,y_0) + iv_x(x_0,y_0),$$

where $u_x(x_0,y_0)$ and $v_x(x_0,y_0)$ are the first partial derivatives with respect to x of the functions u and v at (x_0,y_0).

On the other hand, if $\Delta z = 0 + i\,\Delta y$, the point $z_0 + \Delta z$ is $(x_0, y_0 + \Delta y)$. In this case the existence of $f'(z_0)$ implies that the first partial derivatives $u_y(x_0,y_0)$ and $v_y(x_0,y_0)$ exist and that

$$\text{(6)} \qquad f'(z_0) = v_y(x_0,y_0) - iu_y(x_0,y_0).$$

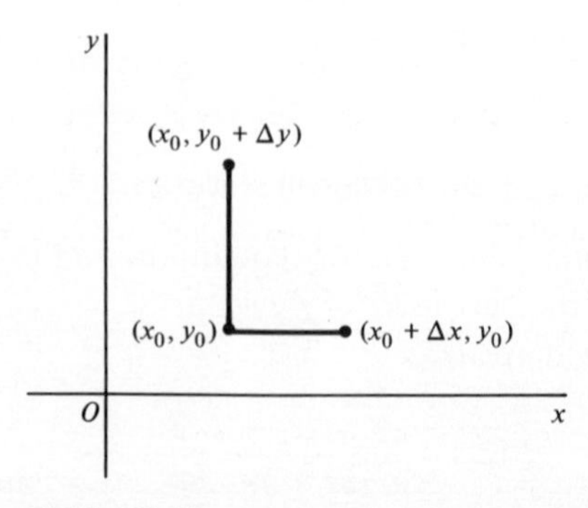

FIGURE 16

Equations (5) and (6) are not only formulas for finding $f'(z_0)$ in terms of partial derivatives of the component functions u and v; they also provide necessary conditions for the existence of $f'(z_0)$. For, equating the real parts and the imaginary parts on the right-hand sides of these equations, we find that the existence of $f'(z_0)$ requires that

$$u_x(x_0,y_0) = v_y(x_0,y_0) \quad \text{and} \quad u_y(x_0,y_0) = -v_x(x_0,y_0). \tag{7}$$

Equations (7) are the *Cauchy-Riemann equations,* so named in honor of the French mathematician A. L. Cauchy (1789–1857), who discovered and used them, and in honor of the German mathematician G. F. B. Riemann (1826–1866) who made them fundamental in his development of the theory of functions of a complex variable.

We summarize the above results as follows.

Theorem. *Suppose that*

$$f(z) = u(x,y) + iv(x,y)$$

and $f'(z_0)$ exists at a point $z_0 = x_0 + iy_0$. Then the first partial derivatives of u and v with respect to x and y exist at (x_0,y_0), and they satisfy the Cauchy-Riemann equations (7) *at that point. Also, $f'(z_0)$ is given in terms of these partial derivatives by either of the equations* (5) *or* (6).

To illustrate the theorem, consider the function

$$f(z) = z^2 = x^2 - y^2 + i2xy.$$

We proved in Sec. 14 that its derivative exists everywhere; in fact, $f'(z) = 2z$. Hence the Cauchy-Riemann equations must be satisfied everywhere. To verify this, we note that $u(x,y) = x^2 - y^2$ and $v(x,y) = 2xy$. Therefore

$$u_x(x,y) = 2x = v_y(x,y), \qquad u_y(x,y) = -2y = -v_x(x,y).$$

Also, according to equations (5) or (6),

$$f'(z) = 2x + i2y = 2z.$$

Since the Cauchy-Riemann equations are necessary for the existence of $f'(z_0)$, they can often be used to locate points where a given function does not have a derivative. Consider, for example, the function $f(z) = |z|^2$ which was also discussed in Sec. 14. In this case $u(x,y) = x^2 + y^2$ and $v(x,y) = 0$, and so $u_x(x,y) = 2x$ and $v_y(x,y) = 0$ while $u_y(x,y) = 2y$ and $v_x(x,y) = 0$. Observe that the Cauchy-Riemann equations are not satisfied unless $x = y = 0$, and $f'(z)$ cannot therefore exist if $z \neq 0$. Note that the above theorem does not guarantee the existence of $f'(0)$; the theorem of the following section does, however, do this.

17. Sufficient Conditions

Satisfaction of the Cauchy-Riemann equations at a point $z_0 = (x_0,y_0)$ is *not* sufficient for the existence of the derivative of f at that point. (See Exercise 6, Sec. 18.) But with certain continuity conditions, we have the following useful theorem.

Theorem. *Let the function*

$$f(z) = u(x,y) + iv(x,y)$$

be defined throughout some ε *neighborhood of the point* $z_0 = x_0 + iy_0$. *Suppose that the first partial derivatives of the functions* u *and* v *with respect to* x *and* y *exist in that neighborhood and are continuous at* (x_0,y_0). *Then if those partial derivatives satisfy the Cauchy-Riemann equations at* (x_0,y_0), *the derivative* $f'(z_0)$ *exists.*

To start the proof, write $\Delta z = \Delta x + i\,\Delta y$, where $0 < |\Delta z| < \varepsilon$, and $\Delta w = f(z_0 + \Delta z) - f(z_0)$. Evidently then,

$$\Delta w = \Delta u + i\,\Delta v$$

where

$$\begin{aligned} \Delta u &= u(x_0 + \Delta x,\ y_0 + \Delta y) - u(x_0,y_0), \\ \Delta v &= v(x_0 + \Delta x,\ y_0 + \Delta y) - v(x_0,y_0). \end{aligned} \tag{1}$$

But, in view of the continuity of the first partial derivatives of u and v at the point (x_0,y_0),

$$\begin{aligned} \Delta u &= u_x(x_0,y_0)\,\Delta x + u_y(x_0,y_0)\,\Delta y + \varepsilon_1\sqrt{(\Delta x)^2 + (\Delta y)^2}, \\ \Delta v &= v_x(x_0,y_0)\,\Delta x + v_y(x_0,y_0)\,\Delta y + \varepsilon_2\sqrt{(\Delta x)^2 + (\Delta y)^2} \end{aligned} \tag{2}$$

where ε_1 and ε_2 tend to 0 as $(\Delta x,\Delta y)$ approaches $(0,0)$. Hence

$$\begin{aligned} \Delta w = {} & u_x(x_0,y_0)\,\Delta x + u_y(x_0,y_0)\,\Delta y + \varepsilon_1\sqrt{(\Delta x)^2 + (\Delta y)^2} \\ & + i[v_x(x_0,y_0)\,\Delta x + v_y(x_0,y_0)\,\Delta y + \varepsilon_2\sqrt{(\Delta x)^2 + (\Delta y)^2}]. \end{aligned} \tag{3}$$

The existence of expressions of type (2) for functions of two real variables with continuous first partial derivatives is established in the calculus of real variables in connection with differentials.[1]

[1] See, for instance, A. E. Taylor and W. R. Mann, "Advanced Calculus," 2d ed., pp. 161–162 and 213–214, 1972.

Since the Cauchy-Riemann equations are satisfied at (x_0,y_0), we can replace $u_y(x_0,y_0)$ by $-v_x(x_0,y_0)$ and $v_y(x_0,y_0)$ by $u_x(x_0,y_0)$ in equation (3) and then divide through by Δz to get

$$\frac{\Delta w}{\Delta z} = u_x(x_0,y_0) + iv_x(x_0,y_0) + (\varepsilon_1 + i\varepsilon_2)\frac{\sqrt{(\Delta x)^2 + (\Delta y)^2}}{\Delta z}. \tag{4}$$

But $\sqrt{(\Delta x)^2 + (\Delta y)^2} = |\Delta z|$, and so

$$\left|\frac{\sqrt{(\Delta x)^2 + (\Delta y)^2}}{\Delta z}\right| = 1.$$

This means that the last term on the right in equation (4) tends to 0 as Δz approaches 0. The limit of the left-hand side of equation (4) therefore exists, and

$$f'(z_0) = u_x(x_0,y_0) + iv_x(x_0,y_0). \tag{5}$$

To illustrate the theorem, let us consider the function

$$f(z) = e^x e^{iy}$$

where $u(x,y) = e^x \cos y$ and $v(x,y) = e^x \sin y$. It is readily seen that the conditions in the theorem are satisfied everywhere in the complex plane. The derivative $f'(z)$ therefore exists everywhere, and

$$f'(z) = e^x \cos y + ie^x \sin y = e^x e^{iy}.$$

Note that $f'(z) = f(z)$.

As another illustration, it follows from the theorem of this section that the function $f(z) = |z|^2$ has a derivative at $z = 0$; in fact, $f'(0) = 0 + i0 = 0$. We saw in Secs. 14 and 16 that this function does not have a derivative anywhere else in the complex plane.

18. The Cauchy-Riemann Equations in Polar Form

When $z_0 \neq 0$, the main results of the previous two sections are exhibited in the context of polar coordinates by means of the coordinate transformation

$$x = r\cos\theta, \qquad y = r\sin\theta. \tag{1}$$

Suppose that $w = f(z)$. Depending on whether we write $z = x + iy$ or $z = re^{i\theta}$, the real and imaginary parts of $w = u + iv$ are expressed in terms of either the variables x and y or r and θ. Using the chain rule for differentiating real-valued functions of two real variables, we obtain the fact that the first partial derivatives of u and v with respect to x and y are continuous functions of (x,y) at a nonzero point if their first partial derivatives with respect to r and θ are continuous functions of (r,θ) at that point, and conversely. Moreover, the Cauchy-Riemann equations (7), Sec. 16, take the form

$$u_r(r_0,\theta_0) = \frac{1}{r_0} v_\theta(r_0,\theta_0), \qquad \frac{1}{r_0} u_\theta(r_0,\theta_0) = -v_r(r_0,\theta_0) \tag{2}$$

in polar coordinates; and if f is differentiable at (r_0,θ_0),

$$\begin{aligned} f'(z_0) &= e^{-i\theta_0}[u_r(r_0,\theta_0) + iv_r(r_0,\theta_0)] \\ &= \frac{1}{r_0} e^{-i\theta_0}[v_\theta(r_0,\theta_0) - iu_\theta(r_0,\theta_0)]. \end{aligned} \tag{3}$$

Verification of these facts forms the contents of Exercises 7, 8, and 9 of this section.

We now state the following alternate form of the theorem in Sec. 17 when $z_0 \neq 0$.

Theorem. *Let the function*

$$f(z) = u(r,\theta) + iv(r,\theta)$$

be defined throughout some ε neighborhood of the nonzero point $z_0 = r_0 e^{i\theta_0}$. Suppose that the first partial derivatives of the functions u and v with respect to r and θ exist in that neighborhood and are continuous functions of (r,θ) at (r_0,θ_0). Then if those partial dervatives satisfy the Cauchy-Riemann equations in polar form at (r_0,θ_0), the derivative $f'(z_0)$ exists.

To illustrate these results, consider the function

$$f(z) = \frac{1}{z} = \frac{1}{re^{i\theta}}.$$

Observe that $u(r,\theta) = \cos\theta/r$ and $v(r,\theta) = -\sin\theta/r$ and that the conditions in the theorem are satisfied at any nonzero point $z = re^{i\theta}$ in the plane. Hence the derivative exists there; and, according to the first of formulas (3),

$$f'(z) = e^{-i\theta}\left(-\frac{\cos\theta}{r^2} + i\frac{\sin\theta}{r^2}\right) = -\frac{1}{(re^{i\theta})^2} = -\frac{1}{z^2}.$$

EXERCISES

1. Use the theorem in Sec. 16 to show that $f'(z)$ does not exist at any point if $f(z)$ is (*a*) $\bar{z}$; (*b*) $z - \bar{z}$; (*c*) $2x + ixy^2$; (*d*) $e^x e^{-iy}$.

2. Using the theorem in Sec. 17 and formula (5) of that section, show that $f'(z)$ and its derivative $f''(z)$ exist everywhere and find $f'(z)$ and $f''(z)$ when
(*a*) $f(z) = iz + 2$; (*b*) $f(z) = e^{-x}e^{-iy}$; (*c*) $f(z) = z^3$;
(*d*) $f(z) = \cos x \cosh y - i \sin x \sinh y$.

Ans. (*b*) $f'(z) = -f(z)$, $f''(z) = f(z)$; (*d*) $f''(z) = -f(z)$.

3. From results given in Secs. 16 and 17 determine where $f'(z)$ exists and find its value when (*a*) $f(z) = 1/z$; (*b*) $f(z) = x^2 + iy^2$; (*c*) $f(z) = z \operatorname{Im} z$.

Ans. (*a*) $f'(z) = -1/z^2$ $(z \neq 0)$; (*b*) $f'(x + ix) = 2x$; (*c*) $f'(0) = 0$.

4. Using the theorem of Sec. 18, show that the function $g(z) = \sqrt{r}\,e^{i\theta/2}$ $(r > 0, -\pi < \theta < \pi)$ has a derivative at each point in its domain of definition and that $g'(z) = 1/[2g(z)]$.

5. If $f(z) = x^3 - i(y-1)^3$, then $u_x(x,y) + iv_x(x,y) = 3x^2$. Why is it true that $f'(z) = 3x^2$ only at the point $z = i$?

6. Show that the function

$$f(z) = \begin{cases} \dfrac{(\bar{z})^2}{z} & (z \neq 0), \\ 0 & (z = 0), \end{cases}$$

is not differentiable at $z = 0$ but that the Cauchy-Riemann equations are satisfied there.

Suggestion: Using definition (1), Sec. 14, of derivative, let z approach 0 along one of the axes and then let it approach 0 along the line $y = x$.

7. Use coordinate transformation (1), Sec. 18, or its inverse, and the chain rule for differentiating real-valued functions of two real variables to obtain the formulas

$$u_x = u_r \cos\theta - \frac{1}{r} u_\theta \sin\theta,$$

$$u_y = u_r \sin\theta + \frac{1}{r} u_\theta \cos\theta$$

as well as similar formulas for v_x and v_y. Conclude that the first partial derivatives of u and v with respect to x and y are continuous functions of (x,y) at a nonzero point if and only if the first partial derivatives of u and v with respect to r and θ are continuous functions of (r,θ) at that point.

8. Using results in Exercise 7, obtain the polar form (2), Sec. 18, of the Cauchy-Riemann equations (7), Sec. 16.

Suggestion: Transform the equations $u_x = v_y$, $u_y = -v_x$ into

$$\left(u_r - \frac{1}{r} v_\theta\right) \cos\theta = \left(\frac{1}{r} u_\theta + v_r\right) \sin\theta,$$

$$\left(u_r - \frac{1}{r} v_\theta\right) \sin\theta = -\left(\frac{1}{r} u_\theta + v_r\right) \cos\theta.$$

9. Using results in Exercises 7 and 8, obtain formulas (3), Sec. 18, for $f'(z_0)$ in polar coordinates from formulas (5) or (6), Sec. 16.

19. Analytic Functions

We are now ready to introduce the concept of an analytic function. A function f of the complex variable z is *analytic* at a point z_0 if its derivative exists not only at z_0 but at each point z in some neighborhood of z_0. It is analytic in a region R if it is analytic at every point in R. The term *holomorphic* is also used in the literature to denote analyticity.

The function $f(z) = |z|^2$, for instance, is not analytic at any point since its derivative exists only at $z = 0$ and not throughout any neighborhood. (See Sec. 14.)

If a function f is analytic in a region R, then about each point z of R there must be a neighborhood on which f is defined. This means that z must be an interior point of the domain of definition, and so analytic functions are usually defined on domains. If, however, we should speak of an analytic function f on the closed disk $|z| \leqq 1$, for example, it is to be understood that f is analytic throughout some domain containing that disk.

An *entire* function is a function that is analytic at each point in the entire plane. Since the derivative of a polynomial exists everywhere, it follows that *every polynomial is an entire function.*

If a function fails to be analytic at a point z_0 but is analytic at some point in every neighborhood of z_0, then z_0 is called a *singular point*, or singularity, of the function. Note, for example, that if

$$f(z) = \frac{1}{z}, \qquad \text{then} \qquad f'(z) = -\frac{1}{z^2} \qquad (z \neq 0).$$

Hence f is analytic at every point except for $z = 0$ where it is not even defined. The point $z = 0$ is therefore a singular point. On the other hand, the function $f(z) = |z|^2$ has no singular points since it is nowhere analytic.

A necessary, but by no means sufficient, condition for a function f to be analytic in a domain D is clearly the continuity of f throughout D. Satisfaction of the Cauchy-Riemann equations is also necessary, but not sufficient. Sufficient conditions for analyticity in D are provided by the theorems in Secs. 17 and 18.

Other useful sufficient conditions are obtained from the differentiation formulas in Sec. 15. The derivatives of the sum and product of two functions exist wherever the functions themselves have derivatives. Thus, *if two functions are analytic in a domain D, their sum and their product are both analytic in D.* Similarly, *their quotient is analytic in D provided the function in the denominator is not zero at any point of D.* In particular, the quotient $P(z)/Q(z)$ of two polynomials is analytic in any domain throughout which $Q(z) \neq 0$.

From the chain rule for the derivative of a composite function (Sec. 15) we find that *a composition of two analytic functions is analytic.* To be precise, suppose that $f(z)$ is analytic in a domain D and $g(z)$ is analytic in a domain that contains the range of f. Then the composition $g[f(z)]$ is analytic in D.

To illustrate, consider the entire function $f(z) = z^2$. According to Exercise 4, Sec. 18, the function

$$g(z) = \sqrt{r}\, e^{i\theta/2} \qquad (r > 0,\ -\pi < \theta < \pi) \tag{1}$$

is analytic everywhere in the indicated domain of defintion, consisting of all points in the plane except for the origin and points lying on the negative real axis. In order to form the composition $g[f(z)]$, we now restrict the domain of definition of f to a domain D such that the range of f is contained in the domain on which g is defined. The largest possible domain D with this property is $r > 0,\ -\pi/2 < \theta < \pi/2$, the right half plane with the imaginary axis excluded. This is readily seen by considering the polar expression

$$f(z) = r^2 e^{i2\theta} \tag{2}$$

and noting that $-\pi < 2\theta < \pi$ when $-\pi/2 < \theta < \pi/2$ Thus the composition $g[f(z)]$ is analytic at any point z where $\operatorname{Re} z > 0$. Indeed, we see from equations (1) and (2) that $g[f(z)] = z$ for any such z.

20. Harmonic Functions

A real-valued function h of two real variables x and y is said to be *harmonic* in a given domain of the xy plane if throughout that domain it has continuous first and second partial derivatives and satisfies the partial differential equation

$$h_{xx}(x,y) + h_{yy}(x,y) = 0, \tag{1}$$

known as *Laplace's equation*. We now assume that a function $f(z) = u(x,y) + iv(x,y)$ is analytic in a domain D and show that the component functions u and v are harmonic in D.

In order to do this, we need a result which will be proved later in Chap. 5 (Sec. 52). Namely, if a function of a complex variable is analytic at a point, then its real and imaginary parts have continuous partial derivatives of all orders at that point.

Since f is analytic in D, the first partial derivatives of its component functions satisfy the Cauchy-Riemann equations throughout D; that is,

$$u_x = v_y, \qquad u_y = -v_x. \tag{2}$$

By differentiating both sides of these equations with respect to x, we have

$$u_{xx} = v_{yx}, \qquad u_{yx} = -v_{xx}. \tag{3}$$

Likewise, differentiation with respect to y yields

$$u_{xy} = v_{yy}, \qquad u_{yy} = -v_{xy}. \tag{4}$$

Now, by a theorem in the calculus of real variables,[1] the continuity of the partial derivatives guarantees that $v_{yx} = v_{xy}$ and $u_{yx} = u_{xy}$. It then follows from equations (3) and (4) that

$$u_{xx}(x,y) + u_{yy}(x,y) = 0 \qquad \text{and} \qquad v_{xx}(x,y) + v_{yy}(x,y) = 0.$$

Thus, *if $f(z) = u(x,y) + iv(x,y)$ is analytic in a domain D, its component functions u and v are harmonic in D.*

If two given functions u and v are harmonic in a domain D and their first partial derivatives satisfy the Cauchy-Riemann equations throughout D, we say that v is a *harmonic conjugate* of u.

Evidently, then, if a function $f(z) = u(x,y) + iv(x,y)$ is analytic in a domain D, v is a harmonic conjugate of u. Conversely, if v is a harmonic conjugate of u in a domain D, the function $f(z) = u(x,y) + iv(x,y)$ is analytic in D. This follows from the theorem in Sec. 17. Hence *a necessary and sufficient condition for a function $f(z) = u(x,y) + iv(x,y)$ to be analytic in a domain D is that v be a harmonic conjugate of u in D.*

We note that if v is a harmonic conjugate of u in some domain, it is not in general true that u is a harmonic conjugate of v there. To illustrate this, we consider the functions

$$u(x,y) = x^2 - y^2, \qquad v(x,y) = 2xy.$$

Since they are the real and imaginary parts, respectively, of the entire function $f(z) = z^2$, v is a harmonic conjugate of u throughout the plane. But u cannot be a harmonic conjugate of v since, according to the theorem in Sec. 16, the function $2xy + i(x^2 - y^2)$ is not analytic anywhere. It is left as an exercise to show that if two functions u and v are to be harmonic conjugates of each other, both u and v must be constant functions.

It is, however, true that if v is a harmonic conjugate of u in a domain D, then $-u$ is a harmonic conjugate of v in D, and conversely. This is because the function $f(z) = u(x,y) + iv(x,y)$ is analytic in D if and only if the function $-if(z) = v(x,y) - iu(x,y)$ is analytic there.

In Chap. 8 (Sec. 78) we shall show that a function u which is harmonic in a domain of a certain type always has a harmonic conjugate. Thus, in such domains every harmonic function is the real part of an analytic function. It is

[1] See, for instance, A. E. Taylor and W. R. Mann, "Advanced Calculus," 2d ed., pp. 214–216, 1972.

also true that a harmonic conjugate, when it exists, is unique except for an additive constant.

We now illustrate one method of obtaining a harmonic conjugate of a given harmonic function. The function

$$u(x,y) = y^3 - 3x^2y \tag{5}$$

is readily seen to be harmonic throughout the entire xy plane. In order to find a harmonic conjugate $v(x,y)$, we note that

$$u_x(x,y) = -6xy.$$

So, in view of the condition $u_x = v_y$, we may conclude that

$$v_y(x,y) = -6xy.$$

Holding x fixed and integrating both sides of this equation with respect to y, we find that

$$v(x,y) = -3xy^2 + \phi(x) \tag{6}$$

where ϕ is at present an arbitrary function of x. Since the condition $u_y = -v_x$ must hold, it follows from equations (5) and (6) that

$$3y^2 - 3x^2 = 3y^2 - \phi'(x).$$

So $\phi'(x) = 3x^2$, or $\phi(x) = x^3 + c$ where c is an arbitrary real number. Hence the function

$$v(x,y) = x^3 - 3xy^2 + c$$

is a harmonic conjugate of $u(x,y)$.

The corresponding analytic function is

$$f(z) = y^3 - 3x^2y + i(x^3 - 3xy^2 + c). \tag{7}$$

It is easily verified that

$$f(z) = i(z^3 + c).$$

This form is suggested by noting that when $y = 0$, equation (7) becomes

$$f(x) = i(x^3 + c).$$

EXERCISES

1. Prove that each of these functions is entire: (*a*) $f(z) = 3x + y + i(3y - x)$; (*b*) $f(z) = \sin x \cosh y + i \cos x \sinh y$; (*c*) $f(z) = e^{-y}e^{ix}$; (*d*) $f(z) = (z^2 - 2)e^{-x}e^{-iy}$.

2. Show why each of these functions is nowhere analytic: (*a*) $f(z) = xy + iy$; (*b*) $f(z) = e^y e^{ix}$.

3. In each case determine the singular points of the function, and state why the function is analytic everywhere except at those points:

$$(a)\ \frac{2z+1}{z(z^2+1)};\qquad (b)\ \frac{z^3+i}{z^2-3z+2};\qquad (c)\ (z+2)^{-1}(z^2+2z+2)^{-1}.$$

Ans. (*a*) $z = 0, \pm i$; (*c*) $z = -2, -1 \pm i$.

4. Show that the function $g(z) = \sqrt{r}\, e^{i\theta/2}$ $(r > 0, 0 < \theta < \pi)$ is analytic in the indicated domain of definition. Then show that the composite function $g(z^2 + 1)$ is analytic in the quadrant $x > 0$, $y > 0$.

5. Show that the function $g(z) = \text{Log}\, r + i\theta$ $(r > 0, -\pi/2 < \theta < \pi/2)$, where the natural logarithm is used, is analytic in the indicated domain of definition, and show that $g'(z) = 1/z$ there. Then show why the composite function $g(2z - 2 + i)$ is an analytic function of z in the domain $x > 1$.

6. State why a composition of two entire functions is entire. Also state why a *linear combination* $cf(z) + dg(z)$ of two entire functions, where c and d are complex constants, is entire.

7. Show that u is harmonic in some domain and find a harmonic conjugate v when (*a*) $u(x,y) = 2x(1 - y)$; (*b*) $u(x,y) = 2x - x^3 + 3xy^2$; (*c*) $u(x,y) = \sinh x \sin y$; (*d*) $u(x,y) = y/(x^2 + y^2)$.

Ans. (*a*) $v(x,y) = x^2 - y^2 + 2y$; (*c*) $v(x,y) = -\cosh x \cos y$.

8. Show that if in some domain v is a harmonic conjugate of u and u is a harmonic conjugate of v, then u and v must be constant functions.

9. Let a function f be analytic in a domain D. Show that f is a constant function if (*a*) the function $\overline{f(z)}$ is also analytic in D; (*b*) $|f(z)|$ is constant for all z in D; (*c*) f is real-valued for all z in D.

10. Show that if v and V are harmonic conjugates of u in a domain D, then $v(x,y)$ and $V(x,y)$ can differ at most by an arbitrary additive constant.

11. Let the function $f(z) = u(r,\theta) + iv(r,\theta)$ be analytic in a domain D that does not include the point $z = 0$. Using the Cauchy-Riemann equations in polar coordinates (Sec. 18), show that the function u satisfies the polar form

$$r^2 u_{rr}(r,\theta) + r u_r(r,\theta) + u_{\theta\theta}(r,\theta) = 0$$

of Laplace's equation throughout D. Show that the function v also satisfies the polar form of Laplace's equation there.

12. Show that the function $u(r,\theta) = \text{Log}\, r$, where the natural logarithm is used, is harmonic in the domain $r > 0$, $0 < \theta < 2\pi$. (See Exercise 11.) Then find a harmonic conjugate v.

Ans. $v(r,\theta) = \theta$.

13. Let the function $f(z) = u(x,y) + iv(x,y)$ be analytic in a domain D, and consider the families of *level curves* $u(x,y) = c_1$ and $v(x,y) = c_2$ where c_1 and c_2 are arbitrary constants. Prove that these families are orthogonal. More precisely,

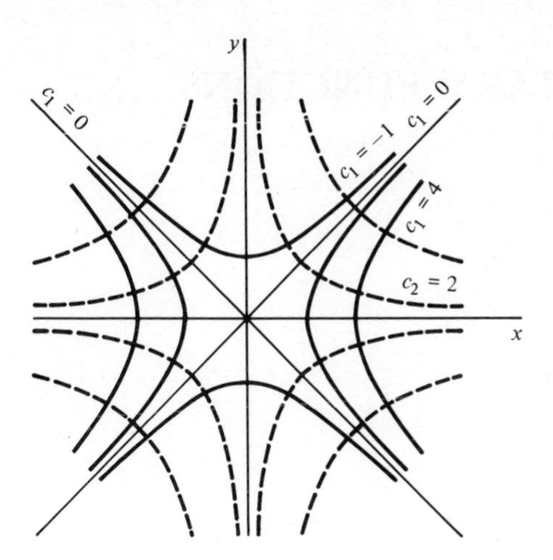

FIGURE 17

show that if $z_0 = (x_0,y_0)$ is a point common to two particular curves $u(x,y) = c_1$ and $v(x,y) = c_2$ and if $f'(z_0) \neq 0$, then the lines tangent to those curves at (x_0,y_0) are perpendicular.

14. Show that when $f(z) = z^2$, the level curves $u(x,y) = c_1$ and $v(x,y) = c_2$ of the component functions are those shown in Fig. 17. Note the orthogonality of the two families, as proved in Exercise 13. The curves $u(x,y) = 0$ and $v(x,y) = 0$ intersect at the origin but are not, however, orthogonal to each other. Why is this fact in agreement with the result of Exercise 13?

15. Sketch the families of level curves of the component functions u and v when $f(z) = 1/z$, and note the orthogonality proved in Exercise 13.

16. Do Exercise 15 using polar coordinates.

17. Sketch the families of level curves of the component functions u and v when $f(z) = (z - 1)/(z + 1)$, and note how the results of Exercise 13 are illustrated here.

3

ELEMENTARY FUNCTIONS

We now consider various elementary functions studied in the calculus of real variables and define corresponding functions of a complex variable. To be specific, we define analytic functions of a complex variable z which reduce to those elementary functions when $z = x + i0$. We start by defining the complex exponential function and then use it to define the others.

21. The Exponential Function

If a function f of the complex variable $z = x + iy$ is to reduce to the familiar exponential function when z is real, we must require that

$$f(x + i0) = e^x \tag{1}$$

for all real numbers x. Inasmuch as $d(e^x)/dx = e^x$ for all real x, it is natural to add the following conditions:

$$f \text{ is entire} \qquad \text{and} \qquad f'(z) = f(z) \text{ for all } z. \tag{2}$$

The function f defined on the entire complex plane by the equation

$$f(z) = e^x \cos y + ie^x \sin y \tag{3}$$

clearly satisfies conditions (1) and (2). When evaluating $\cos y$ and $\sin y$ here, we agree that y is to be taken in radians. It can be shown (Exercise 14, Sec. 22) that the function defined by equation (3) is the only one satisfying conditions (1) and (2), and we write $f(z) = e^z$.

The exponential function of complex analysis is thus defined for all z by the equation

$$e^z = e^x(\cos y + i \sin y). \tag{4}$$

As we have noted, it reduces to the exponential function in the calculus of real variables when $y = 0$, it is entire, and it satisfies the differentiation formula

$$\frac{d}{dz} e^z = e^z \tag{5}$$

for all z.

Note that when z is the pure imaginary number $i\theta$, equation (4) becomes

$$e^{i\theta} = \cos \theta + i \sin \theta.$$

This is Euler's formula which was introduced earlier in Sec. 5. The definition of the symbol $e^{i\theta}$ given there is thus consistent with definition (4).

We agree that when $z = 1/n$, the value of e^z is the positive nth root of e given by equation (4); that is, $e^{1/n}$ is to be read $\sqrt[n]{e}$. This is an exception to the convention (Sec. 6) which would ordinarily require us to interpret $e^{1/n}$ as any one of the nth roots.

Finally, it should be noted that, for convenience, $\exp z$ is often written in place of e^z.

22. Other Properties of $\exp z$

Our definition

$$e^z = e^x(\cos y + i \sin y) \tag{1}$$

furnishes the complex number e^z in the polar form

$$e^z = \rho(\cos \phi + i \sin \phi) \qquad \text{where} \qquad \rho = e^x, \phi = y. \tag{2}$$

Hence the modulus of e^z is e^x and an argument of e^z is y. That is,

$$|e^z| = e^x \qquad \text{and} \qquad \arg e^z = y. \tag{3}$$

With the transformation $w = e^z$, we find from definition (1) that a nonzero point

$$w = \rho(\cos \phi + i \sin \phi) \tag{4}$$

is the image of

$$z = \text{Log}\,\rho + i\phi, \tag{5}$$

where the natural logarithm is used. To illustrate, let us solve for z in the equation $e^z = -1$. Since expression (4) is the polar form of -1 when $\rho = 1$ and $\phi = \pi + 2n\pi$ $(n = 0, \pm 1, \pm 2, \ldots)$, it follows from formula (5) that any number $z = (1 + 2n)\pi i$ is a solution.

Since $|e^z| = e^x$ and $e^x > 0$ for every real number x, we know that $|e^z| > 0$ for every complex number z. Hence

$$e^z \neq 0 \qquad \text{for any complex number } z. \tag{6}$$

This means that the point $w = 0$ cannot be the image of any point in the z plane under the transformation $w = e^z$. We have thus shown that *the range of the exponential function is the entire complex plane except for the origin.*

Any point in the range of the exponential function is actually the image of an infinite number of points in the z plane. This is because, according to definition (1) of the exponential function, any two points in the z plane have the same image whenever their real parts are equal and their imaginary parts differ by an integral multiple of 2π. That is, *the exponential function is periodic with a pure imaginary period of* $2\pi i$:

$$e^{z+2\pi i} = e^z \qquad \text{for all } z. \tag{7}$$

If, however, we restrict the domain of definition to the strip $-\pi < \text{Im}\, z \leq \pi$ (Fig. 18), the mapping $w = e^z$ is *one to one onto* the range. That is, if w is any nonzero point and it is written in the polar form $w = \rho(\cos\Phi + i\sin\Phi)$ where $\Phi = \text{Arg}\, w$, then, in accordance with equations (4) and (5), the point $z = \text{Log}\,\rho + i\Phi$ has w for its image and is the only such point in the strip. Note that the mapping is, in fact, one to one onto the range when the domain of definition is restricted to any strip $y_0 < \text{Im}\, z \leq y_0 + 2\pi$.

The additive property

$$(\exp z_1)(\exp z_2) = \exp(z_1 + z_2) \tag{8}$$

carries over from the real case. To see this, write $z_1 = x_1 + iy_1$, $z_2 = x_2 + iy_2$; then

$$\exp z_1 = \rho_1(\cos\phi_1 + i\sin\phi_1) \qquad \text{where} \qquad \rho_1 = e^{x_1},\ \phi_1 = y_1,$$
$$\exp z_2 = \rho_2(\cos\phi_2 + i\sin\phi_2) \qquad \text{where} \qquad \rho_2 = e^{x_2},\ \phi_2 = y_2.$$

By formula (5), Sec. 5, for the product of two complex numbers in polar form, we have

$$\begin{aligned}(\exp z_1)(\exp z_2) &= \rho_1\rho_2[\cos(\phi_1 + \phi_2) + i\sin(\phi_1 + \phi_2)]\\ &= e^{x_1}e^{x_2}[\cos(y_1 + y_2) + i\sin(y_1 + y_2)].\end{aligned}$$

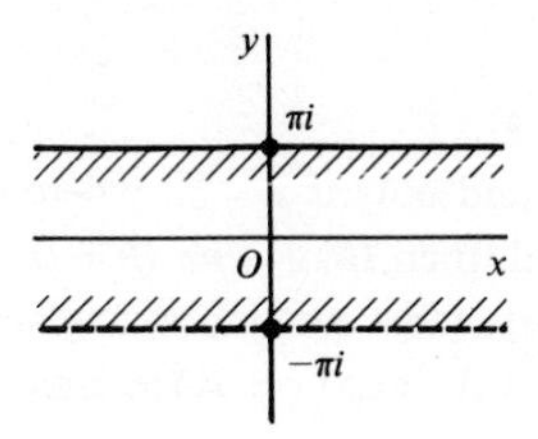

FIGURE 18

But $e^{x_1}e^{x_2} = e^{x_1+x_2}$ and $x_1 + x_2 + i(y_1 + y_2) = z_1 + z_2$, and property (8) is thus established. Note that this property was obtained in Sec. 5 for the case when both z_1 and z_2 are pure imaginary.

In like manner we can derive the identity

$$\frac{\exp z_1}{\exp z_2} = \exp(z_1 - z_2). \tag{9}$$

From this and the fact that $e^0 = 1$, it follows that $1/e^z = e^{-z}$. Another useful identity is

$$(e^z)^n = e^{nz} \qquad (n = 1, 2, \ldots). \tag{10}$$

EXERCISES

1. Show that

(*a*) $\exp(2 \pm 3\pi i) = -e^2$; (*b*) $\exp\frac{2 + \pi i}{4} = \sqrt{e}\,\frac{1 + i}{\sqrt{2}}$; (*c*) $e^{z+\pi i} = -e^z$.

2. State why the function $2z^2 - 3 - ze^z + e^{-z}$ is entire.

3. Find all values of z such that (*a*) $e^z = -2$; (*b*) $e^z = 1 + \sqrt{3}\,i$; (*c*) $\exp(2z - 1) = 1$.

Ans. (*a*) $z = \text{Log}\, 2 + (2n + 1)\pi i$ $(n = 0, \pm 1, \pm 2, \ldots)$;
(*c*) $z = \frac{1}{2} + n\pi i$ $(n = 0, \pm 1, \pm 2, \ldots)$.

4. Write $|\exp(2z + i)|$ and $|\exp(iz^2)|$ in terms of x and y, and then show that $|\exp(2z + i) + \exp(iz^2)| \leqq e^{2x} + e^{-2xy}$.

5. Prove that $|e^{-2z}| < 1$ if and only if $\text{Re}\, z > 0$.

6. Let z be any nonzero complex number. Show that if $z = re^{i\theta}$, then
(*a*) $\bar{z} = re^{-i\theta}$;
(*b*) $\exp(\text{Log}\, r + i\theta) = z$.

7. Derive identities (9) and (10) in Sec. 22.

8. Show that $e^{-nz} = 1/(e^z)^n$ $(n = 1, 2, \ldots)$.

9. Show that

(a) $\exp \bar{z} = \overline{\exp z}$ for all z;

(b) $\exp (i\bar{z}) = \overline{\exp (iz)}$ if and only if $z = n\pi$ where $n = 0, \pm 1, \pm 2, \ldots$.

10. (a) Show that if e^z is real, then $\text{Im } z = n\pi$ $(n = 0, \pm 1, \pm 2, \ldots)$.

(b) If e^z is pure imaginary, what restriction is placed on z?

11. Examine the behavior of (a) $\exp (x + iy)$ as x tends to $-\infty$; (b) $\exp (2 + iy)$ as y tends to ∞.

12. Prove that the function $\exp \bar{z}$ is not analytic anywhere.

13. Show in two ways that the function $\exp (z^2)$ is entire. What is its derivative?

Ans. $2z \exp (z^2)$.

14. Show that if a function $f(z) = u(x,y) + iv(x,y)$ satisfies conditions (1) and (2), Sec. 21, it must be the function defined in equation (3) of that section.

Suggestion: Obtain the equations $u_x = u$, $v_x = v$ and then show that there exist real-valued functions ϕ and ψ of the real variable y such that $u(x,y) = e^x\phi(y)$, $v(x,y) = e^x\psi(y)$. Use the Cauchy-Riemann equations to obtain the differential equation $\phi''(y) + \phi(y) = 0$ whose solution is $\phi(y) = a \cos y + b \sin y$ where a and b are real numbers. Then show that $\psi(y) = a \sin y - b \cos y$, and use the fact that $u(x,0) + iv(x,0) = e^x$ to find a and b.

15. Write $\text{Re}(e^{1/z})$ in terms of x and y. Why is this function harmonic in every domain that does not contain the origin?

16. Let the function $f(z) = u(x,y) + iv(x,y)$ be analytic in some domain D. State why the functions

$$U(x,y) = e^{u(x,y)} \cos [v(x,y)],$$
$$V(x,y) = e^{u(x,y)} \sin [v(x,y)]$$

are harmonic in D and why $V(x,y)$ is, in fact, a harmonic conjugate of $U(x,y)$.

23. Trigonometric Functions

From the formulas

$$e^{ix} = \cos x + i \sin x, \qquad e^{-ix} = \cos x - i \sin x$$

it follows that

$$\sin x = \frac{e^{ix} - e^{-ix}}{2i}, \qquad \cos x = \frac{e^{ix} + e^{-ix}}{2}$$

for every real number x. It is therefore natural to define the sine and cosine functions of a complex variable z by the equations

$$\text{(1)} \qquad \sin z = \frac{e^{iz} - e^{-iz}}{2i}, \qquad \cos z = \frac{e^{iz} + e^{-iz}}{2}.$$

The sine and cosine functions are entire since they are linear combinations (Exercise 6, Sec. 20) of the entire functions e^{iz} and e^{-iz}. Knowing the derivatives of the exponential functions in equations (1), we find that

$$\frac{d}{dz}\sin z = \cos z, \qquad \frac{d}{dz}\cos z = -\sin z. \tag{2}$$

The other four trigonometric functions are defined in terms of the sine and cosine functions by the usual relations:

$$\begin{aligned} \tan z &= \frac{\sin z}{\cos z}, & \cot z &= \frac{\cos z}{\sin z}, \\ \sec z &= \frac{1}{\cos z}, & \csc z &= \frac{1}{\sin z}. \end{aligned} \tag{3}$$

Note that $\tan z$ and $\sec z$ are analytic in any domain where $\cos z \neq 0$, and $\cot z$ and $\csc z$ are analytic in any domain where $\sin z \neq 0$. By differentiating the right-hand sides of equations (3), we obtain the differentiation formulas

$$\begin{aligned} \frac{d}{dz}\tan z &= \sec^2 z, & \frac{d}{dz}\cot z &= -\csc^2 z, \\ \frac{d}{dz}\sec z &= \sec z \tan z, & \frac{d}{dz}\csc z &= -\csc z \cot z. \end{aligned} \tag{4}$$

From the definition of $\sin z$ it follows that

$$\begin{aligned} \sin z = \frac{e^{i(x+iy)} - e^{-i(x+iy)}}{2i} &= (\cos x + i \sin x)\frac{e^{-y}}{2i} - (\cos x - i \sin x)\frac{e^{y}}{2i} \\ &= \sin x\left(\frac{e^{y} + e^{-y}}{2}\right) + i \cos x\left(\frac{e^{y} - e^{-y}}{2}\right), \end{aligned}$$

where $z = x + iy$. Thus the real and imaginary parts of $\sin z$ are displayed as follows:

$$\sin z = \sin x \cosh y + i \cos x \sinh y. \tag{5}$$

In the same way, or by using the first of formulas (2), we find that

$$\cos z = \cos x \cosh y - i \sin x \sinh y. \tag{6}$$

It is evident from these last two formulas that

$$\sin (iy) = i \sinh y, \qquad \cos (iy) = \cosh y \tag{7}$$

and that $\sin \bar{z}$ and $\cos \bar{z}$ are the complex conjugates of $\sin z$ and $\cos z$, respectively.

The periodic character of $\sin z$ and $\cos z$ is also evident from formulas (5) and (6):

$$\sin(z + 2\pi) = \sin z, \qquad \sin(z + \pi) = -\sin z, \tag{8}$$

$$\cos(z + 2\pi) = \cos z, \qquad \cos(z + \pi) = -\cos z. \tag{9}$$

The periodicity of each of the other trigonometric functions follows immediately from identities (8) and (9). For example,

$$\tan(z + \pi) = \tan z. \tag{10}$$

24. Further Properties of Trigonometric Functions

By using either formulas (1) or formulas (5) and (6) of the preceding section, the reader can show that

$$|\sin z|^2 = \sin^2 x + \sinh^2 y, \tag{1}$$

$$|\cos z|^2 = \cos^2 x + \sinh^2 y. \tag{2}$$

It is clear from these two expressions that $\sin z$ and $\cos z$ are not bounded in absolute value, whereas with real variables the absolute values of the sine and cosine are never greater than unity.

The usual trigonometric identities are still valid with complex variables:

$$\sin^2 z + \cos^2 z = 1, \tag{3}$$

$$\sin(z_1 + z_2) = \sin z_1 \cos z_2 + \cos z_1 \sin z_2, \tag{4}$$

$$\cos(z_1 + z_2) = \cos z_1 \cos z_2 - \sin z_1 \sin z_2, \tag{5}$$

$$\sin(-z) = -\sin z, \qquad \cos(-z) = \cos z, \tag{6}$$

$$\sin\left(\frac{\pi}{2} - z\right) = \cos z, \tag{7}$$

$$\sin 2z = 2 \sin z \cos z, \qquad \cos 2z = \cos^2 z - \sin^2 z, \tag{8}$$

etc. Proofs may be based entirely upon properties of the exponential function.

A specific value of z for which $f(z) = 0$ is called a *zero* of a given function f. The zeros of the sine and cosine functions are all real. In fact,

$$\sin z = 0 \quad \text{if and only if} \quad z = n\pi \qquad (n = 0, \pm 1, \pm 2, \ldots) \tag{9}$$

and

$$\cos z = 0 \quad \text{if and only if} \quad z = (n + \tfrac{1}{2})\pi \quad (n = 0, \pm 1, \pm 2, \ldots). \tag{10}$$

To prove statement (9), let us first assume that $\sin z = 0$. According to expression (1), it follows that

$$\sin^2 x + \sinh^2 y = 0;$$

hence x and y must satisfy the equations

$$\sin x = 0 \qquad \text{and} \qquad \sinh y = 0.$$

Evidently, then, $x = n\pi$ $(n = 0, \pm 1, \pm 2, \ldots)$ and $y = 0$; that is, $z = n\pi$. Conversely, if $z = n\pi$, where n is any integer, $\sin z = 0$. We have thus verified statement (9). Statement (10) is proved in a similar fashion.

In view of statement (10), $\tan z$ has singularities at the points $z = (n + \frac{1}{2})\pi$ $(n = 0, \pm 1, \pm 2, \ldots)$ and is analytic everywhere else.

EXERCISES

1. Show that, for any complex number z, $e^{iz} = \cos z + i \sin z$.

2. Establish differentiation formulas (4), Sec. 23.

3. Derive formulas (6) and (7), Sec. 23.

4. Derive formula (1), Sec. 24; then show that $|\sinh y| \leqq |\sin z| \leqq \cosh y$.

5. Derive formula (2), Sec. 24, and show that $|\sinh y| \leqq |\cos z| \leqq \cosh y$.

6. Show that $|\sin z| \geqq |\sin x|$ and $|\cos z| \geqq |\cos x|$.

7. Establish identities (3) and (4), Sec. 24.

8. Prove that (*a*) $1 + \tan^2 z = \sec^2 z$; (*b*) $1 + \cot^2 z = \csc^2 z$.

9. Establish the identities (*a*) $2 \sin (z_1 + z_2) \sin (z_1 - z_2) = \cos 2z_2 - \cos 2z_1$; (*b*) $2 \cos (z_1 + z_2) \sin (z_1 - z_2) = \sin 2z_1 - \sin 2z_2$.

10. Show that $\cos (i\bar{z}) = \overline{\cos (iz)}$ for all z and that $\sin (i\bar{z}) = \overline{\sin (iz)}$ if and only if $z = n\pi i$ $(n = 0, \pm 1, \pm 2, \ldots)$.

11. Prove statement (10), Sec. 24.

12. With the aid of identity (*a*) in Exercise 9, show that if $\cos z_1 = \cos z_2$, then either $z_1 + z_2$ or $z_1 - z_2$ is an integral multiple of 2π. Obtain an analogous result when $\sin z_1 = \sin z_2$.

13. Find all the roots of the equation $\sin z = \cosh 4$ by equating the real parts and the imaginary parts of $\sin z$ and $\cosh 4$.

Ans. $(2n + \frac{1}{2})\pi \pm 4i$ $(n = 0, \pm 1, \pm 2, \ldots)$.

14. Find all the roots of the equation $\cos z = 2$.

Ans. $2n\pi + i \cosh^{-1} 2$; that is, $2n\pi \pm i \operatorname{Log} (2 + \sqrt{3})$ $(n = 0, \pm 1, \pm 2, \ldots)$.

15. Show in two ways that each of the following functions is everywhere harmonic: (*a*) $\sin x \sinh y$; (*b*) $\cos 2x \sinh 2y$.

16. Let the function $f(z)$ be analytic in a domain D. State why the functions $\sin f(z)$ and $\cos f(z)$ are analytic there. Also, write $w = f(z)$ and state why

$$\frac{d}{dz} \sin f(z) = \cos w \frac{dw}{dz}, \qquad \frac{d}{dz} \cos f(z) = -\sin w \frac{dw}{dz}.$$

17. Show that neither (*a*) $\sin \bar{z}$ nor (*b*) $\cos \bar{z}$ is an analytic function of z anywhere.

25. Hyperbolic Functions

The hyperbolic sine and the hyperbolic cosine of a complex variable are defined as they are with a real variable; that is,

$$\sinh z = \frac{e^z - e^{-z}}{2}, \qquad \cosh z = \frac{e^z + e^{-z}}{2}. \tag{1}$$

The hyperbolic tangent of z is defined by the equation

$$\tanh z = \frac{\sinh z}{\cosh z},$$

and then $\coth z$, $\operatorname{sech} z$, and $\operatorname{csch} z$ are defined as the reciprocals of $\tanh z$, $\cosh z$, and $\sinh z$, respectively.

Since e^z and e^{-z} are entire, it follows from definitions (1) that the hyperbolic sine and the hyperbolic cosine functions are entire. The function $\tanh z$ is analytic in every domain in which $\cosh z \neq 0$.

The calculus and algebra of hyperbolic functions can be derived readily from the above definitions. The formulas are the same as those established for the corresponding functions of a real variable; thus

$$\frac{d}{dz} \sinh z = \cosh z, \qquad \frac{d}{dz} \cosh z = \sinh z, \tag{2}$$

$$\frac{d}{dz} \tanh z = \operatorname{sech}^2 z, \qquad \frac{d}{dz} \coth z = -\operatorname{csch}^2 z, \tag{3}$$

$$\frac{d}{dz} \operatorname{sech} z = -\operatorname{sech} z \tanh z, \qquad \frac{d}{dz} \operatorname{csch} z = -\operatorname{csch} z \coth z. \tag{4}$$

Some of the most frequently used identities are

$$\cosh^2 z - \sinh^2 z = 1, \tag{5}$$

$$\sinh (z_1 + z_2) = \sinh z_1 \cosh z_2 + \cosh z_1 \sinh z_2, \tag{6}$$

$$\cosh (z_1 + z_2) = \cosh z_1 \cosh z_2 + \sinh z_1 \sinh z_2, \tag{7}$$

$$\sinh (-z) = -\sinh z, \qquad \cosh (-z) = \cosh z. \tag{8}$$

The hyperbolic functions are, of course, intimately related to the trigonometric functions defined in Sec. 23. In fact, by recalling how all of these functions have been defined in terms of the exponential function, we find such relations as

$$\sinh (iz) = i \sin z, \qquad \cosh (iz) = \cos z, \tag{9}$$

$$\sin (iz) = i \sinh z, \qquad \cos (iz) = \cosh z. \tag{10}$$

The real and imaginary parts of the hyperbolic sine and the hyperbolic cosine functions are shown in the formulas

$$\sinh z = \sinh x \cos y + i \cosh x \sin y, \tag{11}$$

$$\cosh z = \cosh x \cos y + i \sinh x \sin y, \tag{12}$$

where $z = x + iy$. The reader can obtain the expressions

$$|\sinh z|^2 = \sinh^2 x + \sin^2 y, \tag{13}$$

$$|\cosh z|^2 = \sinh^2 x + \cos^2 y \tag{14}$$

in various ways.

The hyperbolic sine and the hyperbolic cosine functions are periodic with period $2\pi i$, and the hyperbolic tangent is periodic with period πi. Also, $\sinh z = 0$ when $z = n\pi i$ and $\cosh z = 0$ when $z = (n + \frac{1}{2})\pi i$ $(n = 0, \pm 1, \pm 2, \ldots)$; in fact, these are the only zeros of $\sinh z$ and $\cosh z$.

EXERCISES

1. Derive differentiation formulas (2) and (4), Sec. 25.
2. Prove identities (5) and (7), Sec. 25.
3. Write $\sinh z = \sinh (x + iy)$ and $\cosh z = \cosh (x + iy)$ and show how formulas (11) and (12) in Sec. 25 follow from identities (6), (7), and (9) in that section.
4. Derive formula (14), Sec. 25; then show that $|\sinh x| \leqq |\cosh z| \leqq \cosh x$.
5. Show that $\sinh (z + \pi i) = -\sinh z$, $\cosh (z + \pi i) = -\cosh z$. Then show that $\tanh (z + \pi i) = \tanh z$.
6. Prove that $\sinh 2z = 2 \sinh z \cosh z$.
7. Using statements (9) and (10) of Sec. 24 together with formulas (10) of Sec. 25, obtain all the zeros of $\sinh z$ and $\cosh z$.
8. Using the results obtained in Exercise 7, locate all the zeros and singularities of the hyperbolic tangent function.
9. Find all the roots of the equations (*a*) $\cosh z = 1/2$; (*b*) $\sinh z = i$; (*c*) $\cosh z = -2$.

   Ans. (*a*) $(2n \pm \frac{1}{3})\pi i$; (*b*) $(2n + \frac{1}{2})\pi i$ $(n = 0, \pm 1, \pm 2, \ldots)$.
10. Why is the function $\sinh (e^z)$ entire? Write its real part as a function of x and y and state why that function must be harmonic everywhere.

26. The Logarithmic Function

Let $\operatorname{Log} r$ denote the natural logarithm of a positive real number r, as defined in the calculus of real variables. We now define the logarithmic function of complex analysis by the equation

$$\log z = \operatorname{Log} r + i\theta \tag{1}$$

where $r = |z|$ and $\theta = \arg z$. It is a multiple-valued function which is defined for all nonzero complex numbers z.

Definition (1) is a natural one in the sense that it is suggested by writing $z = re^{i\theta}$ and using familiar properties of the natural logarithm in elementary calculus to formally expand $\log(re^{i\theta})$.

If Θ denotes the principal value of $\arg z$ $(-\pi < \Theta \leqq \pi)$, we can write

$$\theta = \Theta + 2n\pi \qquad (n = 0, \pm 1, \pm 2, \ldots).$$

Expression (1) then takes the form

$$\log z = \operatorname{Log} r + i(\Theta + 2n\pi) \qquad (n = 0, \pm 1, +2, \ldots). \tag{2}$$

Note that for any particular z in the domain of definition the values of $\log z$ have the same real part, and their imaginary parts differ by integral multiples of 2π.

The *principal value* of $\log z$ is the value obtained from formula (2) when $n = 0$. That value is denoted by $\operatorname{Log} z$ and is given by the equation

$$\operatorname{Log} z = \operatorname{Log} r + i\Theta \qquad (r > 0, \ -\pi < \Theta \leqq \pi). \tag{3}$$

The mapping $w = \operatorname{Log} z$ is single-valued and its domain of definition is the set of all nonzero complex numbers; its range is the strip $-\pi < \operatorname{Im} w \leqq \pi$.

Note that $\operatorname{Log} z$ reduces to the usual natural logarithm of a real variable when the domain of definition is restricted to the positive real axis. For, if z is a positive real number r, $|z| = r$ and $\Theta = 0$; equation (3) then becomes $\operatorname{Log} z = \operatorname{Log} r$.

We saw in Sec. 22, with z and w interchanged, that the equation $z = e^w$ establishes a one to one correspondence between the nonzero points in the z plane and the points in the strip $-\pi < \operatorname{Im} w \leqq \pi$ of the w plane. The point $z = r\exp(i\Theta)$ in the z plane corresponds to the point $w = \operatorname{Log} r + i\Theta$ in the w plane. Therefore, when the domain of definition of the function e^w is restricted to the strip $-\pi < \operatorname{Im} w \leqq \pi$, its inverse is the principal logarithmic function $\operatorname{Log} z$. That is,

$$w = \operatorname{Log} z \qquad \text{if and only if} \qquad z = e^w.$$

The mapping $z = e^w$ also establishes a one to one correspondence between the nonzero points in the z plane and the points in the w plane that lie in the strip $(2k - 1)\pi < \operatorname{Im} w \leqq (2k + 1)\pi$, where k is any fixed integer. When the domain of definition of the function e^w is restricted to that strip, the inverse function is obtained from formula (2) when $n = k$.

27. Branches of $\log z$

The function

$$\operatorname{Log} z = \operatorname{Log} r + i\Theta \qquad (r > 0,\ -\pi < \Theta \leqq \pi) \tag{1}$$

is continuous in the domain $r > 0$, $-\pi < \Theta < \pi$. This is seen by considering the component functions

$$u(r, \Theta) = \operatorname{Log} r \quad \text{and} \quad v(r, \Theta) = \Theta. \tag{2}$$

Each of the component functions, and hence $\operatorname{Log} z$, is continuous in the stated domain. That this is the largest possible domain in which $\operatorname{Log} z$ is continuous is evident from the fact that u is not even defined at the origin and also the fact that the value of the component function v at any point on the negative real axis is π whereas in each neighborhood of that point there are points at which the values of v are near $-\pi$.

The component functions u and v of $\operatorname{Log} z$ possess continuous first partial derivatives with respect to r and Θ in the domain $r > 0$, $-\pi < \Theta < \pi$; and those partial derivatives satisfy the polar form (2), Sec. 18, of the Cauchy-Riemann equations everywhere in that domain. We thus conclude from the theorem in Sec. 18 that *the function* $\operatorname{Log} z$ *is analytic in the domain* $r > 0$, $-\pi < \Theta < \pi$. Furthermore, if $z = re^{i\Theta}$,

$$\frac{d}{dz} \operatorname{Log} z = e^{-i\Theta}\left(\frac{1}{r} + i0\right) = \frac{1}{re^{i\Theta}};$$

that is,

$$\frac{d}{dz} \operatorname{Log} z = \frac{1}{z} \qquad (|z| > 0,\ -\pi < \operatorname{Arg} z < \pi). \tag{3}$$

Since $\operatorname{Log} z$ is not continuous at the origin or on the negative real axis, it cannot be differentiable there.

The notation $\operatorname{Log} z$ is often used to denote both the principal value of $\log z$, as in equation (1), and the analytic function obtained by restricting $\operatorname{Log} z$ to the domain $r > 0$, $-\pi < \Theta < \pi$. The particular usage is clear from the context in which it is found.

If we restrict the value of θ in definition (1) of the preceding section so that $\alpha < \theta < \alpha + 2\pi$ for a fixed but arbitrary α, the function

$$\log z = \operatorname{Log} r + i\theta \qquad (r > 0,\ \alpha < \theta < \alpha + 2\pi) \tag{4}$$

is single-valued and continuous throughout the indicated domain. If $w = \log z$, the range is the horizontal strip $\alpha < \operatorname{Im} \alpha < w + 2\pi$ (Fig. 19).

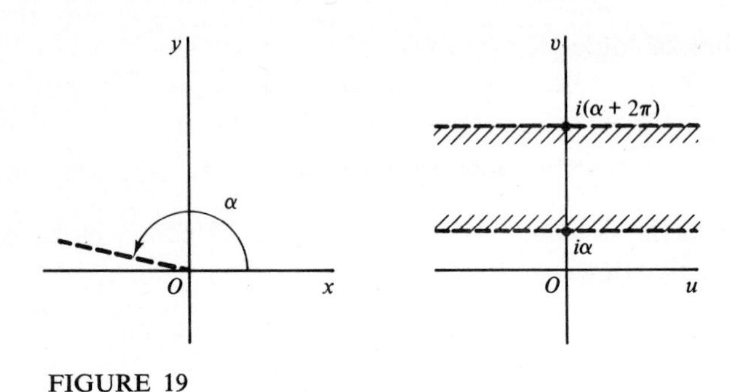

FIGURE 19

Throughout the domain of definition of the function (4) the component functions possess continuous first partial derivatives with respect to the variables r and θ, and at each point in that domain those partial derivatives satisfy the polar form of the Cauchy-Riemann equations. Hence $\log z$ as defined by equation (4) is analytic throughout its domain of definition, and

$$\frac{d}{dz}\log z = \frac{1}{z} \qquad (|z| > 0, \alpha < \arg z < \alpha + 2\pi). \tag{5}$$

A *branch* of a multiple-valued function f is any single-valued function F which is analytic in some domain at each point z of which the value $F(z)$ is one of the values $f(z)$.

In view of this definition, the function $\operatorname{Log} z$ defined at points in the domain $r > 0$, $-\pi < \Theta < \pi$ constitutes a branch of the logarithmic function (1) in the preceding section. This branch is called the *principal branch*. The function (4) is a branch of the same multiple-valued function.

Each point of the negative real axis $\Theta = \pi$, as well as the origin, is a singular point of the principal branch $\operatorname{Log} z$, according to our definition (Sec. 19) of a singular point. The ray $\Theta = \pi$ is called the *branch cut* for the principal branch, a branch cut being a line or curve of singular points introduced in defining a branch of a multiple-valued function. The ray $\theta = \alpha$ is a branch cut for the branch (4) of the logarithmic function. The singular point $z = 0$, common to all branch cuts for that multiple-valued function, is called a *branch point*.

28. Further Properties of Logarithms

Various other properties of logarithms carry over from elementary calculus with certain modifications.

First, let us verify the identity

$$e^{\log z} = z \qquad (z \neq 0). \tag{1}$$

This means that no matter what value of $\log z$ we select, the number $e^{\log z}$ will always be z. To see this, write $z = re^{i\theta}$ and $\log z = \operatorname{Log} r + i\theta$, where θ is any value of $\arg z$. Then

$$e^{\log z} = e^{\operatorname{Log} r + i\theta} = e^{\operatorname{Log} r} e^{i\theta} = re^{i\theta} = z.$$

It is not, however, true that $\log e^z$ is always equal to z. This is evident from the fact that $\log e^z$ is multiple-valued. In fact, if $z = x + iy$,

$$\begin{aligned} \log e^z &= \operatorname{Log} |e^z| + i \arg e^z = x + i(y + 2n\pi) \\ &= z + 2n\pi i \qquad (n = 0, \pm 1, \pm 2, \ldots). \end{aligned} \tag{2}$$

Let z_1 and z_2 be any two nonzero complex numbers, where

$$z_1 = r_1 \exp(i\theta_1), \qquad z_2 = r_2 \exp(i\theta_2).$$

It is easy to show that

$$\log(z_1 z_2) = \log z_1 + \log z_2. \tag{3}$$

This means, of course, that any value of $\log(z_1 z_2)$ can be expressed as some value of $\log z_1$ plus some value of $\log z_2$; and, conversely, any value of $\log z_1$ plus any value of $\log z_2$ is a value of $\log(z_1 z_2)$. Statement (3) follows immediately from the fact that $\operatorname{Log}(r_1 r_2) = \operatorname{Log} r_1 + \operatorname{Log} r_2$ together with the statement that $\arg(z_1 z_2) = \arg z_1 + \arg z_2$, as demonstrated in Sec. 5.

In like manner, it can be shown that

$$\log \frac{z_1}{z_2} = \log z_1 - \log z_2. \tag{4}$$

To illustrate statement (3), write

$$z_1 = z_2 = -1 \qquad \text{and} \qquad \log(z_1 z_2) = \log 1 = 0.$$

Equation (3) is satisfied when $\log z_1 = \pi i$ and $\log z_2 = -\pi i$; but it is not satisfied when $\log z_1 = \log z_2 = \pi i$, for example. It should therefore be noted that statement (3), as well as statement (4), is not in general valid when *log* is replaced everywhere by *Log*.

Let $z = r \exp(i\Theta)$ be a nonzero complex number, where Θ denotes the principal value of its argument, and let n be a positive integer. Recalling the formula (Sec. 6) for the nth roots of a nonzero complex number and the definition of the multiple-valued logarithm, we find that

$$\begin{aligned} \log(z^{1/n}) &= \log\left[\sqrt[n]{r} \exp \frac{i(\Theta + 2k\pi)}{n}\right] \\ &= \log \sqrt[n]{r} + i\left(\frac{\Theta + 2k\pi}{n} + 2p\pi\right) \\ &= \frac{1}{n} \operatorname{Log} r + i\frac{\Theta + 2(pn + k)\pi}{n} \end{aligned}$$

where k is any integer between 0 and $n - 1$ inclusive and p is any integer. On the other hand,

$$\frac{1}{n}\log z = \frac{1}{n}[\operatorname{Log} r + i(\Theta + 2q\pi)]$$

$$= \frac{1}{n}\operatorname{Log} r + i\frac{\Theta + 2q\pi}{n}$$

where q is any integer. Clearly, then, any one of the values of $\log(z^{1/n})$ is a value of $(1/n)\log z$. To prove the converse, recall that the remainder upon dividing an integer by a positive integer n is always an integer between 0 and $n - 1$ inclusive. That is, for any integer q there exists an integer p and an integer k between 0 and $n - 1$ inclusive such that $q = pn + k$. We thus conclude that

$$\log(z^{1/n}) = \frac{1}{n}\log z \qquad (n = 1, 2, \ldots) \tag{5}$$

where, corresponding to a value of $\log(z^{1/n})$ taken on the left, the appropriate value of $\log z$ is to be selected on the right, and conversely.

Observe that statement (5) together with property (1) yields the formula

$$z^{1/n} = \exp\left(\frac{1}{n}\log z\right) \qquad (n = 1, 2, \ldots). \tag{6}$$

For a fixed z the right-hand side of equation (6) takes on just n distinct values, the values of $z^{1/n}$.

In order to further clarify how results such as statement (5), involving multiple-valued logarithms, are to be interpreted in the sense of set equalities, we note that in general

$$\log(z^n) \neq n\log z \qquad (n = 1, 2, \ldots).$$

In the particular case when $z = i$ and $n = 2$, for example, the values of $\log(i^2)$ are the numbers $(2k + 1)\pi i$ $(k = 0, \pm 1, \pm 2, \ldots)$, whereas the values of $2\log i$ are the numbers $(4k + 1)\pi i$ $(k = 0, \pm 1, \pm 2, \ldots)$. Consequently, the set of values of $\log(i^2)$ is not the same as the set of values of $2\log i$; and $\log(i^2) \neq 2\log i$.

The statement $\log(z^n) = n\log z$ may or may not be true for specific values of z and n when the multiple-valued logarithmic function is replaced by a single-valued branch of it. Note, for example, that $\operatorname{Log}[(1 + i)^2] = 2\operatorname{Log}(1 + i)$ while $\operatorname{Log}[(-1 + i)^2] \neq 2\operatorname{Log}(-1 + i)$.

EXERCISES

1. Show that (*a*) $\text{Log}(-ei) = 1 - (\pi/2)i$; (*b*) $\text{Log}(1-i) = \frac{1}{2}\text{Log}\, 2 - (\pi/4)i$.

2. Show that when $n = 0, \pm 1, \pm 2, \ldots$, (*a*) $\log 1 = 2n\pi i$; (*b*) $\log(-1) = (2n+1)\pi i$; (*c*) $\log i = (2n + \frac{1}{2})\pi i$; (*d*) $\log(i^{1/2}) = (n + \frac{1}{4})\pi i$.

3. Find all the roots of the equation $\log z = (\pi/2)i$. *Ans.* $z = i$.

4. Find all the roots of the equation $e^z = -3$.

Ans. $z = \text{Log}\, 3 + (2n+1)\pi i \; (n = 0, \pm 1, \pm 2, \ldots)$.

5. Establish formula (4), Sec. 28.

6. By choosing specific nonzero values of z_1 and z_2, show that formula (4), Sec. 28, is not always valid when *log* is replaced everywhere by *Log*.

7. Show that if $\text{Re}\, z_1 > 0$ and $\text{Re}\, z_2 > 0$, then

$$\text{Log}(z_1 z_2) = \text{Log}\, z_1 + \text{Log}\, z_2.$$

8. Prove that if $z = re^{i\theta}$, then

$$\text{Log}(z^2) = 2\,\text{Log}\, z \qquad (r > 0, -\pi/2 < \theta < \pi/2).$$

9. Show that (*a*) if

$$\log z = \text{Log}\, r + i\theta \qquad (r > 0,\ \pi/4 < \theta < 9\pi/4),$$

then $\log(i^2) = 2 \log i$; (*b*) if

$$\log z = \text{Log}\, r + i\theta \qquad (r > 0,\ 3\pi/4 < \theta < 11\pi/4),$$

then $\log(i^2) \neq 2 \log i$.

10. Prove that if z is any nonzero complex number,

$$z^n = \exp(n \log z) \qquad (n = 1, 2, \ldots).$$

Recalling how in Sec. 6 we wrote $z^0 = 1$ and $z^n = (z^{-1})^{-n}$ when $n = -1, -2, \ldots$, show that this result is actually true when $n = 0, \pm 1, \pm 2, \ldots$.

11. Show that for all points z in the right half plane $x > 0$ the function $\text{Log}\, z$ can be written

$$\text{Log}\, z = \frac{1}{2}\text{Log}(x^2 + y^2) + i \arctan \frac{y}{x}$$

where $-\pi/2 < \arctan t < \pi/2$. Use this expression for $\text{Log}\, z$ together with the theorem in Sec. 17 to give another proof that the principal branch $\text{Log}\, z$ is analytic in the domain $x > 0$ and that formula (3), Sec. 27, is valid there. But note that some complications arise with the inverse tangent and its derivative in the remaining part of the full domain of analyticity, $r > 0$, $-\pi < \Theta < \pi$, of $\text{Log}\, z$, especially on the line $x = 0$.

12. Show in two ways that the function $\text{Log}(x^2 + y^2)$ is harmonic in every domain that does not contain the origin.

13. Show that (*a*) the function $\text{Log}(z - i)$ is analytic everywhere except on the half line $x \leqq 0$, $y = 1$; (*b*) the function

$$\frac{\text{Log}(z+4)}{z^2 + i}$$

is analytic everywhere except at the points $\pm(1-i)/\sqrt{2}$ and on the half line $x \leqq -4$, $y = 0$.

14. Write $z = r \exp(i\theta)$ and show that

$$\text{Re}\,[\log(z-1)] = \tfrac{1}{2}\,\text{Log}\,(1 - 2r\cos\theta + r^2) \qquad (z \neq 1).$$

Why must this function satisfy Laplace's equation when $z \neq 1$?

29. Complex Exponents

Formula (6) of the preceding section and Exercise 10 there suggest that we define z^c, where *the exponent c is any complex number*, by the equation

$$z^c = \exp(c \log z) \qquad (z \neq 0). \tag{1}$$

The exponential function used on the right in equation (1) is, of course, defined according to equation (4), Sec. 21, and $\log z$ denotes the multiple-valued logarithmic function. Definition (1) is thus a consistent definition in the sense that it includes the special cases noted above when $c = 1/n$ $(n = 1, 2, \ldots)$ and $c = n$ $(n = 0, \pm 1, \pm 2, \ldots)$.

Such powers of z are, in general, multiple-valued. For example,

$$i^{-2i} = \exp(-2i \log i) = \exp\left[-2i\left(\frac{\pi}{2} + 2n\pi\right)i\right]$$

$$= \exp[(4n+1)\pi] \qquad (n = 0, \pm 1, \pm 2, \ldots).$$

Note that, in view of the property $e^{-z} = 1/e^z$, the two sets of numbers z^{-c} and $1/z^c$ are the same. We can therefore write

$$z^{-c} = \frac{1}{z^c} \qquad (z \neq 0). \tag{2}$$

Other familiar properties of exponents carry over from the theory of real variables. Suppose, for instance, that $z = re^{i\theta}$ and α is any real number. The function

$$\log z = \text{Log}\, r + i\theta \qquad (r > 0, \alpha < \theta < \alpha + 2\pi) \tag{3}$$

is single-valued and analytic in the domain indicated, as is the composite function $\exp(c \log z)$. Thus the function z^c defined by equation (1), where $\log z$ is given by equation (3), is single-valued and analytic in the domain $r > 0$, $\alpha < \theta < \alpha + 2\pi$. The derivative of that branch of the multiple-valued power function (1) can be written in terms of the logarithmic function defined by equation (3) as follows:

$$\frac{d}{dz} z^c = \frac{d}{dz} \exp(c \log z) = \exp(c \log z)\, \frac{c}{z}$$

$$= c\,\frac{\exp(c \log z)}{\exp(\log z)} = c \exp[(c-1)\log z].$$

The final term here is the single-valued function cz^{c-1}; thus

$$\frac{d}{dz} z^c = cz^{c-1} \qquad (|z| > 0, \alpha < \arg z < \alpha + 2\pi). \tag{4}$$

When $\alpha = -\pi$, so that $-\pi < \arg z < \pi$, the function

$$z^c = \exp (c \operatorname{Log} z) \qquad (z \neq 0) \tag{5}$$

is called the *principal branch* of the multiple-valued power function (1). It is single-valued and analytic in the domain $|z| > 0$, $-\pi < \operatorname{Arg} z < \pi$.

As an example, the principal value of $(-i)^i$ is

$$\exp [i \operatorname{Log} (-i)] = \exp\left[i\left(-i\frac{\pi}{2}\right)\right] = \exp\frac{\pi}{2}.$$

As another example, the principal branch of $z^{2/3}$,

$$\exp (\tfrac{2}{3} \operatorname{Log} z) = \exp (\tfrac{2}{3} \operatorname{Log} r + \tfrac{2}{3} i\Theta) = \sqrt[3]{r^2} \exp (i\tfrac{2}{3}\Theta),$$

is analytic in the domain $r > 0$, $-\pi < \Theta < \pi$, as we can see also from the theorem in Sec. 18.

Note that if $z = e$ in definition (1), the quantity e^c on the left is in general multiple-valued. The usual definition of e^c occurs when the principal branch is taken.

According to definition (1), *the exponential function with base c*, where c is any nonzero complex constant, is written

$$c^z = \exp (z \log c) \qquad (c \neq 0). \tag{6}$$

When a value of $\log c$ is specified, c^z is an entire function of z. It is easily seen that

$$\frac{d}{dz} c^z = c^z \log c \qquad (c \neq 0). \tag{7}$$

30. Inverse Trigonometric Functions

Inverses of the trigonometric and hyperbolic functions can be described in terms of logarithms.

To define the inverse sine function $\sin^{-1} z$, we write $w = \sin^{-1} z$ when $z = \sin w$. That is, $w = \sin^{-1} z$ when

$$z = \frac{e^{iw} - e^{-iw}}{2i}.$$

To express w in terms of z, we first obtain e^{iw} by solving the equation

$$e^{2iw} - 2ize^{iw} - 1 = 0,$$

which is quadratic in e^{iw}. We find that

$$e^{iw} = iz + (1 - z^2)^{1/2}$$

where $(1 - z^2)^{1/2}$ is, as we know, a double-valued function of z. Taking the logarithm of each side and recalling that $w = \sin^{-1} z$, we arrive at the formula

$$\sin^{-1} z = -i \log [iz + (1 - z^2)^{1/2}]. \tag{1}$$

Note that $\sin^{-1} z$ is a multiple-valued function with infinitely many values at each point z. When specific branches of the square root and the logarithm are used, this function becomes single-valued and analytic because it is then a composition of analytic functions.

In a similar fashion, the inverse cosine and the inverse tangent functions can be written

$$\cos^{-1} z = -i \log [z + i(1 - z^2)^{1/2}], \tag{2}$$

$$\tan^{-1} z = \frac{i}{2} \log \frac{i + z}{i - z}. \tag{3}$$

The derivatives of these three functions are readily obtained from the above formulas. The derivatives of the first two depend on the values chosen for the square roots:

$$\frac{d}{dz} \sin^{-1} z = \frac{1}{(1 - z^2)^{1/2}}; \qquad \frac{d}{dz} \cos^{-1} z = \frac{-1}{(1 - z^2)^{1/2}}. \tag{4}$$

The derivative of the last one,

$$\frac{d}{dz} \tan^{-1} z = \frac{1}{1 + z^2}, \tag{5}$$

does not, however, depend on the manner in which the function is made single-valued.

Inverse hyperbolic functions can be treated in a corresponding manner. It turns out that

$$\sinh^{-1} z = \log [z + (z^2 + 1)^{1/2}], \tag{6}$$

$$\cosh^{-1} z = \log [z + (z^2 - 1)^{1/2}], \tag{7}$$

$$\tanh^{-1} z = \frac{1}{2} \log \frac{1 + z}{1 - z}. \tag{8}$$

Finally, we remark that common alternative notation for all these inverse functions is arcsin z, etc.

EXERCISES

1. Show that when $n = 0, \pm 1, \pm 2, \ldots$,
(*a*) $(1+i)^i = \exp(-\pi/4 + 2n\pi) \exp[(i/2) \text{ Log } 2]$;
(*b*) $(-1)^{1/\pi} = \exp[(2n+1)i]$.

2. Find the principal value of (*a*) i^i; (*b*) $[(e/2)(-1-i\sqrt{3})]^{3\pi i}$;
(*c*) $(1-i)^{4i}$.

Ans. (*a*) $\exp(-\pi/2)$; (*b*) $-\exp(2\pi^2)$.

3. Show that if $z \neq 0$ and k is a real number, $|z^k| = \exp(k \text{ Log } |z|) = |z|^k$.

4. Let c, d, and z denote complex numbers, where $z \neq 0$. Prove that if all the powers involved are principal values, (*a*) $z^{-c} = 1/z^c$;
(*b*) $(z^c)^n = z^{cn}$ $(n = 1, 2, \ldots)$; (*c*) $z^c z^d = z^{c+d}$; (*d*) $z^c/z^d = z^{c-d}$.

5. Using the principal branch of z^i, write the functions $u(r,\theta)$ and $v(r,\theta)$ when $z^i = u(r,\theta) + iv(r,\theta)$.

6. Derive formula (7), Sec. 29.

7. Assuming that $f'(z)$ exists, find the formula for $d[c^{f(z)}]/dz$.

8. Find the values of (*a*) $\tan^{-1}(2i)$; (*b*) $\tan^{-1}(1+i)$; (*c*) $\cosh^{-1}(-1)$;
(*d*) $\tanh^{-1} 0$.

Ans. (*a*) $(n+\frac{1}{2})\pi + \dfrac{i}{2} \text{Log } 3$ $(n = 0, \pm 1, \pm 2, \ldots)$;

(*d*) $n\pi i$ $(n = 0, \pm 1, \pm 2, \ldots)$.

9. Let c be a fixed nonzero complex number and note that i^c is multiple-valued. What restriction must be placed on the constant c in order that the values of $|i^c|$ be all the same. *Ans.* c is real.

10. Solve the equation $\sin z = 2$ for z (*a*) by identifying real parts and imaginary parts; (*b*) by using formula (1), Sec. 30.

Ans. $(2n+\frac{1}{2})\pi \pm i \text{ Log } (2+\sqrt{3})$ $(n = 0, \pm 1, \pm 2, \ldots)$.

11. Solve the equation $\cos z = \sqrt{2}$ for z.

12. Derive formulas (2) and (4), Sec. 30.

13. Derive formulas (3) and (5), Sec. 30.

14. Derive formulas (6) and (8), Sec. 30.

4

MAPPING BY ELEMENTARY FUNCTIONS

The geometric interpretation of a function of a complex variable as a mapping, or transformation, was introduced in Sec. 10. We pointed out there that the nature of such a function may be displayed graphically, to some extent, by the manner in which it maps certain curves and regions.

In this chapter we shall see how various curves and regions are mapped by elementary analytic functions. Applications of such results to physical problems will be illustrated later on in Chaps. 9 and 10.

31. Linear Functions

The mapping of the z plane onto the w plane by means of the equation

$$w = z + C, \tag{1}$$

where C is a complex constant, is a translation by means of the vector representing C. That is, if

$$z = x + iy \qquad \text{and} \qquad C = C_1 + iC_2,$$

then the image of any point (x,y) in the z plane is the point

$$(x + C_1, y + C_2)$$

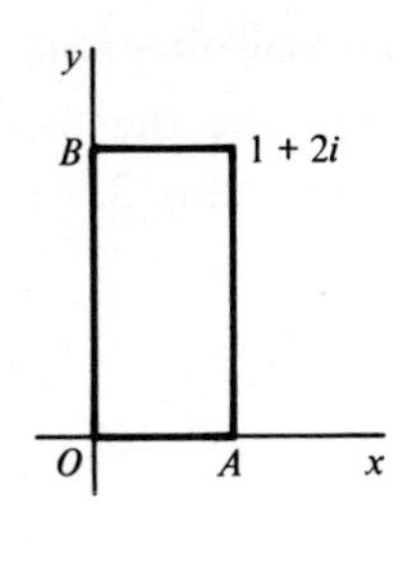

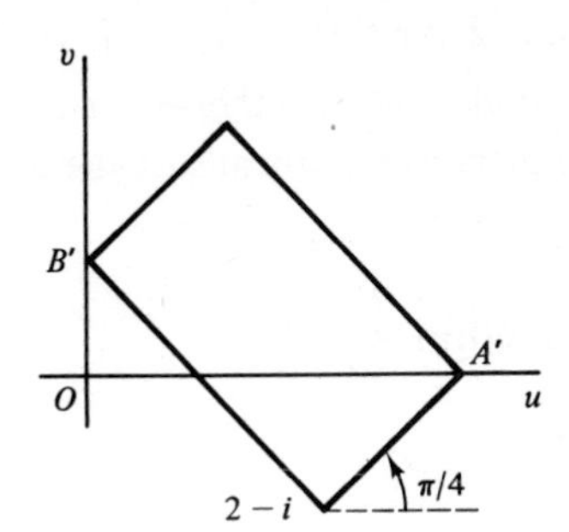

FIGURE 20

in the w plane. Since each point in any given region of the z plane is mapped into the w plane in this manner, the image region is geometrically congruent to the original one.

Properties of the mapping defined by the equation

$$w = Bz \qquad (B \neq 0), \tag{2}$$

where B is complex, are readily obtained by using the polar forms of B and z. For if $B = be^{i\beta}$ and $z = re^{i\theta}$, then

$$w = bre^{i(\beta+\theta)}.$$

Thus, the transformation defined by equation (2) maps any nonzero point z with polar coordinates (r,θ) onto the nonzero point whose polar coordinates are $(br, \beta + \theta)$. This mapping consists of both a rotation of the vector representing z about the origin through an angle $\beta = \arg B$ and also an expansion or contraction of the vector by the factor $b = |B|$. The image of a given region in the z plane is, therefore, geometrically similar to that region.

By applying transformation (1) to the variable w in equation (2), we obtain the general *linear transformation*

$$w = Bz + C \qquad (B \neq 0) \tag{3}$$

which consists of a rotation and an expansion or contraction followed by a translation.

As an illustration, the transformation

$$w = (1 + i)z + 2 - i \tag{4}$$

maps the rectangular region shown in the z plane of Fig. 20 onto the rectangular region shown in the w plane there. This is seen by noting that transformation (4) is a composition of the transformations $Z = (1 + i)z$ and $w = Z + 2 - i$.

Since $1 + i = \sqrt{2}\exp(i\pi/4)$, the first of these transformations is a rotation through the angle $\pi/4$ together with an expansion by the factor $\sqrt{2}$. The second is a translation by means of the vector representing $2 - i$.

32. The Function $\dfrac{1}{z}$

The equation

$$w = \frac{1}{z} \tag{1}$$

establishes a one to one correspondence between the nonzero points of the z and the w planes. Since $z\bar{z} = |z|^2$, the mapping can be described by means of the consecutive transformations

$$Z = \frac{1}{|z|^2}\,z, \qquad w = \bar{Z}. \tag{2}$$

The first of these transformations is an inversion with respect to the circle $|z| = 1$. That is, the image of a nonzero point z is the point Z with the properties

$$|Z| = \frac{1}{|z|} \qquad \text{and} \qquad \arg Z = \arg z.$$

Thus the points exterior to the circle $|z| = 1$ are mapped onto the nonzero points interior to it, and conversely (Fig. 21). Any point on the circle is mapped onto itself. The second of transformations (2) is simply a reflection in the real axis.

Observe that the image of a circle $|z| = \varepsilon$ is the circle $|w| = 1/\varepsilon$. Also, an ε neighborhood $|z| < \varepsilon$ of the origin with the origin excluded corresponds to an ε neighborhood $|w| > 1/\varepsilon$ of the point at infinity (Sec. 8). It is therefore natural to define a transformation T on the extended complex plane by writing $T(0) = \infty$, $T(\infty) = 0$, and $T(z) = 1/z$ for the remaining values of z. The transformation T is then a one to one continuous mapping of the extended complex plane onto itself. The continuity of this transformation throughout the extended plane was established in Exercise 12(*a*), Sec. 15.

If a, b, c, and d are real numbers, the equation

$$a(x^2 + y^2) + bx + cy + d = 0$$

represents any circle or line, depending on whether $a \neq 0$ or $a = 0$, respectively. When $w = 1/z$, that equation becomes

$$d(u^2 + v^2) + bu - cv + a = 0.$$

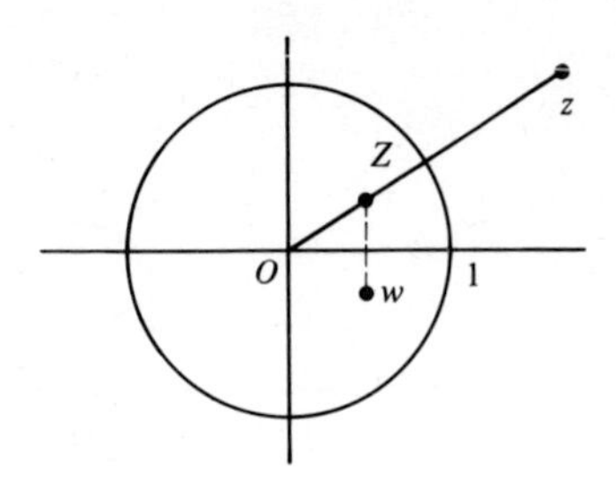

FIGURE 21
$w = 1/z$.

This is readily seen by using cartesian coordinates and noting that the equation

$$u + iv = \frac{1}{x + iy}$$

yields the relations

$$u = \frac{x}{x^2 + y^2}, \qquad v = -\frac{y}{x^2 + y^2},$$

or

$$x = \frac{u}{u^2 + v^2}, \qquad y = -\frac{v}{u^2 + v^2}.$$

Thus, a circle ($a \neq 0$) not passing through the origin ($d \neq 0$) in the z plane is transformed into a circle not passing through the origin in the w plane. A circle through the origin in the z plane is transformed into a line which does not pass through the origin in the w plane. A line not passing through the origin in the z plane is transformed into a circle through the origin in the w plane, and a line through the origin in the z plane is transformed into a line through the origin in the w plane. If we consider lines in the extended plane as circles through the point at infinity, we can say that the function T defined above always transforms circles into circles.

Note that the line $x = c_1$ ($c_1 \neq 0$) is transformed into the circle

$$u^2 + v^2 - \frac{u}{c_1} = 0 \tag{3}$$

which is tangent to the v axis at the origin; and the line $y = c_2$ ($c_2 \neq 0$) is transformed into the circle

$$u^2 + v^2 + \frac{v}{c_2} = 0 \tag{4}$$

which is tangent to the u axis at the origin. (See Fig. 22.)

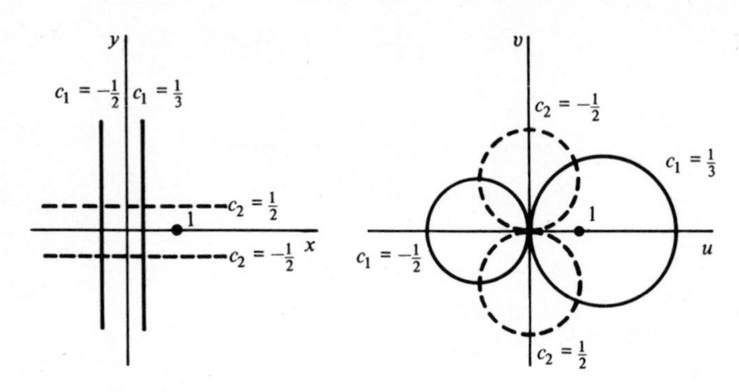

FIGURE 22

The half plane $x > c_1$ $(c_1 > 0)$ has for its image the region

$$\frac{u}{u^2 + v^2} > c_1, \tag{5}$$

or

$$\left(u - \frac{1}{2c_1}\right)^2 + v^2 < \left(\frac{1}{2c_1}\right)^2;$$

so the image of any point in that half plane lies inside the circle given by equation (3). Conversely, every point inside the circle satisfies inequality (5) and is therefore the image of a point in the half plane. Hence the image of the half plane is the entire circular region.

The function $1/z$ is useful in studying properties of a function f when the point at infinity is involved. If the limit of $f(z)$ as z tends to infinity is a number w_0, we can define f at ∞ to be w_0 so that it is continuous there; we then write $f(\infty) = w_0$. The number w_0 can also be determined by evaluating the limit of $f(1/z)$ as z approaches 0. For, according to the definitions of limits (Sec. 11),

$$\lim_{z\to\infty} f(z) = w_0 \qquad \text{if and only if} \qquad \lim_{z\to 0} f\left(\frac{1}{z}\right) = w_0. \tag{6}$$

In like manner, we can make f continuous at a point z_0 by writing $f(z_0) = \infty$ when the limit of $1/f(z)$ as z approaches z_0 is 0. Observe that this is all in agreement with the way in which we defined the function T earlier in this section by extending the domain of definition of the function $1/z$.

For an illustration, consider the function

$$f(z) = \frac{4z^2}{(1 - z)^2}.$$

In order to make f continuous at ∞, we write $f(\infty) = 4$ since the limit of

$$f\left(\frac{1}{z}\right) = \frac{4}{(z-1)^2}$$

as z approaches 0 is 4. We can also make f continuous at the point $z = 1$ by writing $f(1) = \infty$ since the limit of $1/f(z)$ as z approaches 1 is 0.

Finally, we can write $f(\infty) = \infty$ if the limit of the function $1/f(1/z)$ as z approaches 0 is 0.

EXERCISES

1. Show that the transformation $w = iz$ is a rotation of the z plane through the angle $\pi/2$. Find the image of the infinite strip $0 < x < 1$.

   Ans. $0 < v < 1$.

2. Show that the transformation $w = iz + i$ maps the half plane $x > 0$ onto the half plane $v > 1$.
3. Find the region onto which the half plane $y > 0$ is mapped by the transformation $w = (1 + i)z$ (*a*) by using polar coordinates; (*b*) by using cartesian coordinates. Sketch this region.

   Ans. $v > u$.

4. Find the image of the region $y > 1$ under the transformation $w = (1 - i)z$.
5. Find the image of the semi-infinite strip $x > 0, 0 < y < 2$ under the transformation $w = iz + 1$. Sketch the strip and its image.

   Ans. $-1 < u < 1, v > 0$.

6. Give a geometric description of the transformation $w = B(z + C)$ where B and C are complex constants and $B \neq 0$.
7. Show that if $c_1 < 0$, the image of the half plane $x < c_1$ under the transformation $w = 1/z$ is the interior of a circle. What is the image when $c_1 = 0$?
8. Show that the image of the half plane $y > c_2$ under the transformation $w = 1/z$ is the interior of a circle, provided $c_2 > 0$. Find the image when $c_2 < 0$; also find it when $c_2 = 0$.
9. Find the image of the infinite strip $0 < y < 1/(2c)$ under the transformation $w = 1/z$. Sketch the strip and its image.

   Ans. $u^2 + (v + c)^2 > c^2, v < 0$.

10. Find the image of the quadrant $x > 1, y > 0$ under the transformation $w = 1/z$.

    Ans. $|w - \frac{1}{2}| < \frac{1}{2}, v < 0$.

11. Verify the mapping, where $w = 1/z$, of the regions and parts of the boundaries indicated in (*a*) Fig. 4, Appendix 2; (*b*) Fig. 5, Appendix 2.
12. Describe geometrically the transformation $w = 1/(z - 1)$.
13. Describe geometrically the transformation $w = i/z$; also show that it transforms circles and lines into circles and lines.

14. Find the image of the semi-infinite strip $x \geqq 0$, $0 \leqq y \leqq 1$ under the transformation $w = i/z$. Sketch the strip and its image.

Ans. $0 \leqq \phi \leqq \pi/2$, $\rho \geqq \cos \phi$, where $w = \rho e^{i\phi}$.

15. Find the image of the hyperbola $x^2 - y^2 = 1$ under the transformation $w = 1/z$.

Ans. $\rho^2 = \cos 2\phi$, where $w = \rho e^{i\phi}$.

16. Let the circle $|z| = 1$ have a positive, or counterclockwise, orientation. Determine the orientation of its image under the transformation $w = 1/z$.

17. Show that when a circle is transformed into a circle under the transformation $w = 1/z$, the center of the original circle is never mapped onto the center of the image circle.

18. Prove statement (6), Sec. 32.

19. Show that

$$\lim_{z \to \infty} f(z) = \infty \qquad \text{if and only if} \qquad \lim_{z \to 0} \frac{1}{f(1/z)} = 0.$$

20. Determine $f(-1)$ and $f(\infty)$ so that if

$$f(z) = \frac{-z + 1}{z + 1} \qquad (z \neq -1, \infty),$$

the function f is continuous in the extended plane.

Ans. $f(-1) = \infty, f(\infty) = -1$.

33. Linear Fractional Transformations

The transformation

$$w = \frac{az + b}{cz + d} \qquad (ad - bc \neq 0), \tag{1}$$

where a, b, c, and d are complex constants, is called a *linear fractional transformation*, or Möbius transformation. When $c = 0$, it is simply a linear transformation (Sec. 31). When $c \neq 0$, equation (1) can be written

$$w = \frac{a}{c} + \frac{bc - ad}{c} \frac{1}{cz + d}, \tag{2}$$

and it is clear that the condition $ad - bc \neq 0$ ensures that a linear fractional tranformation is not a constant function.

When cleared of fractions, equation (1) takes the form

$$Azw + Bz + Cw + D = 0 \tag{3}$$

which is linear in z and linear in w, or bilinear in z and w. Hence another name for a linear fractional transformation is *bilinear transformation*.

Solving equation (1) for z, we find that

$$z = \frac{-dw + b}{cw - a}. \tag{4}$$

Hence, if $c = 0$, each point in the w plane is the image of one and only one point in the z plane; the same is true if $c \neq 0$, except for the case $w = a/c$. We now extend the domain of definition of transformation (1) in order to define a linear fractional transformation T on the extended z plane. We first write

$$T(z) = \frac{az + b}{cz + d}. \tag{5}$$

We then write $T(\infty) = \infty$ if $c = 0$; and we write $T(-d/c) = \infty$ and $T(\infty) = a/c$ if $c \neq 0$. This agrees with the convention established at the end of the previous section.

When its domain of definition is extended in this way, the linear fractional transformation (1) is a *one to one* mapping of the extended complex plane *onto* itself. That is, $T(z_1) \neq T(z_2)$ whenever $z_1 \neq z_2$, and for each point w in the extended plane there is a point z in that plane such that $T(z) = w$. Hence, associated with the transformation T, there is an *inverse transformation* T^{-1} defined on the extended plane as follows:

$$T^{-1}(w) = z \qquad \text{if and only if} \qquad T(z) = w.$$

If we interchange z and w in this definition and in equation (4), we find that

$$T^{-1}(z) = \frac{-dz + b}{cz - a} \qquad (ad - bc \neq 0).$$

Thus, T^{-1} is itself a linear fractional transformation where $T^{-1}(\infty) = \infty$ if $c = 0$, and $T^{-1}(a/c) = \infty$ and $T^{-1}(\infty) = -d/c$ if $c \neq 0$.

If T and S are two linear fractional transformations, then so is the composition $S[T(z)]$. This can be verified by combining expressions of the type (5).

We have already observed that if $c = 0$, the linear fractional transformation (1) is of the special type $w = Bz + C$ $(B \neq 0)$. If, on the other hand, $c \neq 0$, form (2) of equation (1) reveals that the linear fractional transformation is a composition of the special types

$$Z = cz + d, \qquad W = \frac{1}{Z}, \qquad w = \frac{a}{c} + \frac{bc - ad}{c} W.$$

It thus follows that *a linear fractional transformation always transforms circles into circles* because these special linear fractional transformations do this. (See Secs. 31 and 32.) Here lines in the extended plane are considered as circles through the point at infinity.

There is just one linear fractional transformation that maps three given distinct points z_1, z_2, and z_3 onto three specified distinct points w_1, w_2, and w_3, respectively. Indeed, the equation

$$\frac{(w - w_1)(w_2 - w_3)}{(w - w_3)(w_2 - w_1)} = \frac{(z - z_1)(z_2 - z_3)}{(z - z_3)(z_2 - z_1)} \tag{6}$$

defines such a transformation. To see this, note first that equation (6) can be written

$$(z - z_3)(w - w_1)(z_2 - z_1)(w_2 - w_3) = (z - z_1)(w - w_3)(z_2 - z_3)(w_2 - w_1), \tag{7}$$

which in turn can be put in form (3) by expanding the products. Next observe that if $z = z_1$, the right-hand side of equation (7) is zero and it follows that $w = w_1$. Similarly, if $z = z_3$, the left-hand side is zero and consequently $w = w_3$. If $z = z_2$, we obtain the linear equation

$$(w - w_1)(w_2 - w_3) = (w - w_3)(w_2 - w_1)$$

whose unique solution is $w = w_2$. It is left to the exercises to show that equation (6) defines the only linear fractional transformation mapping the points z_1, z_2, and z_3 onto w_1, w_2, and w_3, respectively.

In equation (6) the point at infinity can be introduced as one of the prescribed points in either the z or the w plane. For example, if w_2 is to be ∞, replace w_2 by $1/w_2$ in that equation and then, after clearing fractions in the numerator and denominator on the left-hand side, write $w_2 = 0$. This yields the equation

$$\frac{w - w_1}{w - w_3} = \frac{(z - z_1)(z_2 - z_3)}{(z - z_3)(z_2 - z_1)}. \tag{8}$$

To illustrate, we determine the linear fractional transformation T that maps $z_1 = 1$, $z_2 = 0$, $z_3 = -1$ onto $w_1 = i$, $w_2 = \infty$, $w_3 = 1$, respectively. Substituting the given values of z_1, z_2, z_3 and w_1, w_3 into equation (8), we find that

$$w = \frac{(1 + i)z + (i - 1)}{2z}.$$

It is readily verified that the given points in the z plane map onto the specified points in the w plane.

34. Special Linear Fractional Transformations

Let us determine all linear fractional transformations that map the upper half plane $\operatorname{Im} z \geqq 0$ onto the unit disk $|w| \leqq 1$.

Since a linear fractional transformation

$$w = \frac{az + b}{cz + d} \qquad (ad - bc \neq 0) \tag{1}$$

transforms lines in the z plane into circles or lines in the w plane, we know that the boundary $\operatorname{Im} z = 0$ of the half plane is transformed into either a circle or a line. The boundary $\operatorname{Im} z = 0$ must, in fact, be transformed into a circle because its image lies in the disk $|w| \leqq 1$ and is thus bounded. Suppose now that there are points on this circle which lie in the interior of the disk. Then, since equation (1) defines a continuous function of z, there would be a point just below the x axis which is mapped onto a point near that circle and inside the disk. This image point would also be the image of a point on or above the x axis since the mapping of the half plane is *onto* the disk. This, however, contradicts the fact that a linear fractional transformation defined on the entire plane is one to one. Thus, the image of the boundary $\operatorname{Im} z = 0$ of the half plane must be the boundary $|w| = 1$ of the disk.

Now a linear fractional transformation of the line $\operatorname{Im} z = 0$ onto the circle $|w| = 1$ is uniquely determined if we require it to map three given points of the line onto three specified points of the circle. Let us select the points $z = 0$, $z = 1$, and $z = \infty$ on the line and determine all transformations of the form (1) which map those points onto points lying on the circle.

In view of equation (1), the requirement that $|w| = 1$ when $z = 0$ and $z = \infty$ leads to the equations

$$|d| = |b|, \tag{2}$$

$$|c| = |a|. \tag{3}$$

From the second of these equations and the fact that $ad - bc \neq 0$, it follows that $a \neq 0$ and $c \neq 0$. Hence

$$w = \frac{a}{c}\,\frac{z + b/a}{z + d/c},$$

or, since $|a/c| = 1$,

$$w = e^{i\alpha}\,\frac{z - z_0}{z - z_1} \tag{4}$$

where α is any real constant and z_0, z_1 are complex constants. Since $|d/c| = |b/a|$ according to equations (2) and (3), we know that $|z_1| = |z_0|$.

We now impose on equation (4) the condition that $|w| = 1$ when $z = 1$. This yields

$$|1 - z_1| = |1 - z_0|,$$

or

$$(1 - z_1)(1 - \bar{z}_1) = (1 - z_0)(1 - \bar{z}_0).$$

But $z_1\bar{z}_1 = z_0\bar{z}_0$ since $|z_1| = |z_0|$, and the above relation reduces to

$$z_1 + \bar{z}_1 = z_0 + \bar{z}_0,$$

or $\operatorname{Re} z_1 = \operatorname{Re} z_0$. It follows that either $z_1 = z_0$ or $z_1 = \bar{z}_0$, again since $|z_1| = |z_0|$. The condition $z_1 = z_0$ leads to the mapping $w = \exp(i\alpha)$ of the z plane onto a single point; hence $z_1 = \bar{z}_0$.

The required transformation must therefore have the form

$$w = e^{i\alpha}\frac{z - z_0}{z - \bar{z}_0}. \tag{5}$$

Observe that the point $w = 0$ is the image of z_0. So the point z_0 must lie above the real axis, or

$$\operatorname{Im} z_0 > 0. \tag{6}$$

It remains to show that, with condition (6), transformation (5) actually maps the half plane $\operatorname{Im} z \geqq 0$ *onto* the disk $|w| \leqq 1$. We do this by interpreting the equation

$$|w| = \frac{|z - z_0|}{|z - \bar{z}_0|}$$

geometrically. If a point z lies above the real axis, both it and the point z_0 lie on the same side of that axis which is the perpendicular bisector of the line segment joining z_0 and $\bar{z}_0$. It follows that the distance $|z - z_0|$ is less than the distance $|z - \bar{z}_0|$ (Fig. 23); that is, $|w| < 1$. Likewise, if z lies below the real axis, the distance $|z - z_0|$ is greater than the distance $|z - \bar{z}_0|$, and so $|w| > 1$. Since any linear fractional transformation is a one to one mapping of the extended plane onto itself, every point w such that $|w| < 1$ must therefore be the image of exactly one point z above the real axis.

We conclude that any linear fractional transformation of the form (5), where the real number α is arbitrary and the imaginary part of the complex number z_0 is positive, maps the half plane $\operatorname{Im} z \geqq 0$ onto the disk $|w| \leqq 1$ in a one to one manner.

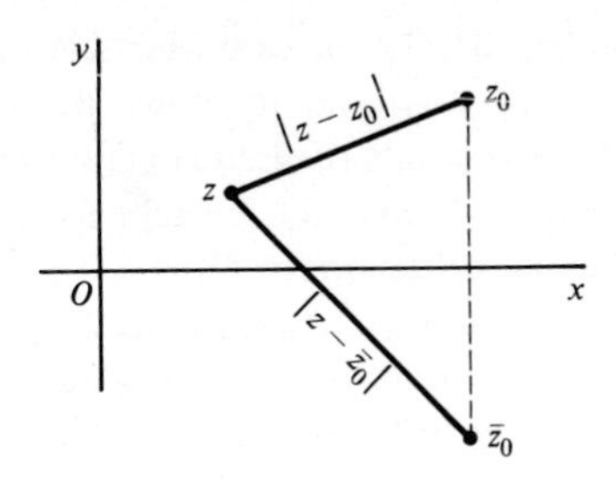

FIGURE 23

This result can be used to illustrate how the identity transformation is not necessarily the only one that maps a given region onto itself. In fact, any linear fractional transformation of the form

$$w = e^{i\alpha} \frac{z - z_0}{\bar{z}_0 z - 1}, \tag{7}$$

where α is real and $|z_0| < 1$, maps the disk $|z| \leqq 1$ onto the disk $|w| \leqq 1$ in a one to one manner. The proof of this is left as an exercise.

EXERCISES

1. Find the linear fractional transformation that maps the points $z_1 = 2$, $z_2 = i$, $z_3 = -2$ onto the points $w_1 = 1$, $w_2 = i$, $w_3 = -1$.

   Ans. $w = (3z + 2i)/(iz + 6)$.

2. Find the linear fractional transformation that maps the points $z_1 = -i$, $z_2 = 0$, $z_3 = i$ onto the points $w_1 = -1$, $w_2 = i$, $w_3 = 1$. Into what curve is the imaginary axis transformed?

3. Find the bilinear transformation that maps the points $z_1 = \infty$, $z_2 = i$, $z_3 = 0$ onto the points $w_1 = 0$, $w_2 = i$, $w_3 = \infty$.

   Ans. $w = -1/z$.

4. Find the bilinear transformation that maps the points z_1, z_2, z_3 onto the points $w_1 = 0$, $w_2 = 1$, $w_3 = \infty$.

   Ans. $w = [(z - z_1)(z_2 - z_3)]/[(z - z_3)(z_2 - z_1)]$.

5. Show that a composition of two linear fractional transformations is again a linear fractional transformation.

6. A *fixed point* of a transformation $w = f(z)$ is a point z_0 such that $z_0 = f(z_0)$. Show that every bilinear transformation, with the exception of the identity transformation $w = z$, has at most two fixed points in the extended plane.

7. Find the fixed points (Exercise 6) of the transformations
   (*a*) $w = (z - 1)/(z + 1)$; (*b*) $w = (6z - 9)/z$.

   Ans. (*a*) $z = \pm i$; (*b*) $z = 3$.

8. Modify equation (6), Sec. 33, for the case when both z_2 and w_2 are the point at infinity. Then show that any linear fractional transformation must be of the form $w = az$ when its fixed points (Exercise 6) are 0 and ∞.
9. Prove that if the origin is a fixed point (Exercise 6) of a bilinear transformation, the transformation can be written in the form $w = z/(cz + d)$.
10. Verify the mapping shown in Fig. 12, Appendix 2, where $w = (z - 1)/(z + 1)$.
11. Determine the constants in transformation (5), Sec. 34, such that it maps the half plane $\text{Im } z \geqq 0$ onto the disk $|w| \leqq 1$ where the images of the points $z = \infty$, $z = 0$, and $z = 1$ are the points $w = -1$, $w = 1$, and $w = i$, respectively. Show that the image of the positive x axis is the upper half of the circle $|w| = 1$. Then verify the mapping illustrated in Fig. 13, Appendix 2.
12. Use the transformation $w = (i - z)/(i + z)$, illustrated in Fig. 13, Appendix 2, to show that the disk $|z - 1| \leqq 1$ is mapped onto the half plane $\text{Re } w \leqq 0$ by the linear fractional transformation $w = (z - 2)/z$.

    Suggestion: First translate the given disk one unit to the left. Then apply the inverse of the transformation in Appendix 2, and follow this with a rotation of $\pi/2$ radians.
13. Transformation (5), Sec. 34, with condition (6) there, maps the point $z = \infty$ onto the point $w = \exp(i\alpha)$ which lies on the boundary of the disk $|w| \leqq 1$. Show that if $0 < \alpha < 2\pi$ and the points $z = 0$ and $z = 1$ are to be mapped onto the points $w = 1$ and $w = \exp(i\alpha/2)$, respectively, the transformation can be written

$$w = \exp(i\alpha)\,\frac{z + \exp(-i\alpha/2)}{z + \exp(i\alpha/2)}.$$

14. Note that when $\alpha = \pi/2$, the transformation in Exercise 13 becomes

$$w = \frac{iz + \exp(i\pi/4)}{z + \exp(i\pi/4)}.$$

    Verify that this special case maps the half plane $\text{Im } z > 0$ and segments of its boundary in the manner indicated in Fig. 24.
15. Show that when $\text{Im } z_0 < 0$, transformation (5), Sec. 34, maps the lower half plane $\text{Im } z \leqq 0$ onto the unit disk $|w| \leqq 1$.
16. Derive transformation (7), Sec. 34.

    Suggestion: Successive linear fractional transformations may be used to first map the disk $|z| \leqq 1$ onto the half plane $\text{Im } Z \geqq 0$ and then to map that half plane onto the disk $|w| \leqq 1$.
17. Show that when $z_0 = 0$, transformation (7), Sec. 34, is a rotation about the origin through the angle $\alpha + \pi$.
18. Show that there is no linear fractional transformation of the form (7), Sec. 34, which maps the disk $|z| \leqq 1$ onto the disk $|w| \leqq 1$ such that the points $z_1 = 1$, $z_2 = i$, $z_3 = -1$ have the images $w_1 = 1$, $w_2 = -i$, $w_3 = -1$.

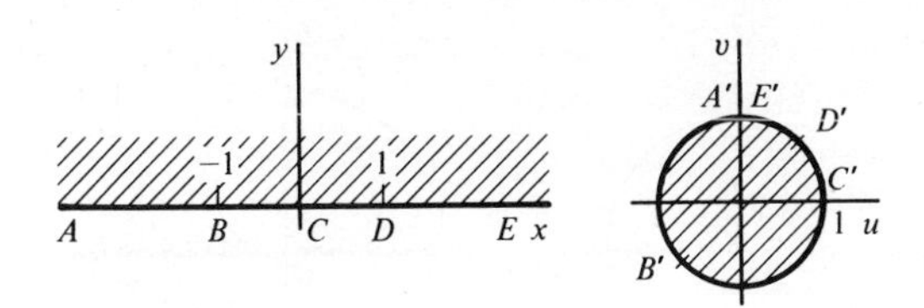

FIGURE 24

$$w = \frac{iz + e^{i\pi/4}}{z + e^{i\pi/4}}.$$

19. Show that equation (6), Sec. 33, defines the only linear fractional transformation that maps three given distinct points z_1, z_2, and z_3 onto three specified distinct points w_1, w_2, and w_3, respectively.

Suggestion: Let T and S be two such linear fractional transformations. Then, using the result of Exercise 6, show that $w = S^{-1}[T(z)]$ is the identity transformation $w = z$.

20. Prove that if a bilinear transformation maps each point of the x axis into the u axis, then the coefficients in the transformation are all real, except possibly for a common complex factor. The converse statement is evident.

35. The Function z^n

First let us consider the transformation

$$w = z^2 \tag{1}$$

which can be described easily in terms of polar coordinates. If $z = re^{i\theta}$ and $w = \rho e^{i\phi}$, then

$$\rho e^{i\phi} = r^2 e^{i2\theta}.$$

Hence the image of any nonzero point z is found by squaring the modulus and doubling an argument of z; that is, $|w| = |z|^2$ and $\arg w = 2 \arg z$.

Observe that transformation (1) maps the entire z plane onto the entire w plane. The transformation is a one to one mapping of the first quadrant $r \geqq 0$, $0 \leqq \theta \leqq \pi/2$ in the z plane onto the upper half $\rho \geqq 0$, $0 \leqq \phi \leqq \pi$ of the w plane (Fig. 25). It is also a mapping of the upper half plane $r \geqq 0$, $0 \leqq \theta \leqq \pi$ onto the entire w plane. However, in this case the transformation is not one to one since both the positive and negative real axes in the z plane are mapped onto the positive real axis in the w plane.

A circle $r = r_0$ is transformed into the circle $\rho = r_0^2$, and the sector $r \leqq r_0$, $0 \leqq \theta \leqq \pi/2$ is mapped in a one to one manner onto the semicircular region $\rho \leqq r_0^2$, $0 \leqq \phi \leqq \pi$ (Fig. 25).

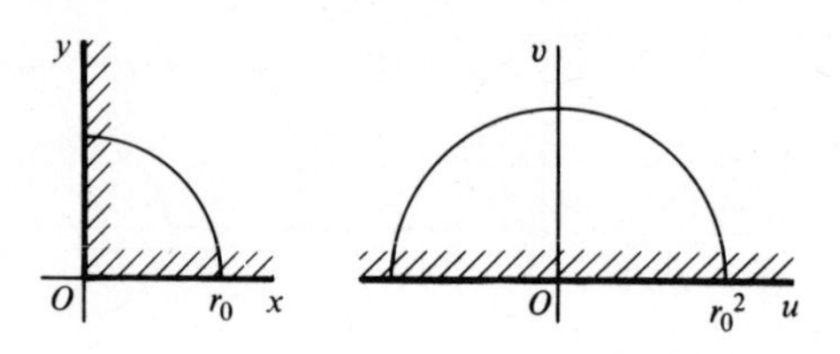

FIGURE 25
$w = z^2$.

In cartesian coordinates the transformation $w = z^2$ becomes

$$u + iv = x^2 - y^2 + i2xy.$$

Hence the image of the hyperbola $x^2 - y^2 = c_1$ $(c_1 \neq 0)$ is the line $u = c_1$ and the image of the hyperbola $2xy = c_2$ $(c_2 \neq 0)$ is the line $v = c_2$. These hyperbolas are illustrated in Chap. 2 (Fig. 17).

Two nonzero points z and $-z$ always have the same image, and each point on the line $u = c_1$ is the image of exactly two such points in the z plane where those two points lie on different branches of the hyperbola $x^2 - y^2 = c_1$. Thus, the points on a particular branch of this hyperbola are in a one to one correspondence with the points on the line $u = c_1$.

Likewise, the mapping of the hyperbola $2xy = c_2$ onto the line $v = c_2$ is such that each branch is mapped in a one to one manner onto that line.

Images of regions whose boundaries involve such hyperbolas are easily obtained. Note, for example, that the domain $x > 0$, $y > 0$, $xy < 1$ consists of all points lying on the upper branches of hyperbolas from the family $xy = c$ where $0 < c < 1$. The image of the domain therefore consists of all points lying on the lines $v = 2c$. That is, the image of the domain is the horizontal strip $0 < v < 2$.

When n is a positive integer, the transformation

$$w = z^n, \qquad \text{or} \qquad \rho e^{i\phi} = r^n e^{in\theta}, \tag{2}$$

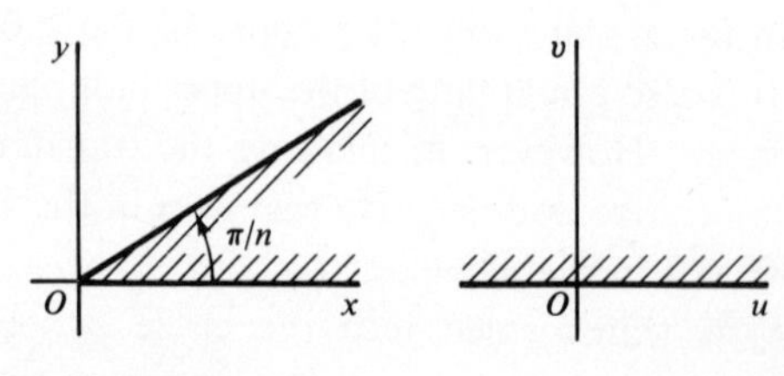

FIGURE 26
$w = z^n$.

maps the region $r \geqq 0$, $0 \leqq \theta \leqq \pi/n$ onto the upper half plane $\rho \geqq 0$, $0 \leqq \phi \leqq \pi$ (Fig. 26). The entire z plane is mapped onto the entire w plane, where each nonzero point in the w plane is the image of n distinct points in the z plane. The circle $r = r_0$ is mapped onto the circle $\rho = r_0^n$; and the arc $r = r_0$, $0 \leqq \theta < 2\pi/n$ is mapped in a one to one manner onto the circle $\rho = r_0^n$.

36. The Function $z^{1/2}$

According to Sec. 6, the values of $z^{1/2}$ are the two square roots of z when $z \neq 0$. In Sec. 28 we saw that this multiple-valued function could also be written

$$z^{1/2} = \exp\left(\tfrac{1}{2}\log z\right) \qquad (z \neq 0). \tag{1}$$

If we use polar coordinates and the fact that $\log z = \text{Log}\, r + i(\Theta + 2k\pi)$, where $r = |z|$, $\Theta = \text{Arg}\, z$, and $k = 0, \pm 1, \pm 2, \ldots$, we can write equation (1) in the form

$$z^{1/2} = \sqrt{r} \exp \frac{i(\Theta + 2k\pi)}{2} \qquad (r > 0, k = 0, 1). \tag{2}$$

Since the complex exponential function has period $2\pi i$, this gives the two values of $z^{1/2}$ for each nonzero complex number z when $k = 0$ and $k = 1$.

The *principal branch* F_0 of the multiple-valued function $z^{1/2}$ is the analytic function obtained from equation (1) by using the principal branch of $\log z$. Hence, if we set k equal to zero in equation (2), we find that

$$F_0(z) = \sqrt{r} \exp \frac{i\Theta}{2} \qquad (r > 0, -\pi < \Theta < \pi). \tag{3}$$

The ray $\Theta = \pi$ is the branch cut for F_0, and the point $z = 0$ is the branch point. Note that the right-hand side of equation (3) is defined for points on the branch cut of F_0; however, the function obtained by this extension is not even continuous there. This is due to the fact that there are values of Θ close to π as well as values close to $-\pi$ in any neighborhood of a point on the negative real axis.

The transformation $w = F_0(z)$ maps the domain $r > 0$, $-\pi < \Theta < \pi$ onto the right-hand half $\rho > 0$, $-\pi/2 < \phi < \pi/2$ of the w plane, where $w = \rho e^{i\phi}$. It also maps the region $0 < r \leqq r_0$, $-\pi < \Theta < \pi$ onto the half disk $0 < \rho \leqq \sqrt{r_0}$, $-\pi/2 < \phi < \pi/2$ (Fig. 27) since $\rho = \sqrt{r}$ and $\phi = \Theta/2$.

When $k = 1$ in equation (2), we have the branch

$$F_1(z) = \sqrt{r} \exp \frac{i(\Theta + 2\pi)}{2} \qquad (r > 0, -\pi < \Theta < \pi). \tag{4}$$

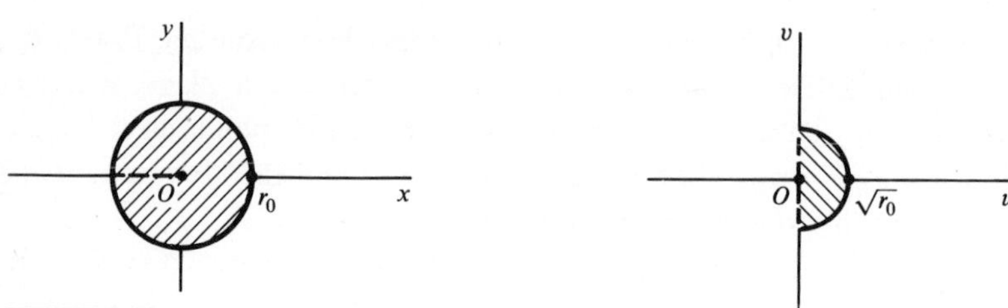

FIGURE 27
$w = F_0(z) = \sqrt{r}\, e^{i\Theta/2}$ $(r > 0, -\pi < \Theta < \pi)$.

Since $\exp(i\pi) = -1$, it follows that $F_1(z) = -F_0(z)$. The values $\pm F_0(z)$ thus represent the totality of values of $z^{1/2}$ at all points in the domain $r > 0$, $-\pi < \Theta < \pi$. If, by means of expression (3), we extend the domain of definition of F_0 to include the ray $\Theta = \pi$ and if we write $F_0(0) = 0$, the values $\pm F_0(z)$ represent the totality of values of $z^{1/2}$ in the entire z plane.

Further branches of $z^{1/2}$ are obtained by using various other branches of $\log z$ in expression (1). A branch where the ray $\theta = \alpha$ is the branch cut is given by

$$f_\alpha(z) = \sqrt{r} \exp \frac{i\theta}{2} \qquad (r > 0, \alpha < \theta < \alpha + 2\pi). \tag{5}$$

Observe that when $\alpha = -\pi$ we have the branch $F_0(z)$ and that when $\alpha = \pi$ we have the branch $F_1(z)$. Just as in the cases of F_0 and F_1, the domain of definition of f_α can be extended to the entire complex plane by using expression (5) to define f_α at the nonzero points on the branch cut and by writing $f_\alpha(0) = 0$. Such extensions are, of course, not continuous in the entire complex plane.

It is sometimes useful to note that if $w = f_\alpha(z)$, then $z = w^2$. For example, we know (Sec. 35) that the function $w = z^2$ is a one to one mapping of the branch of the hyperbola $2xy = 1$ which lies in the first quadrant of the z plane onto the line $v = 1$ in the w plane. Hence, after interchanging the z and the w planes, we find that the branch f_0, with branch cut $\theta = 0$, maps the line $y = 1$ in the z plane in a one to one manner onto the branch of the hyperbola $2uv = 1$ lying in the first quadrant of the w plane.

37. Other Irrational Functions

Let n be any positive integer greater than one. The values of $z^{1/n}$ are the nth roots of z when $z \neq 0$; and, according to Sec. 28, the multiple-valued function $z^{1/n}$ can be written

$$z^{1/n} = \exp\left(\frac{1}{n} \log z\right) \qquad (z \neq 0).$$

Since $\log z = \text{Log}\, r + i(\Theta + 2k\pi)$, where $r = |z|$, $\Theta = \text{Arg}\, z$, and $k = 0, \pm 1, \pm 2, \ldots$, we find that

$$(1) \qquad z^{1/n} = \sqrt[n]{r} \exp \frac{i(\Theta + 2k\pi)}{n} \qquad (k = 0, 1, 2, \ldots, n-1).$$

The case $n = 2$ was considered in the preceding section. In the general case, each of the n functions

$$(2) \qquad F_k(z) = \sqrt[n]{r} \exp \frac{i(\Theta + 2k\pi)}{n}$$

$$(r > 0, -\pi < \Theta < \pi, k = 0, 1, 2, \ldots, n-1)$$

is a branch of $z^{1/n}$. The transformation $w = F_k(z)$ is a one to one mapping of the domain $r > 0$, $-\pi < \Theta < \pi$ onto the domain $\rho > 0$, $(2k-1)\pi/n < \phi < (2k+1)\pi/n$, where $w = \rho e^{i\phi}$. These n branches of $z^{1/n}$ yield the n distinct nth roots of z at any point z in the domain $r > 0$, $-\pi < \Theta < \pi$. The principal branch occurs when $k = 0$, and further branches of the type (5) in the preceding section are readily constructed.

Branches of the double-valued function $(z - z_0)^{1/2}$ can be obtained by noting that it is a composition of the translation $Z = z - z_0$ with the double-valued function $Z^{1/2}$. Each branch of $Z^{1/2}$ yields a branch of $(z - z_0)^{1/2}$. To obtain expressions for branches of $(z - z_0)^{1/2}$, we write $r_0 = |z - z_0|$ and $\theta_0 = \arg(z - z_0)$, as well as $\Theta_0 = \text{Arg}(z - z_0)$. Two such branches are

$$(3) \qquad G_0(z) = \sqrt{r_0} \exp \frac{i\Theta_0}{2} \qquad (r_0 > 0, -\pi < \Theta_0 < \pi)$$

and

$$(4) \qquad g_0(z) = \sqrt{r_0} \exp \frac{i\theta_0}{2} \qquad (r_0 > 0, 0 < \theta_0 < 2\pi).$$

The branch of $Z^{1/2}$ used in writing $G_0(z)$ is defined at all points in the Z plane except for the origin and points on the half line $\text{Arg}\, Z = \pi$. The transformation $w = G_0(z)$ is therefore a one to one mapping of the domain $|z - z_0| > 0$, $-\pi < \text{Arg}(z - z_0) < \pi$ onto the right half $\text{Re}\, w > 0$ of the w plane (Fig. 28). The transformation $w = g_0(z)$ maps the domain $|z - z_0| > 0$, $0 < \arg(z - z_0) < 2\pi$ in a one to one manner onto the upper half plane $\text{Im}\, w > 0$.

For an instructive but less elementary example, we now consider the double-valued function $(z^2 - 1)^{1/2}$. Using established properties of logarithms, we write

$$(5) \qquad (z^2 - 1)^{1/2} = \exp\left[\tfrac{1}{2} \log(z^2 - 1)\right] = \exp\left[\tfrac{1}{2} \log(z - 1) + \tfrac{1}{2} \log(z + 1)\right]$$

$$= (z - 1)^{1/2}(z + 1)^{1/2} \qquad (z \neq \pm 1).$$

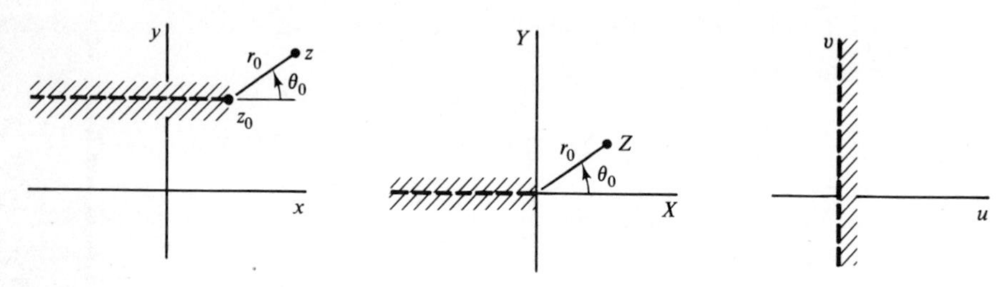

FIGURE 28

Thus, if $f_1(z)$ is a branch of $(z-1)^{1/2}$ defined on a domain D_1 and $f_2(z)$ is a branch of $(z+1)^{1/2}$ defined on a domain D_2, then the product $f_1(z)f_2(z)$ is a branch of $(z^2-1)^{1/2}$ defined at all points lying in both D_1 and D_2.

To obtain a specific branch of $(z^2-1)^{1/2}$, we use the branch of $(z-1)^{1/2}$ and the branch of $(z+1)^{1/2}$ given by equation (4). If we write $r_1 = |z-1|$ and $\theta_1 = \arg(z-1)$, that branch of $(z-1)^{1/2}$ is

$$\sqrt{r_1}\exp\frac{i\theta_1}{2} \qquad (r_1 > 0, 0 < \theta_1 < 2\pi).$$

The branch of $(z+1)^{1/2}$ given by equation (4) is

$$\sqrt{r_2}\exp\frac{i\theta_2}{2} \qquad (r_2 > 0, 0 < \theta_2 < 2\pi)$$

where $r_2 = |z+1|$ and $\theta_2 = \arg(z+1)$. The product of these two branches is therefore the branch f of $(z^2-1)^{1/2}$ defined by the equation

$$f(z) = \sqrt{r_1 r_2}\exp\frac{i(\theta_1+\theta_2)}{2} \qquad (r_1 > 0, r_2 > 0, 0 < \theta_k < 2\pi, k = 1, 2). \tag{6}$$

As illustrated in Fig. 29, the branch f is defined everywhere in the z plane except on the half line $x \geqq -1$, $y = 0$.

The branch f of $(z^2-1)^{1/2}$ given in equation (6) can be extended to a function

$$F(z) = \sqrt{r_1 r_2}\exp\frac{i(\theta_1+\theta_2)}{2} \tag{7}$$

$$(r_1 > 0, r_2 > 0, r_1 + r_2 > 2, 0 \leqq \theta_k < 2\pi, k = 1, 2),$$

which is analytic everywhere in the z plane except on the line segment $-1 \leqq x \leqq 1$, $y = 0$. Since $F(z) = f(z)$ for all z in the domain of definition of F except on the half line $x > 1$, $y = 0$, we need only show that F is analytic on that half line. To do this, we form the product of the branches of $(z-1)^{1/2}$ and

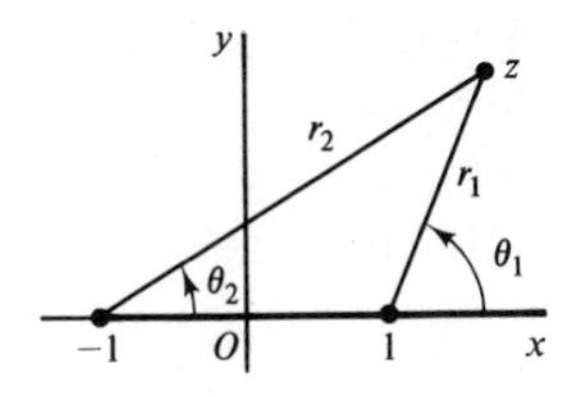

FIGURE 29

$(z+1)^{1/2}$ which are given by equation (3). That is, we consider the function

$$G(z) = \left(\sqrt{r_1}\exp\frac{i\Theta_1}{2}\right)\left(\sqrt{r_2}\exp\frac{i\Theta_2}{2}\right)$$

$$(r_1 > 0, r_2 > 0, -\pi < \Theta_k < \pi, k = 1, 2),$$

where $r_1 = |z-1|$, $r_2 = |z+1|$, $\Theta_1 = \text{Arg}(z-1)$, and $\Theta_2 = \text{Arg}(z+1)$. Observe that G is analytic in the entire z plane except for the half line $x \leqq 1$, $y = 0$. Now $F(z) = G(z)$ when the point z lies above or on the half line $x > 1$, $y = 0$; for then $\theta_k = \Theta_k$ $(k = 1, 2)$. When z lies below that half line, $\theta_k = \Theta_k + 2\pi$ $(k = 1, 2)$. Consequently, $\exp(i\theta_k/2) = -\exp(i\Theta_k/2)$ and again $F(z) = G(z)$. Since $F(z) = G(z)$ in a domain containing the half line $x > 1$, $y = 0$ and since $G(z)$ is analytic in that domain, $F(z)$ is analytic there. Hence $F(z)$ is analytic everywhere except on the line segment $-1 \leqq x \leqq 1$, $y = 0$.

The function F defined in equation (7) cannot itself be extended to a function which is analytic at points on the line segment $-1 \leqq x \leqq 1$, $y = 0$. For the value on the right of equation (7) jumps from $i\sqrt{r_1 r_2}$ to numbers near $-i\sqrt{r_1 r_2}$ as the point z moves downward across that line segment. Hence the extension would not even be continuous there.

The transformation $w = F(z)$ is a one to one mapping of the domain D_z consisting of all points in the z plane except those on the line segment $-1 \leqq x \leqq 1$, $y = 0$ onto the domain D_w consisting of the entire w plane with the exception of the line segment $u = 0$, $-1 \leqq v \leqq 1$ (Fig. 30).

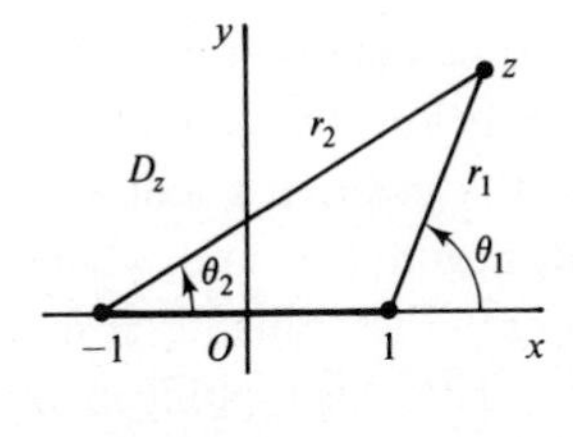

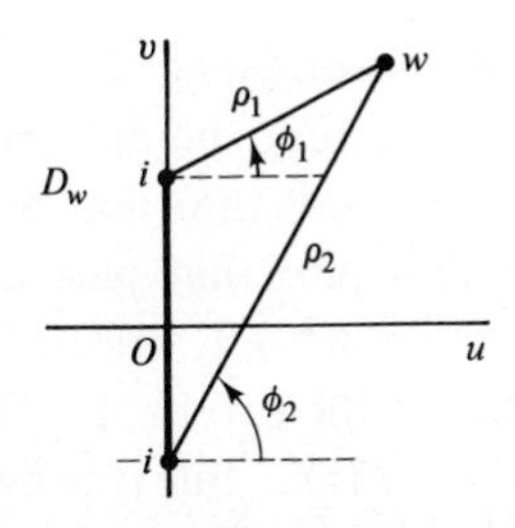

FIGURE 30 $w = F(z)$.

Before showing this, we note that when $z = iy$ $(y > 0)$, $\theta_1 + \theta_2 = \pi$ and $r_1 = r_2 > 1$; hence the positive y axis is mapped onto that part of the v axis for which $v > 1$. The negative y axis is, moreover, mapped onto that part of the v axis for which $v < -1$. Each point in the upper half $y > 0$ of the domain D_z is mapped into the upper half $v > 0$ of the w plane, and each point in the lower half $y < 0$ of the domain D_z is mapped into the lower half $v < 0$ of the w plane. The half line $x > 1$, $y = 0$ is mapped onto the positive real axis in the w plane, and the half line $x < -1$, $y = 0$ is mapped onto the negative real axis there.

To show that the transformation $w = F(z)$ is *one to one*, we note that if $F(z_1) = F(z_2)$, then $z_1^2 - 1 = z_2^2 - 1$. From this it follows that $z_1 = z_2$ or $z_1 = -z_2$. However, because of the manner in which F maps the upper and lower halves of the domain D_z, as well as the portions of the real axis lying in D_z, the case $z_1 = -z_2$ is impossible. Thus, if $F(z_1) = F(z_2)$, then $z_1 = z_2$; and F is one to one.

We show that F maps the domain D_z *onto* the domain D_w by finding a function H mapping D_w into D_z with the property that if $z = H(w)$, then $w = F(z)$. This will show that for any point w in D_w, there exists a point z in D_z such that $F(z) = w$; and the mapping is therefore onto. The mapping H will be the inverse of F.

To find H, we first note that if w is a value of $(z^2 - 1)^{1/2}$ for a specific z, then z is a value of $(w^2 + 1)^{1/2}$ for that w. The function H will be a branch of the multiple-valued function

$$(w^2 + 1)^{1/2} = (w - i)^{1/2}(w + i)^{1/2}.$$

Following our procedure for obtaining the branch $F(z)$ of $(z^2 - 1)^{1/2}$, we write $w - i = \rho_1 \exp(i\phi_1)$ and $w + i = \rho_2 \exp(i\phi_2)$. With the restrictions $\rho_1 > 0$, $\rho_2 > 0$, $\rho_1 + \rho_2 > 2$ and $-\pi/2 \leqq \phi_k < 3\pi/2$, $k = 1, 2$, we then write

$$H(w) = \sqrt{\rho_1 \rho_2} \exp \frac{i(\phi_1 + \phi_2)}{2}, \tag{8}$$

the domain of definition being D_w. This branch maps points of D_w lying above or below the u axis onto points above or below the x axis, respectively. It maps the positive u axis into that part of the positive x axis where $x > 1$ and the negative u axis into that part of the negative x axis where $x < -1$. If $z = H(w)$, then $z^2 = w^2 + 1$, and so $w^2 = z^2 - 1$. Since z is in D_z and since $F(z)$ and $-F(z)$ are the two values of $(z^2 - 1)^{1/2}$ for a point in D_z, we see that $w = F(z)$ or $w = -F(z)$. But it is evident from the manner in which F and H

map the upper and lower halves of their domains of definition, including the portions of the real axes lying in those domains, that $w = F(z)$.

Branches of double-valued functions

$$w = (z^2 + Az + B)^{1/2} = [(z - z_0)^2 - z_1{}^2]^{1/2} \qquad (z_1 \neq 0), \tag{9}$$

where $A = -2z_0$ and $B = z_0{}^2 - z_1{}^2$, and mappings by those branches, can be treated with the aid of the results found for F above and the successive transformations

$$Z = \frac{z - z_0}{z_1}, \qquad W = (Z^2 - 1)^{1/2}, \qquad w = z_1 W. \tag{10}$$

EXERCISES

1. Describe the region onto which the sector $r < 1$, $0 < \theta < \pi/4$ is mapped by the transformation (*a*) $w = z^2$; (*b*) $w = z^3$; (*c*) $w = z^4$.
2. Show that the transformation $w = z^2$ is a one to one mapping of the lines $x = c$ $(c \neq 0)$ onto the parabolas $v^2 = -4c^2(u - c^2)$, and the lines $y = d\ (d \neq 0)$ onto the parabolas $v^2 = 4d^2(u + d^2)$. Note that all these parabolas have foci at the point $w = 0$.
3. Find a region in the z plane whose image under the transformation $w = z^2$ is the rectangular domain in the w plane bounded by the lines $u = 1$, $u = 2$, $v = 1$, $v = 2$.
4. Show that the principal branch of the function $z^{1/2}$ maps the domain bounded by the y axis and the parabola $y^2 = -4(x - 1)$ onto the triangular domain bounded by the lines $v = u$, $v = -u$, and $u = 1$. Show corresponding parts of the boundaries of the two regions. See Exercise 2.
5. Show that the branch $w = \sqrt{r} \exp(i\theta/2)$, $r > 0$, $0 < \theta < 2\pi$, of the multiple-valued function $z^{1/2}$ is a one to one mapping of the domain between the two parabolas
$$r = \frac{2a^2}{1 - \cos\theta}, \qquad r = \frac{2b^2}{1 - \cos\theta}$$
onto the strip $a < v < b$ where $b > a > 0$. See Exercise 2.
6. Determine the image of the domain $r > 0$, $-\pi < \theta < \pi$ in the z plane under each of the transformations defined by the four branches of $z^{1/4}$ given by equation (2), Sec. 37, when $n = 4$. Use these four branches of $z^{1/4}$ to determine the fourth roots of i.
7. The branch F of $(z^2 - 1)^{1/2}$ in Sec. 37 was defined in terms of the coordinates $r_1, r_2, \theta_1, \theta_2$. Explain geometrically why the conditions $r_1 > 0$, $0 < \theta_1 + \theta_2 < \pi$ describe the quadrant $x > 0$, $y > 0$ of the z plane. Then show that the transfor-

mation $w = F(z)$ maps that quadrant onto the quadrant $u > 0$, $v > 0$ in the w plane.

Suggestion: To show that the quadrant $x > 0$, $y > 0$ in the z plane is described, note that $\theta_1 + \theta_2 = \pi$ at each point on the positive y axis and that $\theta_1 + \theta_2$ decreases as a point z moves to the right along a ray $\theta_2 = c$ $(0 < c < \pi/2)$.

8. For the transformation $w = F(z)$ of the first quadrant of the z plane onto the first quadrant of the w plane (Exercise 7), show that

$$u = \frac{1}{\sqrt{2}} \sqrt{r_1 r_2 + x^2 - y^2 - 1}, \qquad v = \frac{1}{\sqrt{2}} \sqrt{r_1 r_2 - x^2 + y^2 + 1},$$

where $r_1{}^2 r_2{}^2 = (x^2 + y^2 + 1)^2 - 4x^2$, and that the image of the portion of the hyperbola $x^2 - y^2 = 1$ in the first quadrant is the ray $u > 0$, $v = u$.

9. Show that in Exercise 8 the domain D that lies under the hyperbola and in the first quadrant of the z plane is described by the conditions $r_1 > 0$, $0 < \theta_1 + \theta_2 < \pi/2$. Then show that the image of D is the octant $0 < v < u$. Sketch the domains.

10. Let F be the branch of $(z^2 - 1)^{1/2}$ defined in Sec. 37, and let $z_0 = r_0 \exp(i\theta_0)$ be a fixed point, where $r_0 > 0$ and $0 \leqq \theta_0 < 2\pi$. Show that a branch F_0 of $(z^2 - z_0{}^2)^{1/2}$, whose branch cut is the line segment between the points z_0 and $-z_0$, is given by the formula $F_0(z) = z_0 F(Z)$, where $Z = z/z_0$.

11. Write $z - 1 = r_1 \exp(i\theta_1)$ and $z + 1 = r_2 \exp(i\Theta_2)$, where $0 < \theta_1 < 2\pi$ and $-\pi < \Theta_2 < \pi$, to define a branch of the function

(*a*) $(z^2 - 1)^{1/2}$; (*b*) $\left(\dfrac{z-1}{z+1}\right)^{1/2}$.

In each case the branch cut should consist of the two rays $\theta_1 = 0$ and $\Theta_2 = \pi$.

12. Using the notation of Sec. 37, show that the function

$$\left(\frac{z-1}{z+1}\right)^{1/2} = \sqrt{\frac{r_1}{r_2}} \exp \frac{i(\theta_1 - \theta_2)}{2}$$

is a branch with the same domain of definition D_z and the same branch cut as F. Show that this transformation maps D_z onto the right half plane $\rho > 0$, $-\pi/2 < \phi < \pi/2$, where the point $w = 1$ is the image of the point $z = \infty$. The inverse transformation is

$$z = \frac{1 + w^2}{1 - w^2} \qquad (u > 0).$$

13. Show that the transformation in Exercise 12 maps the region outside the unit circle $|z| = 1$ in the upper half of the z plane onto the region in the first quadrant between the line $v = u$ and the u axis. Sketch the regions.

14. Write $z = r \exp(i\Theta)$, $z - 1 = r_1 \exp(i\Theta_1)$, $z + 1 = r_2 \exp(i\Theta_2)$, where the values of all three angles lie between $-\pi$ and π, and define a branch of the function $[z(z^2 - 1)]^{1/2}$ whose branch cut consists of the two segments $x \leqq -1$, $y = 0$ and $0 \leqq x \leqq 1$, $y = 0$ of the x axis.

38. The Transformation $w = \exp z$

The transformation

$$w = e^z, \qquad \text{or} \qquad \rho e^{i\phi} = e^x e^{iy},$$

where $z = x + iy$ and $w = \rho \exp(i\phi)$, can be written

$$\rho = e^x, \qquad \phi = y.$$

This transformation is a one to one mapping of the line $y = c$ onto the ray $\phi = c$, the origin being excluded from that ray. The line $x = c$ is mapped onto the circle $\rho = e^c$. Observe, however, that there is an infinite number of points on the line $x = c$ with the same image point.

The rectangular region $a \leqq x \leqq b$, $c \leqq y \leqq d$ is mapped onto the region

$$e^a \leqq \rho \leqq e^b, \qquad c \leqq \phi \leqq d$$

bounded by portions of circles and rays. This mapping is one to one if $d - c < 2\pi$. The two regions and corresponding parts of their boundaries are shown in Fig. 31. In particular, if $c = 0$ and $d = \pi$, then $0 \leqq y \leqq \pi$ and the rectangle is mapped onto half of a circular ring, as shown in Fig. 8, Appendix 2.

According to Sec. 22, the transformation $w = e^z$ is a one to one mapping of the strip $(2n - 1)\pi < y \leqq (2n + 1)\pi$, where n is any integer, onto the set of nonzero points in the w plane. From Sec. 26 we know that if z lies in the strip and $w = e^z$, then

$$z = \text{Log}\, w + 2n\pi i.$$

The infinite strip $0 < y < \pi$ is mapped onto the upper half $\rho > 0, 0 < \phi < \pi$ of the w plane. Corresponding parts of the boundaries of the two regions are shown in Fig. 6 of Appendix 2. This mapping of a strip onto a half plane is especially useful in applications.

The semi-infinite strip $x \leqq 0$, $0 \leqq y \leqq \pi$ is mapped onto the semicircular region $0 < \rho \leqq 1$, $0 \leqq \phi \leqq \pi$ (Fig. 7, Appendix 2). Note that the point $w = 0$ is not included in the image region since e^z is never zero.

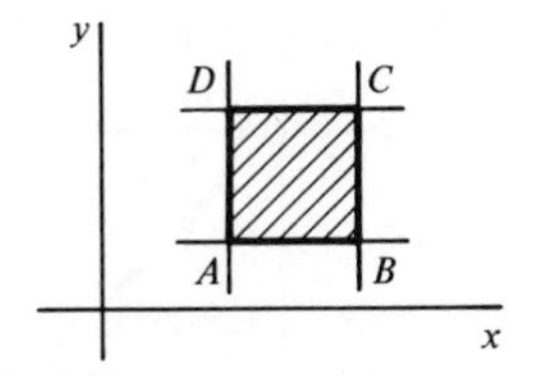

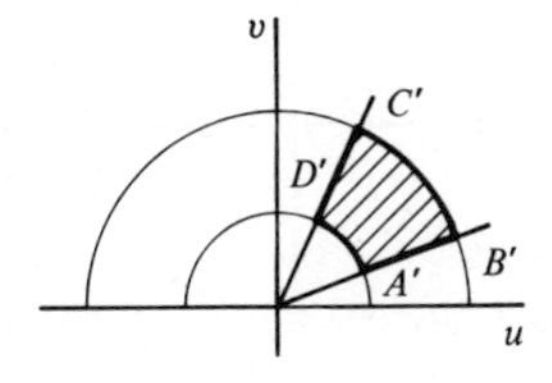

FIGURE 31
$w = e^z$.

39. The Transformation $w = \sin z$

Since

$$\sin z = \sin x \cosh y + i \cos x \sinh y,$$

the transformation $w = \sin z$ can be written

$$u = \sin x \cosh y, \qquad v = \cos x \sinh y. \tag{1}$$

Of particular importance in the applications is the fact that the transformation $w = \sin z$ is a one to one mapping of the strip

$$-\frac{\pi}{2} \leqq x \leqq \frac{\pi}{2}, \qquad y \geqq 0$$

in the z plane onto the upper half $v \geqq 0$ of the w plane, as shown in Fig. 9 of Appendix 2. We now verify this by considering images of vertical lines $x = c$ where $-\pi/2 \leqq c \leqq \pi/2$.

The y axis ($x = 0$) is mapped onto the v axis ($u = 0$) in a one to one manner, and the lower portions of these axes correspond to each other, as do the upper portions. For the image of a point $(0,y)$ on the y axis is the point $(0, \sinh y)$ on the v axis.

If $0 < c < \pi/2$, the line $x = c$ is mapped in a one to one manner onto the curve

$$u = \sin c \cosh y, \qquad v = \cos c \sinh y \tag{2}$$

which is the right-hand branch of the hyperbola

$$\frac{u^2}{\sin^2 c} - \frac{v^2}{\cos^2 c} = 1 \tag{3}$$

with foci $(\pm 1,0)$. See Fig. 32.

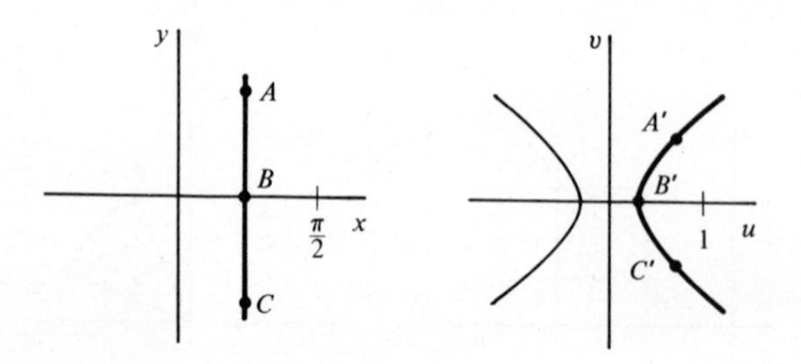

FIGURE 32
$w = \sin z$.

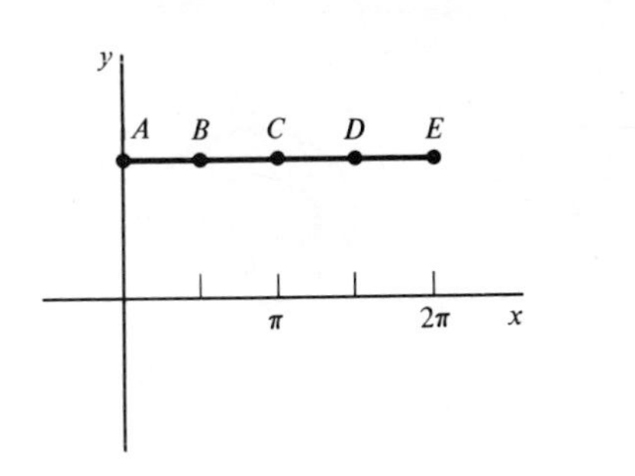

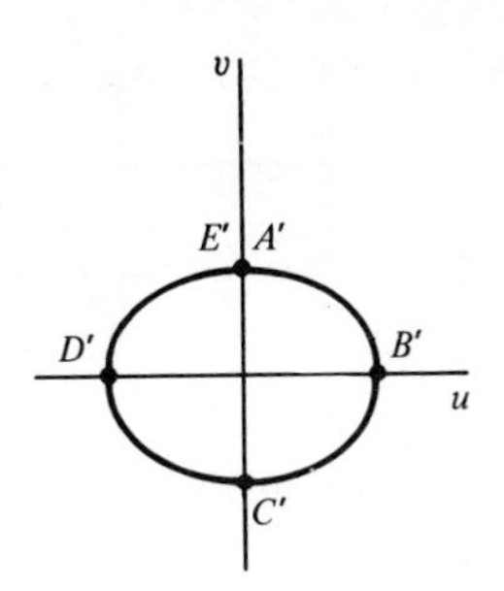

FIGURE 33
$w = \sin z$.

The image of the line $x = \pi/2$, obtained by writing $x = \pi/2$ in equations (1), consists of those points $(u,0)$ on the u axis where $u \geqq 1$. The mapping of the line $x = \pi/2$ is not one to one since the points $\pi/2 + iy$ and $\pi/2 - iy$ have the same image. The mapping is, however, one to one on either the upper or the lower half of that line.

The image of a line $x = c$, $-\pi/2 \leqq c < 0$, is readily obtained from the results above by means of the identity $\sin z = -\sin(-z)$. The image is the left-hand branch of hyperbola (3), except when $c = -\pi/2$. In that case the image consists of the points $(u,0)$ where $u \leqq -1$.

It is now evident, by noting the images of just the upper halves of all these lines, that the transformation $w = \sin z$ maps the semi-infinite strip $-\pi/2 \leqq x \leqq \pi/2$, $y \geqq 0$ in a one to one manner onto the upper half $v \geqq 0$ of the w plane. The right-hand half of the strip is mapped onto the first quadrant of the w plane as shown in Fig. 10, Appendix 2.

The transformation $w = \sin z$ maps the horizontal line $y = c$ onto the curve

$$u = \sin x \cosh c, \qquad v = \cos x \sinh c. \tag{4}$$

If $c \neq 0$, this curve is the ellipse

$$\frac{u^2}{\cosh^2 c} + \frac{v^2}{\sinh^2 c} = 1. \tag{5}$$

The line segment $0 \leqq x \leqq \pi$, $y = c$ is mapped in a one to one manner onto the right half of that ellipse, while the segment $\pi \leqq x \leqq 2\pi$, $y = c$ is mapped in a one to one manner onto the left half. See Fig. 33. The image of the x axis, obtained by writing $c = 0$ in equations (4), is the portion of the u axis where $-1 \leqq u \leqq 1$.

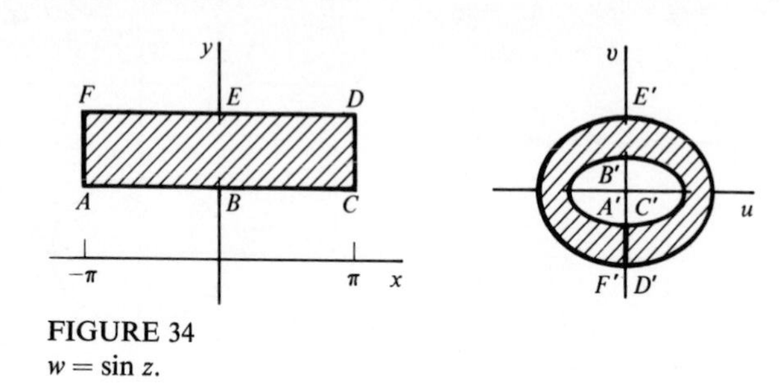

FIGURE 34
$w = \sin z$.

The rectangular region $-\pi \leqq x \leqq \pi$, $c_1 \leqq y \leqq c_2$ is mapped onto the region bounded by two confocal ellipses, as shown in Fig. 34. Both of the sides $x = \pm\pi$, $c_1 \leqq y \leqq c_2$ are mapped onto the line segment $u = 0$, $-\sinh c_2 \leqq v \leqq -\sinh c_1$. Thus, if $c_1 > 0$, the image of the rectangular region is the elliptic ring with a cut along the negative v axis. As the point z describes the boundary, its image makes a circuit around one ellipse, then along the cut and around the other ellipse, and back again along the cut to the starting point.

The rectangular region $-\pi/2 \leqq x \leqq \pi/2$, $0 \leqq y \leqq c$ is mapped in a one to one manner onto a semielliptic region as shown in Fig. 11 of Appendix 2.

40. Successive Transformations

Since $\cos z = \sin(z + \pi/2)$, the transformation

$$w = \cos z$$

can be written successively as

$$Z = z + \frac{\pi}{2}, \qquad w = \sin Z.$$

Hence this transformation is the same as the sine transformation preceded by a translation to the right through $\pi/2$ units.

The transformation

$$w = \sinh z$$

can be written $w = -i \sin(iz)$, or

$$Z = iz, \qquad W = \sin Z, \qquad w = -iW.$$

It is therefore a combination of the sine transformation with rotations through right angles. The transformation

$$w = \cosh z$$

is, likewise, essentially a cosine transformation.

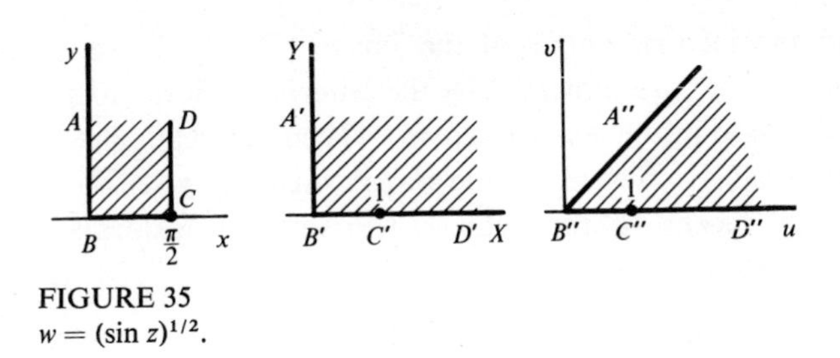

FIGURE 35
$w = (\sin z)^{1/2}$.

The transformation

$$w = (\sin z)^{1/2},$$

where the fractional power denotes the principal branch, can be written as the successive transformations

$$Z = \sin z, \qquad w = Z^{1/2}.$$

As noted in the preceding section, the first transformation maps the semi-infinite strip $0 \leqq x \leqq \pi/2$, $y \geqq 0$ onto the first quadrant $X \geqq 0$, $Y \geqq 0$ in the Z plane; the second transforms that quadrant into an octant in the w plane. The successive transformations are illustrated in Fig. 35.

For another illustration of successive transformations, consider first the linear fractional transformation

$$Z = \frac{z-1}{z+1}.$$

It is readily shown that this transformation maps the half plane $y > 0$ onto the half plane $Y > 0$ in the Z plane. Next observe that the transformation $w = \text{Log}\, Z$ maps the half plane $Y > 0$ onto the strip $0 < v < \pi$ in the w plane. We conclude then that the transformation

$$w = \text{Log}\frac{z-1}{z+1}$$

maps the half plane $y > 0$ onto the strip $0 < v < \pi$. Corresponding points on the boundaries are shown in Fig. 19, Appendix 2.

41. Table of Transformations of Regions

Appendix 2 consists of a set of figures showing transformations of a number of simple and useful regions by means of various elementary functions. In each case there is a one to one correspondence between the interior points of the given region and the interior points of the image of that region. Corresponding parts of boundaries are indicated by the lettering. Some mappings that have not been discussed in the text are shown. Their verification can be left as exercises for the reader. Several of the transformations given in Appendix 2 can be derived by means of the Schwarz-Christoffel transformation, to be developed in Chap. 10.

EXERCISES

1. Show that the lines $ky = x$ are mapped onto the spirals $\rho = \exp(k\phi)$ under the transformation $w = \exp z$, where $w = \rho \exp(i\phi)$.

2. Verify the mapping of the region and boundaries shown in Fig. 7 of Appendix 2, where the transformation is $w = e^z$.

3. Find the image of the semi-infinite strip $x \geqq 0$, $0 \leqq y \leqq \pi$ under the transformation $w = \exp z$, and exhibit corresponding portions of the boundaries.

4. Define a branch of $\log(z - 1)$ that maps the cut z plane consisting of all points except those on the segment $x \geqq 1$, $y = 0$ of the real axis onto the strip $0 < v < 2\pi$ in the w plane.

5. Verify that the transformation $w = \sin z$ maps the line $x = c$ $(0 < c < \pi/2)$ in a one to one manner onto the right-hand branch of the hyperbola given in equation (3), Sec. 39.

6. Show that under the transformation $w = \sin z$ the line $x = c$ $(\pi/2 < c < \pi)$ is mapped onto the right-hand branch of the hyperbola given in equation (3), Sec. 39. Note that the mapping is one to one and that the upper and lower halves of the line are mapped onto the lower and upper half of the branch, respectively.

7. Determine the image of the line $x = c$ $(-\pi < c < -\pi/2)$ under the transformation $w = \sin z$.

8. Verify the mapping by $\sin z$ shown in Fig. 10, Appendix 2.

9. Show that under the transformation $w = \sin z$ the images of the line segments forming the boundary of the rectangular region $0 \leqq x \leqq \pi/2$, $0 \leqq y \leqq 1$ are the line segments and the arc $D'E'$ indicated in Fig. 36. The arc $D'E'$ is a quarter of the ellipse $(u/\cosh 1)^2 + (v/\sinh 1)^2 = 1$.

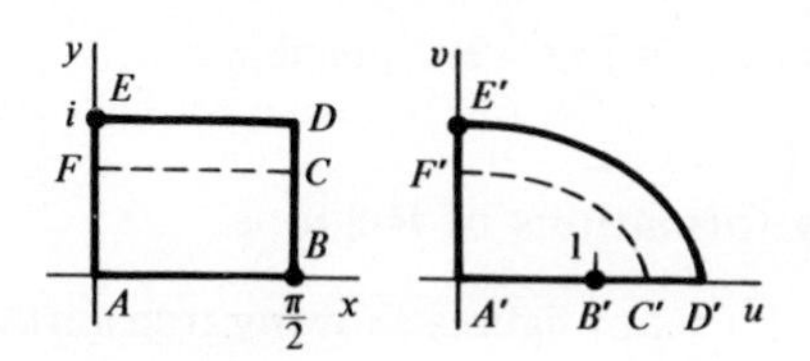

FIGURE 36
$w = \sin z$.

10. Complete the mapping indicated in Fig. 36 by using the mapping of line segments $0 \leqq x \leqq \pi/2$, $y = c$ to prove that the transformation $w = \sin z$ establishes a one to one correspondence between points of the rectangular region and points of the region $A'B'D'E'$.

11. Verify the mapping by $\sin z$ shown in Fig. 11, Appendix 2.

12. Show that the transformation $w = \cosh z$ maps the points $z = iy$ $(0 \leqq y \leqq \pi/2)$ into the segment $0 \leqq u \leqq 1$, $v = 0$ of the u axis.

13. Show that under the transformation $w = \cosh z$ the image of the semi-infinite strip $x \geqq 0$, $0 \leqq y \leqq \pi/2$ is the first quadrant of the w plane. Indicate corresponding parts of the boundaries of the regions.

14. Describe the transformation $w = \cosh z$ in terms of the transformation $w = \sin z$ and rotations and translations.

15. Show that the transformation $w = \sin^2 z$ maps the region $0 \leqq x \leqq \pi/2$, $y \geqq 0$ onto the region $v \geqq 0$, and indicate corresponding parts of the boundaries.

16. Show that under the transformation $w = (\sin z)^{1/4}$ the semi-infinite strip $-\pi/2 \leqq x \leqq \pi/2$, $y \geqq 0$ is mapped onto the part of the first quadrant lying below the line $v = u$, and determine corresponding parts of the boundaries.

17. Show that the linear fractional transformation $Z = (z - 1)/(z + 1)$ maps the x axis onto the X axis, the segment $-1 < x < 1$, $y = 0$ of the x axis onto the negative half of the X axis, and the half planes $y > 0$ and $y < 0$ onto the half planes $Y > 0$ and $Y < 0$, respectively. Show that when the principal branch is used, the composite function

$$w = Z^{1/2} = \left(\frac{z-1}{z+1}\right)^{1/2}$$

maps the z plane, except for the segment $-1 \leqq x \leqq 1$, $y = 0$ of the x axis, onto the half plane $u > 0$. (Compare Exercise 12, Sec. 37.)

18. Using the polar representation of z, show that the transformation

$$w = z + \frac{1}{z}$$

maps both the upper and the lower half of the circle $r = 1$ onto the line segment $-2 \leqq u \leqq 2$, $v = 0$.

19. Show that the transformation $w = z + 1/z$ maps the circle $r = c$ onto the ellipse

$$u = \left(c + \frac{1}{c}\right) \cos \theta, \qquad v = \left(c - \frac{1}{c}\right) \sin \theta.$$

20. Verify the mapping indicated in Fig. 16, Appendix 2, where $w = z + 1/z$.

21. Describe the mapping $w = \cosh z$ in terms of the transformations

$$Z = e^z, \qquad w = \frac{1}{2}\left(Z + \frac{1}{Z}\right).$$

5

INTEGRALS

The reader may pass directly to Chap. 8 and continue with mappings and their applications to physical problems. It would seem natural to present that material next since we have just completed a study of mapping by elementary functions. However, we have not yet established the continuity of the first and second partial derivatives of the component functions of an analytic function. If the reader takes up Chap. 8 at this time, he will have to assume that continuity since some of the theory of integration in the present chapter is needed to establish it.

Integrals are extremely important in the study of functions of a complex variable. The theory of integration is noted for its mathematical elegance. The theorems are generally concise and powerful, and most of the proofs are simple. The theory is, however, also noted for its great utility in applied mathematics.

42. Definite Integrals

In order to introduce integrals of $f(z)$ in a fairly simple way, we first define the definite integral of *a complex-valued function F of a real variable t.*

We write

$$F(t) = U(t) + iV(t) \qquad (a \leqq t \leqq b). \tag{1}$$

Here U and V are real-valued piecewise continuous, or sectionally continuous, functions of t defined on a closed bounded interval $a \leqq t \leqq b$. That is, each of those two functions is real-valued and continuous everywhere in the given interval except possibly for a finite number of points where the function, although discontinuous, has both left-hand and right-hand limits. Only the right-hand limit is, of course, required at a, and only the left-hand limit is required at b. We then say that F is *piecewise continuous* and define the definite integral of F on the interval $a \leqq t \leqq b$ in terms of two definite integrals of the type encountered in the calculus of real variables:

$$\int_a^b F(t)\,dt = \int_a^b U(t)\,dt + i\int_a^b V(t)\,dt. \tag{2}$$

The above conditions on the functions U and V are sufficient to ensure the existence of their integrals. An improper integral of F on an unbounded interval is defined in a like manner, and it exists when the improper integrals of U and V both converge.

From definition (2) it follows that

$$\operatorname{Re}\int_a^b F(t)\,dt = \int_a^b \operatorname{Re}[F(t)]\,dt. \tag{3}$$

Furthermore, for each complex constant $\gamma = c_1 + ic_2$,

$$\begin{aligned}\int_a^b \gamma F\,dt &= \int_a^b (c_1 U - c_2 V)\,dt + i\int_a^b (c_2 U + c_1 V)\,dt\\ &= (c_1 + ic_2)\left(\int_a^b U\,dt + i\int_a^b V\,dt\right);\end{aligned}$$

that is,

$$\int_a^b \gamma F(t)\,dt = \gamma\int_a^b F(t)\,dt. \tag{4}$$

Rules such as those for integrating a sum and for interchanging the limits of integration also hold true just as they do for real-valued functions of t.

To establish another basic property, let us assume that the value of the integral defined in equation (2) is a nonzero complex number. If r_0 is the modulus and θ_0 is an argument of that number, then

$$r_0 e^{i\theta_0} = \int_a^b F\,dt.$$

Solving for r_0 and using property (4), we obtain the equation

$$r_0 = \int_a^b e^{-i\theta_0}F\,dt.$$

Observe that each side of this equation is a real number; and recall that when a complex number is real, it is the same as its real part. Property (3) then allows us to write

$$r_0 = \int_a^b \operatorname{Re}\,(e^{-i\theta_0}F)\,dt. \tag{5}$$

But

$$\operatorname{Re}\,(e^{-i\theta_0}F) \leqq |e^{-i\theta_0}F| = |e^{-i\theta_0}|\,|F| = |F|;$$

and so

$$r_0 \leqq \int_a^b |F|\,dt,$$

provided $a < b$. That is,

$$\left|\int_a^b F(t)\,dt\right| \leqq \int_a^b |F(t)|\,dt \qquad (a \leqq b). \tag{6}$$

This inequality is clearly also valid when the value of the integral on the left is zero.

With only minor modifications, the above discussion als o yields inequalities such as

$$\left|\int_a^\infty F(t)\,dt\right| \leqq \int_a^\infty |F(t)|\,dt \tag{7}$$

when both integrals exist.

43. Contours

Classes of curves that are adequate for the study of integrals of a function $f(z)$ of a complex variable will now be introduced.

An *arc* C is a set of points $z = (x,y)$ in the complex plane such that

$$x = x(t), \qquad y = y(t) \qquad (a \leqq t \leqq b), \tag{1}$$

where $x(t)$ and $y(t)$ are continuous functions of the real parameter t. This definition establishes a continuous mapping of the interval $a \leqq t \leqq b$ into the xy plane, and the image points are ordered according to increasing values of t. It is convenient to describe the points of C by the equation

$$z = z(t) = x(t) + iy(t) \qquad (a \leqq t \leqq b), \tag{2}$$

and we say that $z(t)$ is continuous since $x(t)$ and $y(t)$ are both continuous.

The arc C is a *simple arc*, or a Jordan arc, if it does not cross itself; that is, C is simple if $z(t_1) \neq z(t_2)$ when $t_1 \neq t_2$. When the arc C is simple except for the fact that $z(b) = z(a)$, we say that C is a *simple closed curve*, or a Jordan curve. The polygonal line

$$z(t) = \begin{cases} t + it & (0 \leqq t \leqq 1), \\ t + i & (1 \leqq t \leqq 2), \end{cases} \tag{3}$$

consisting of a line segment from 0 to $1 + i$ followed by one from $1 + i$ to $2 + i$, is an example of a simple arc. The unit circle

$$z(t) = \cos t + i \sin t \qquad (0 \leqq t \leqq 2\pi) \tag{4}$$

about the origin is, on the other hand, a simple closed curve.

The complex-valued function $z(t)$ given by equation (2) is said to be a differentiable function of the real parameter t if both $x(t)$ and $y(t)$ are differentiable functions of t. The derivative $z'(t)$, or $d[z(t)]/dt$, is defined as

$$z'(t) = x'(t) + iy'(t) \qquad (a \leqq t \leqq b). \tag{5}$$

The derivatives of $x(t)$ and $y(t)$ are, of course, the right-hand and left-hand derivatives of those functions at the end points $t = a$ and $t = b$, respectively.

An arc C, as described by equation (2), is called *smooth* if $z'(t)$ exists and is continuous in the interval $a \leqq t \leqq b$ and if $z'(t)$ is never zero there. If $x'(t) = 0$ at a point t, the vector $z'(t) = iy'(t)$ is vertical; and if $x'(t) \neq 0$, the slope of the vector $z'(t)$ is $y'(t)/x'(t)$ which is equal to the slope dy/dx of the line tangent to the arc C at the point corresponding to t. Thus the angle of inclination of the tangent line is given by $\arg z'(t)$. Moreover, since $z'(t)$ is continuous in the interval $a \leqq t \leqq b$, we conclude that a smooth arc has a continuously turning tangent.

In view of the identity

$$|z'(t)| = \sqrt{[x'(t)]^2 + [y'(t)]^2},$$

we may express the length of a smooth arc by the formula

$$L = \int_a^b |z'(t)| \, dt. \tag{6}$$

To examine the effect on formula (6) of changing the parameter for C, write

$$t = \phi(r) \qquad (c \leqq r \leqq d) \tag{7}$$

where ϕ is a real-valued function mapping the interval $c \leqq r \leqq d$ onto the interval $a \leqq t \leqq b$. We assume that ϕ is continuous with a continuous derivative

We also assume that $\phi'(r) > 0$ for each r; this ensures that t increases with r. Observe that, with the change of variable indicated in equation (7), formula (6) for the length of C takes the form

$$L = \int_c^d |z'[\phi(r)]|\,\phi'(r)\,dr.$$

But C may be described in terms of the new parametric representation

$$z = Z(r) = z[\phi(r)] \qquad (c \leqq r \leqq d) \tag{8}$$

Then, because (see Exercise 6 of this section)

$$Z'(r) = z'[\phi(r)]\phi'(r), \tag{9}$$

we have the result

$$L = \int_c^d |Z'(r)|\,dr.$$

Thus, the number L given by formula (6) is invariant under such changes in the parametric representation of C.

A *contour*, or piecewise smooth arc, is an arc consisting of a finite number of smooth arcs joined end to end. If equation (2) represents a contour, then $x(t)$ and $y(t)$ are continuous, whereas their first derivatives are piecewise continuous. The polygonal line (3) is, for example, a contour. When only the initial and final values of $z(t)$ are the same, a contour C is called a *simple closed contour*. Examples are the circle (4) as well as the boundary of a triangle or a rectangle taken in a specific direction. The length of a contour or a simple closed contour is the sum of the lengths of the smooth arcs which make up the contour.

Associated with any simple closed curve or simple closed contour C are two domains each of which has the points of C as its only boundary points. One of these domains, called the interior of C, is bounded; the other, the exterior of C, is unbounded. It will be convenient to accept this statement, known as the Jordan curve theorem, as geometrically evident; the proof is not simple.[1]

EXERCISES

1. Evaluate

(*a*) $\displaystyle\int_0^{\pi/4} e^{it}\,dt$; (*b*) $\displaystyle\int_0^{\infty} e^{-zt}\,dt \; (\text{Re}\, z > 0)$; (*c*) $\dfrac{d}{dt}e^{it}$.

Ans. (*a*) $1/\sqrt{2} + i(1 - 1/\sqrt{2})$; (*b*) $1/z$; (*c*) ie^{it}.

[1] See Sec. 13 of the book by Thron cited in Appendix 1.

2. Show that if F is a function of the type (1), Sec. 42, then

$$\int_a^b F(t)\,dt = \int_{-b}^{-a} F(-t)\,dt.$$

3. Show that if m and n are integers,

$$\int_0^{2\pi} e^{imt}e^{-int}\,dt = \begin{cases} 0 & (m \neq n), \\ 2\pi & (m = n). \end{cases}$$

4. Let F be a continuous complex-valued function of t defined on the interval $a \leqq t \leqq b$. Give an example where there is no number c between a and b such that $F(c)$ times $b - a$ equals the value of the definite integral of F on that interval. Thus, show that the mean-value theorem for integrals in elementary calculus does *not* apply to such functions.

   Suggestion: Use a special case of the result in Exercise 3.

5. Let $f(t) = u(t) + iv(t)$ be a piecewise continuous complex-valued function of a real variable t defined on an interval $a \leqq t \leqq b$. Show that if $F(t) = U(t) + iV(t)$ is a function such that $F'(t) = f(t)$, then

$$\int_a^b f(t)\,dt = F(b) - F(a).$$

6. Derive formula (9), Sec. 43.

   Suggestion: Write $Z(r) = x[\phi(r)] + iy[\phi(r)]$ and apply the chain rule for real-valued functions of a real variable.

7. Let the function $z(t) = x(t) + iy(t)$, $a \leqq t \leqq b$, represent a smooth arc, and suppose that each value of this function lies in the domain of definition of an analytic function $w = f(z)$. Show that if $w(t) = f[z(t)]$, then

$$w'(t) = f'[z(t)]z'(t).$$

   Suggestion: Write $w(t) = u[x(t), y(t)] + iv[x(t), y(t)]$ and apply the chain rule from the calculus of real variables.

44. Line Integrals

We now define the definite integral of a complex-valued function f of the complex variable z. This integral is defined in terms of the values $f(z)$ along a given contour C extending from a point $z = \alpha$ to a point $z = \beta$ in the complex plane. It is therefore a line integral, and its value depends in general upon the contour C as well as the function f. Such an integral is written

$$\int_C f(z)\,dz \qquad \text{or} \qquad \int_\alpha^\beta f(z)\,dz;$$

the latter notation is often used when the value of the integral is independent of the choice of the contour taken between the two end points. While a line integral may be defined directly as a limit of a sum, we choose to define it in terms of a definite integral of the type introduced in Sec. 42.

Let C be the contour represented by the equation

$$z(t) = x(t) + iy(t) \qquad (a \leqq t \leqq b) \tag{1}$$

and extending from the point $\alpha = z(a)$ to the point $\beta = z(b)$. Let the function $f(z) = u(x,y) + iv(x,y)$ be *piecewise continuous* on C; that is, the real and imaginary parts

$$u[x(t), y(t)] \qquad \text{and} \qquad v[x(t), y(t)]$$

of $f[z(t)]$ are piecewise continuous functions of t. We define the line integral of f along C as follows:

$$\int_C f(z)\,dz = \int_a^b f[z(t)]z'(t)\,dt. \tag{2}$$

Since

$$f[z(t)]z'(t) = \{u[x(t),y(t)] + iv[x(t),y(t)]\}[x'(t) + iy'(t)],$$

definition (2) can be written in terms of integrals of real-valued functions of a real variable. That is, in accordance with expression (2), Sec. 42,

$$\int_C f(z)\,dz = \int_a^b (ux' - vy')\,dt + i\int_a^b (vx' + uy')\,dt. \tag{3}$$

Note that since C is a contour, the functions x' and y' are, in addition to the functions u and v, piecewise continuous functions of t. Hence the integrals on the right in equation (3) exist, and the existence of the integral defined by equation (2) is thus ensured.

In terms of line integrals of real-valued functions of two real variables,

$$\int_C f(z)\,dz = \int_C u\,dx - v\,dy + i\int_C v\,dx + u\,dy. \tag{4}$$

Observe that expression (4) can be written formally by replacing f by $u + iv$ and dz by $dx + i\,dy$ and then expanding the product.

Unless it is otherwise indicated, we agree that paths of integration are to be contours and that integrands are to be piecewise continuous functions on those contours.

Associated with the contour C of integral (2) is the contour $-C$ consisting of the same set of points but with the order reversed so that the new contour extends from the point β to the point α. The contour $-C$ is described by the equation $z = z(-t)$ where $-b \leqq t \leqq -a$. Thus

$$\int_{-C} f(z)\,dz = \int_{-b}^{-a} f[z(-t)][-z'(-t)]\,dt,$$

where $z'(-t)$ denotes the derivative of $z(t)$ with respect to t evaluated at $-t$.

After a change of variable in the last integral (see Exercise 2, Sec. 43), we find that

$$\int_{-C} f(z)\,dz = -\int_C f(z)\,dz. \tag{5}$$

We note three further properties of line integrals which follow immediately from either expression (2) or (3). Namely,

$$\int_C \gamma f(z)\,dz = \gamma \int_C f(z)\,dz, \tag{6}$$

for any complex constant γ, and

$$\int_C [f(z) + g(z)]\,dz = \int_C f(z)\,dz + \int_C g(z)\,dz. \tag{7}$$

Also, if C consists of a contour C_1 from α_1 to β_1 and a contour C_2 from α_2 to β_2 where $\beta_1 = \alpha_2$, then

$$\int_C f(z)\,dz = \int_{C_1} f(z)\,dz + \int_{C_2} f(z)\,dz. \tag{8}$$

According to definition (2) above and property (6), Sec. 42,

$$\left|\int_C f(z)\,dz\right| \leqq \int_a^b |f[z(t)]z'(t)|\,dt.$$

So for any constant M such that $|f(z)| \leqq M$ whenever z is on the contour C,

$$\left|\int_C f(z)\,dz\right| \leqq M \int_a^b |z'(t)|\,dt.$$

Now the integral on the right here represents the length L of the contour. Hence the modulus of the value of the integral of f along C does not exceed ML; that is,

$$\left|\int_C f(z)\,dz\right| \leqq ML. \tag{9}$$

The above inequality is, of course, a strict inequality if $|f(z)| < M$ for all points z on C.

It should be noted that a number M such as that appearing in inequality (9) always exists for the arcs and functions considered here. For if a function f is defined on an arc C, we can write

$$|f(z)| = |f[z(t)]| \qquad (a \leqq t \leqq b)$$

when the point z lies on C. If, moreover, f is continuous on C, $|f[z(t)]|$ represents a continuous real-valued function defined on a finite closed interval, and such a function always reaches a maximum value M on that interval.[1] These remarks can now be extended immediately to the case when f is piecewise continuous on C.

The value of a line integral is invariant under a change in the parametric representation of its contour when the change is of the type described by equation (7), Sec. 43. This is seen by writing the integrals on the right in equation (3) in terms of the new parameter r and following the procedure used in Sec. 43 to show the invariance of arc length.

Definite integrals in elementary calculus can be interpreted as areas, and they have other interpretations as well. Except in special cases, no corresponding helpful interpretation, geometrical or physical, is available for integrals in the complex plane. Nonetheless, as noted earlier, the theory of integration in the complex plane is remarkably useful in both pure and applied mathematics.

45. Examples

Let us find the value of the integral

$$I_1 = \int_{C_1} z^2\, dz$$

when C_1 is the line segment OB from $z = 0$ to $z = 2 + i$ (Fig. 37). Observe that points of C_1 lie on the line $x = 2y$. Hence, if the coordinate y is used as the parameter, a parametric equation for C_1 is

$$z(y) = 2y + iy \qquad (0 \leqq y \leqq 1).$$

Also, on C_1 the integrand z^2 becomes

$$z^2 = x^2 - y^2 + i2xy = 3y^2 + i4y^2.$$

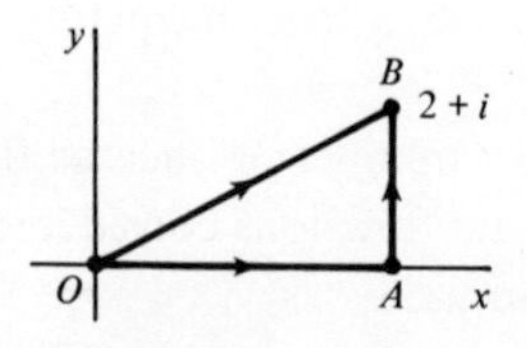

FIGURE 37

[1] See, for instance, A. E. Taylor and W. R. Mann, "Advanced Calculus," 2d ed., pp. 92–96, 1972.

Therefore,

$$I_1 = \int_0^1 (3y^2 + i4y^2)(2 + i)\, dy$$
$$= (3 + 4i)(2 + i)\int_0^1 y^2\, dy = \tfrac{2}{3} + \tfrac{11}{3}i.$$

We next take the path of integration as C_2, the contour OAB shown in Fig. 37, and evaluate the integral

$$I_2 = \int_{C_2} z^2\, dz = \int_{OA} z^2\, dz + \int_{AB} z^2\, dz.$$

A parametric equation for the arc OA is $z(x) = x\ (0 \leqq x \leqq 2)$, and for the arc AB we can write $z(y) = 2 + iy\ (0 \leqq y \leqq 1)$. Hence

$$I_2 = \int_0^2 x^2\, dx + \int_0^1 (2 + iy)^2 i\, dy$$
$$= \tfrac{8}{3} + i\left[\int_0^1 (4 - y^2)\, dy + 4i\int_0^1 y\, dy\right] = \tfrac{2}{3} + \tfrac{11}{3}i.$$

Incidentally, the equation of the contour OAB here can be written in the form

$$z(t) = \begin{cases} t & (0 \leqq t \leqq 2), \\ 2 + i(t - 2) & (2 \leqq t \leqq 3). \end{cases}$$

We note that $I_2 = I_1$. Thus the integral of z^2 over the simple closed contour $OABO$ has the value $I_2 - I_1 = 0$; we shall soon see that this is a consequence of the fact that the integrand z^2 is analytic interior to and on the contour.

For a third example, we let the integrand be the function

$$f(z) = \bar{z},$$

and we take the path of integration C_3 as the upper half of the circle $|z| = 1$ from $z = -1$ to $z = 1$ (Fig. 38). A parametric equation for $-C_3$ is $z(\theta) = \cos\theta + i\sin\theta$, or

$$z(\theta) = e^{i\theta} \qquad (0 \leqq \theta \leqq \pi).$$

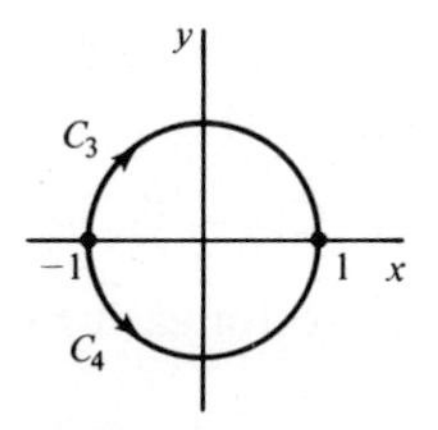

FIGURE 38

Hence

$$I_3 = \int_{C_3} \bar{z}\,dz = -\int_{-C_3} \bar{z}\,dz = -\int_0^\pi e^{-i\theta} ie^{i\theta}\,d\theta = -\pi i.$$

The integral I_4 between the same two points along the lower semicircle C_4, represented by the equation

$$z(\theta) = e^{i\theta} \qquad (\pi \leqq \theta \leqq 2\pi),$$

is readily evaluated:

$$I_4 = \int_{C_4} \bar{z}\,dz = \int_\pi^{2\pi} e^{-i\theta} ie^{i\theta}\,d\theta = \pi i.$$

Note that $I_4 \neq I_3$ and that the integral I_C of $\bar{z}$ around the entire circle C in the counterclockwise direction is not zero:

$$I_C = \int_C \bar{z}\,dz = I_4 - I_3 = 2\pi i.$$

When z is on the unit circle C,

$$\frac{1}{z} = \frac{\bar{z}}{|z|^2} = \bar{z};$$

thus the integrands of the integrals I_3, I_4, and I_C can be replaced by $1/z$. In particular,

$$I_C = \int_C \frac{dz}{z} = 2\pi i.$$

For our final example we let C_5 denote the line segment from $z = i$ to $z = 1$. Without evaluating the integral

$$I_5 = \int_{C_5} \frac{dz}{z^4},$$

let us determine an upper bound for its absolute value. The path of integration is a segment of the line $y = 1 - x$. Hence, if z is a point on C_5, we have

$$|z^4| = (x^2 + y^2)^2 = [x^2 + (1 - x)^2]^2 = (2x^2 - 2x + 1)^2$$

which means that

$$|z^4| = [2(x - \tfrac{1}{2})^2 + \tfrac{1}{2}]^2 \geqq \tfrac{1}{4},$$

since $(x - \frac{1}{2})^2 \geqq 0$. Consequently, for each z on C_5,

$$\left|\frac{1}{z^4}\right| \leqq 4.$$

This inequality is also evident from a figure. We may now write $M = 4$ in inequality (9), Sec. 44; and since the length L of C_5 is $\sqrt{2}$, it follows that

$$|I_5| \leqq 4\sqrt{2}.$$

EXERCISES

For each arc C and function f in Exercises 1 through 5, find the value of

$$\int_C f(z)\,dz$$

after observing that C is a contour and f is piecewise continuous on C.

1. $f(z) = y - x - 3x^2 i$ and C
(*a*) is the line segment from $z = 0$ to $z = 1 + i$;
(*b*) consists of two line segments, one from $z = 0$ to $z = i$ and the other from $z = i$ to $z = 1 + i$.

Ans. (*a*) $1 - i$; (*b*) $(1 - i)/2$.

2. $f(z) = (z + 2)/z$ and C is
(*a*) the semicircle $z = 2e^{i\theta}$ $(0 \leqq \theta \leqq \pi)$;
(*b*) the semicircle $z = 2e^{i\theta}$ $(\pi \leqq \theta \leqq 2\pi)$;
(*c*) the circle $z = 2e^{i\theta}$ $(-\pi \leqq \theta \leqq \pi)$.

Ans. (*a*) $-4 + 2\pi i$; (*b*) $4 + 2\pi i$; (*c*) $4\pi i$.

3. $f(z) = z - 1$ and C is the arc from $z = 0$ to $z = 2$ consisting of
(*a*) the semicircle $z - 1 = e^{i\theta}$ $(0 \leqq \theta \leqq \pi)$;
(*b*) the segment $0 \leqq x \leqq 2$, $y = 0$ of the real axis.

Ans. (*a*) 0; (*b*) 0.

4. C is the arc from $z = -1 - i$ to $z = 1 + i$ along the curve $y = x^3$, and

$$f(z) = \begin{cases} 4y & (y > 0), \\ 1 & (y < 0). \end{cases}$$

Ans. $2 + 3i$.

5. $f(z) = e^z$ and C is the arc from $z = \pi i$ to $z = 1$ consisting of
(*a*) the line segment joining those points;
(*b*) the portion of the coordinate axes joining those points.

Ans. (*a*) $1 + e$; (*b*) $1 + e$.

6. Evaluate the integral

$$\int_C z^m \bar{z}^n\,dz$$

where m and n are integers and C is the circle $|z| = 1$ taken counterclockwise. (See Exercise 3, Sec. 43.)

7. Show that if C is the boundary of the square with vertices at the points $z = 0$, $z = 1$, $z = 1 + i$, $z = i$ and the orientation of C is counterclockwise, then

$$\int_C (3z + 1)\, dz = 0.$$

8. Let C be the contour in Exercise 7 and evaluate

$$\int_C \pi \exp(\pi \bar{z})\, dz.$$

Ans. $4(e^\pi - 1)$.

9. Evaluate the integral I_3 in Sec. 45 using this representation for C_3:

$$z(t) = t + i\sqrt{1 - t^2} \qquad (-1 \leq t \leq 1).$$

10. Let C be the arc of the circle $|z| = 2$ from $z = 2$ to $z = 2i$ that lies in the first quadrant. Without actually evaluating the integral, show that

$$\left| \int_C \frac{dz}{z^2 + 1} \right| \leq \frac{\pi}{3}.$$

11. Show that if C is the boundary of the triangle with vertices at the points $z = 0$, $z = 3i$, $z = -4$ and the orientation of C is counterclockwise, then

$$\left| \int_C (e^z - \bar{z})\, dz \right| \leq 60.$$

*12. Let C be the circle $|z| = R$ taken in the counterclockwise direction, where $R > 1$. Show that

$$\left| \int_C \frac{\operatorname{Log} z}{z^2}\, dz \right| < 2\pi \frac{\pi + \operatorname{Log} R}{R}$$

and hence that the value of the integral approaches zero as R tends to ∞.

13. By writing the integral in terms of integrals of real-valued functions of a real variable, prove that

$$\int_\alpha^\beta dz = \beta - \alpha$$

whenever the path of integration from the point $z = \alpha$ to the point $z = \beta$ is (*a*) a smooth arc; (*b*) a contour.

14. Prove that

$$\int_\alpha^\beta z\, dz = \tfrac{1}{2}(\beta^2 - \alpha^2)$$

whenever the path of integration is (*a*) a smooth arc; (*b*) a contour.

15. Show that if C_0 is the circle

$$z - z_0 = r_0 e^{i\theta} \qquad (0 \leq \theta \leq 2\pi),$$

described counterclockwise, and f is continuous on C_0, then

$$\int_{C_0} f(z)\, dz = ir_0 \int_0^{2\pi} f(z_0 + r_0 e^{i\theta}) e^{i\theta}\, d\theta.$$

16. Obtain the following special cases of the result in Exercise 15:

$$\int_{C_0} \frac{dz}{z - z_0} = 2\pi i, \qquad \int_{C_0} (z - z_0)^{n-1}\, dz = 0 \qquad (n = \pm 1, \pm 2, \ldots).$$

46. The Cauchy-Goursat Theorem

Suppose that two real-valued functions $P(x,y)$ and $Q(x,y)$, together with their first partial derivatives, are continuous throughout a closed region R consisting of points within and on a simple closed contour C. The contour is described in the *positive sense* (counterclockwise) so that the interior points of R lie to the left of C. According to Green's theorem for line integrals in the calculus of real variables,

$$\int_C P\, dx + Q\, dy = \iint_R (Q_x - P_y)\, dx\, dy.$$

We now consider a function

$$f(z) = u(x,y) + iv(x,y)$$

which is analytic throughout such a region R in the z plane. We assume, moreover, that $f'(z)$ is *continuous* there. The component functions u and v, along with their first partial derivatives, are thus continuous in R; consequently,

$$\int_C u\, dx - v\, dy = -\iint_R (v_x + u_y)\, dx\, dy,$$

$$\int_C v\, dx + u\, dy = \iint_R (u_x - v_y)\, dx\, dy.$$

In view of the Cauchy-Riemann equations, the integrands of these two double integrals are zero throughout R; and according to equation (4), Sec. 44, the line integrals on the left represent the real and imaginary parts of the value of the integral of $f(z)$ along C. So we have the result

$$\int_C f(z)\, dz = 0$$

which was obtained by Cauchy in the early part of the last century.

For elementary examples, we note that if C is a simple closed contour,

$$\int_C dz = 0, \qquad \int_C z\, dz = 0, \qquad \int_C z^2\, dz = 0$$

because the functions 1, z, and z^2 are entire and their derivatives are everywhere continuous.

Goursat[1] was the first to prove that the condition of continuity on $f'(z)$ can be omitted. The removal of that condition is important. One of the consequences will be, for example, that derivatives of analytic functions are also analytic. The revised form of Cauchy's result is the following theorem, known as the *Cauchy-Goursat theorem.*

Theorem. *If a function f is analytic at all points within and on a simple closed contour C, then*

$$\int_C f(z)\,dz = 0.$$

The proof is presented in the next two sections where, to be specific, we assume that C is positively oriented. It will be a simple matter to extend the theorem to more general paths such as the entire boundary of the region between two concentric circles.

47. A Preliminary Lemma

We start by forming subsets of the region R which consists of the points on C together with the points interior to C. To do this, we draw equally spaced lines parallel to the real and imaginary axes such that the distance between adjacent vertical lines is the same as that between adjacent horizontal lines. We thus form a finite number of closed square subregions where each point of R lies in at least one subregion.

We shall refer to such square subregions simply as squares, always keeping in mind that by a square we mean a boundary together with the points interior to it. If a particular square contains points that are not in R, we remove those points and call what remains a partial square. We thus *cover* the region R with a finite number of squares and partial squares (Fig. 39), and our proof of the following lemma starts with this covering.

Lemma. *Let f be analytic at all points of a closed region R consisting of the interior of a simple closed contour C together with the points on C itself. For any positive number ε, it is always possible to cover R with a finite number (n) of squares and partial squares, whose boundaries will be denoted by C_j, such that a point z_j exists within or on each C_j for which the inequality*

$$\left| \frac{f(z) - f(z_j)}{z - z_j} - f'(z_j) \right| < \varepsilon \qquad (j = 1, 2, \ldots, n) \tag{1}$$

is satisfied by all points z distinct from z_j within or on C_j.

[1] E. Goursat (1858–1936), pronounced *gour-sah'*.

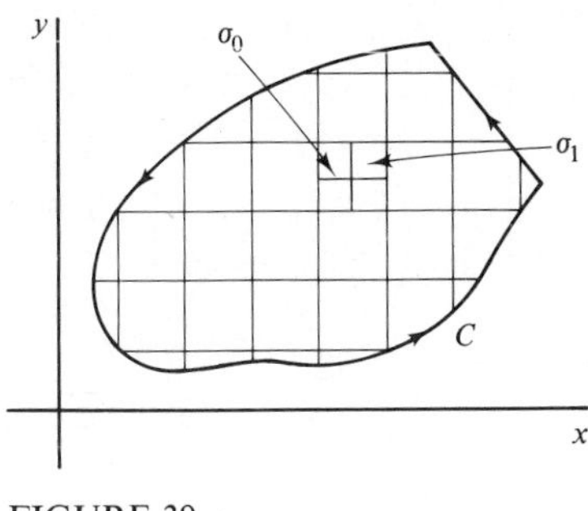

FIGURE 39

Suppose that in the covering constructed just prior to the statement of the lemma there is some square or partial square with boundary C_j such that no point z_j exists for which inequality (1) holds. If that subregion is a square, construct four smaller squares by drawing line segments joining the midpoints of its opposite sides (Fig. 39). If the subregion is a partial square, treat the whole square in the same manner and then let the portions that lie outside of R be discarded. If in any one of these smaller subregions no point z_j exists for which inequality (1) holds, construct still smaller squares and partial squares, etc.

Doing this to each of the original subregions that requires it, we may find that after a finite number of steps we can cover the region R with a collection of squares and partial squares such that inequality (1) is true for all of them. In that case the lemma is true.

Suppose, on the other hand, that points z_j do not exist such that condition (1) is satisfied after subdividing one of the original subregions a finite number of times. Let σ_0 denote that subregion if it is a square; if it is a partial square, let σ_0 denote the entire square of which it is a part. After we subdivide σ_0, at least one of the four smaller squares, denoted by σ_1, must contain points of R but no appropriate point z_j. We then subdivide σ_1 and continue in this manner. It may be that after a square σ_{k-1} $(k = 1,2,\ldots)$ has been selected, more than one of the four smaller squares constructed from it can be chosen. In order to make a specific choice, we take σ_k to be the one lowest and then furthest to the left.

In view of the manner in which the nested infinite sequence

$$\sigma_0, \sigma_1, \sigma_2, \ldots, \sigma_{k-1}, \sigma_k, \ldots$$

of squares is constructed, it is easily shown (Exercise 13, Sec. 50) that there is a point z_0 common to each σ_k; also, each of these squares contains points of R. Recall how the sizes of the squares in the sequence are decreasing, and note that

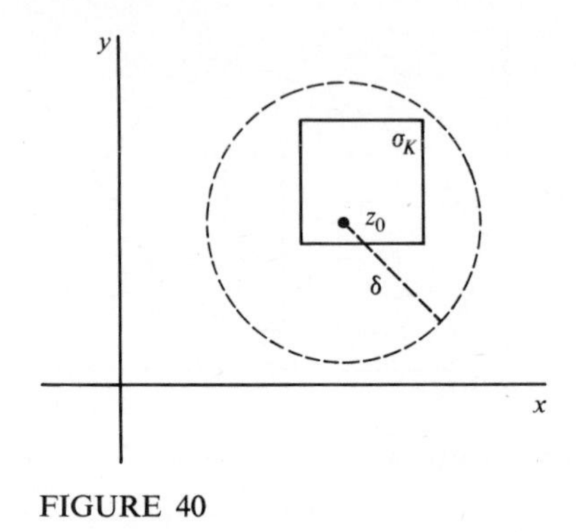

FIGURE 40

any δ neighborhood $|z - z_0| < \delta$ of z_0 contains such squares when their diagonals have lengths less than δ. Every δ neighborhood $|z - z_0| < \delta$ therefore contains points of R distinct from z_0, and z_0 must therefore be an accumulation point of R. Since the region R is a closed set, it follows that z_0 is a point in R.

The function f is analytic throughout R, and we now see that, in particular, it is analytic at z_0. Consequently, $f'(z_0)$ exists. According to the definition of derivative, there is for each positive number ε a δ neighborhood $|z - z_0| < \delta$ such that the inequality

$$\left| \frac{f(z) - f(z_0)}{z - z_0} - f'(z_0) \right| < \varepsilon$$

is satisfied by all points distinct from z_0 in that neighborhood. But the neighborhood $|z - z_0| < \delta$ contains a square σ_K when K is large enough so that the length of a diagonal of that square is less than δ (Fig. 40). Consequently, z_0 serves as the point z_j such that inequality (1) is satisfied in the subregion consisting of the square σ_K or a part of σ_K. Contrary to our hypothesis then, it is not necessary to subdivide σ_K. We have thus arrived at a contradiction, and the proof of the lemma is complete.

48. Proof of the Cauchy-Goursat Theorem

We now show that the inequality

$$\left| \int_C f(z)\, dz \right| < \gamma \tag{1}$$

is true for every positive number γ and conclude that the value of the integral must therefore be zero.

Given an arbitrary positive number ε, let C_j $(j = 1, 2, \ldots, n)$ denote the boundaries of a collection of squares and partial squares covering R in accordance with the lemma of the preceding section. We can state inequality (1) of that lemma in the following form. Each function

$$\delta_j(z) = \begin{cases} \dfrac{f(z) - f(z_j)}{z - z_j} - f'(z_j) & (z \neq z_j) \\ 0 & (z = z_j) \end{cases} \tag{2}$$

defined on the jth square or partial square satisfies the inequality

$$|\delta_j(z)| < \varepsilon \tag{3}$$

at all points z in its domain of definition. Note that each of these functions is continuous throughout the region on which it is defined.

Next observe that the value of f at a point z on any particular C_j can be written

$$f(z) = f(z_j) - z_j f'(z_j) + f'(z_j)z + (z - z_j)\delta_j(z). \tag{4}$$

Hence, integrating $f(z)$ counterclockwise around C_j and recalling (Sec. 46) that

$$\int_{C_j} dz = 0, \qquad \int_{C_j} z\, dz = 0,$$

we see that

$$\int_{C_j} f(z)\, dz = \int_{C_j} (z - z_j)\delta_j(z)\, dz. \tag{5}$$

If we add all n integrals on the left in equation (5), we find that

$$\sum_{j=1}^{n} \int_{C_j} f(z)\, dz = \int_C f(z)\, dz.$$

This is true because the integrals along the common boundary of every pair of adjacent subregions cancel each other; the integral is taken in one sense along that line segment in one subregion and in the opposite sense in the other (Fig. 41). Only the integrals along the arcs that are parts of C remain. Thus, in view of equation (5),

$$\int_C f(z)\, dz = \sum_{j=1}^{n} \int_{C_j} (z - z_j)\, \delta_j(z)\, dz,$$

or

$$\left|\int_C f(z)\, dz\right| \leqq \sum_{j=1}^{n} \left|\int_{C_j} (z - z_j)\, \delta_j(z)\, dz\right|. \tag{6}$$

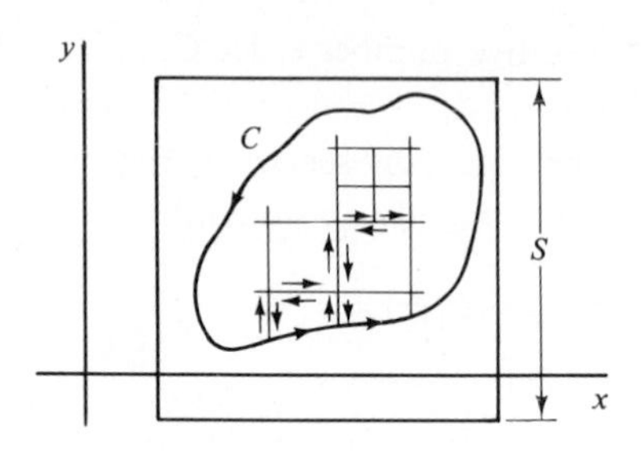

FIGURE 41

Let us now use property (9), Sec. 44, to find an upper bound for each integral on the right in inequality (6). To do this, we first recall that each C_j coincides either entirely or partially with the boundary of a square. In either case, we let s_j denote the length of a side of the square. Now in the jth integral on the right in inequality (6) both the variable z and the point z_j lie in that square; hence

$$|z - z_j| \leqq \sqrt{2}s_j.$$

In view of inequality (3) we therefore know that each integrand on the right in inequality (6) satisfies the condition

$$|(z - z_j)\delta_j(z)| < \sqrt{2}s_j\varepsilon. \tag{7}$$

As for the length of the path C_j, it is $4s_j$ if C_j is the boundary of a square. In that case we let A_j denote the area of the square and observe that

$$\left|\int_{C_j}(z - z_j)\delta_j(z)\,dz\right| < \sqrt{2}s_j\varepsilon 4s_j = 4\sqrt{2}A_j\varepsilon. \tag{8}$$

If C_j is the boundary of a partial square, its length does not exceed $4s_j + L_j$ where L_j is the length of that part of C_j which is also a part of C. Again letting A_j denote the area of the full square, we find that

$$\left|\int_{C_j}(z - z_j)\delta_j(z)\,dz\right| < \sqrt{2}\,s_j\varepsilon(4s_j + L_j) < 4\sqrt{2}\,A_j\varepsilon + \sqrt{2}\,SL_j\varepsilon \tag{9}$$

where S is the length of a side of some square that encloses the entire contour C as well as all the squares originally used in covering C (Fig. 41). Note that the sum of all the A_j's does not exceed S^2.

If L denotes the length of C, it now follows from inequalities (6), (8), and (9) that

$$\left|\int_C f(z)\,dz\right| < (4\sqrt{2}\,S^2 + \sqrt{2}\,SL)\varepsilon.$$

For each positive number γ, the right-hand member here can be made equal to γ by assigning the proper value to the positive number ε. Hence inequality (1) is established, and the Cauchy-Goursat theorem is proved.

49. Simply and Multiply Connected Domains

A *simply connected domain* D is a domain such that every simple closed contour within it encloses only points of D. The set of points interior to a simple closed contour is an example. The annular domain between two concentric circles is, however, not simply connected. A domain that is not simply connected is said to be a *multiply connected domain.*

The Cauchy-Goursat theorem can be stated in the following alternate form.

If a function f is analytic throughout a simply connected domain D, then for every simple closed contour C within D

$$\int_C f(z)\,dz = 0. \tag{1}$$

The simple closed contour here can be replaced by an arbitrary closed contour C which is not necessarily simple. For, if C intersects itself only a finite number of times, then it consists of a finite number of simple closed contours. By applying the Cauchy-Goursat theorem to each of these simple closed contours, we obtain the desired result for C. Also, a portion of C may be traversed twice in opposite directions since the integrals along that portion in the two directions cancel each other. Subtleties arise if the closed contour has an infinite number of self-intersection points.[1]

The Cauchy-Goursat theorem can also be modified in the following manner.

Theorem. *Let C be a simple closed contour and let C_j $(j = 1, 2, \ldots, n)$ be a finite number of simple closed contours interior to C such that the sets interior to each C_j have no points in common. Let R be the closed region consisting of all points within and on C except for points interior to each C_j (Fig. 42). Let B denote the entire oriented boundary of R consisting of C and all the C_j described in a direction such that the points of R lie to the left of B. Then, if f is analytic in R,*

$$\int_B f(z)\,dz = 0. \tag{2}$$

[1] For a proof of the above theorem involving more general paths, see, for example, Secs. 63–65 in vol. I of the books by Markushevich cited in Appendix 1.

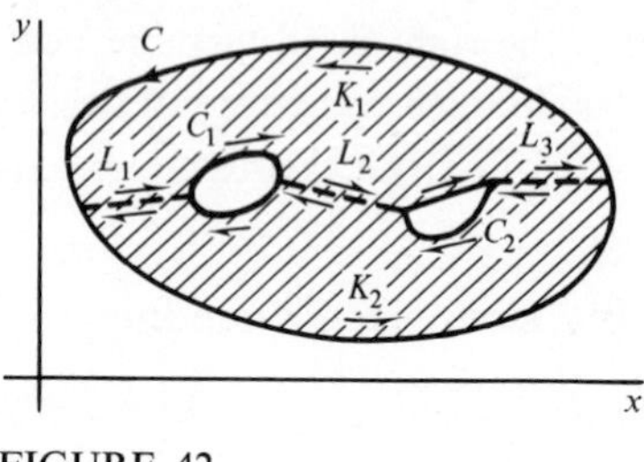

FIGURE 42

To establish this result, we introduce a polygonal path L_1, consisting of a finite number of line segments joined end to end, to connect the outer contour C to the inner contour C_1. We introduce another polygonal path L_2 which connects C_1 to C_2; and we continue in this manner, with L_{n+1} connecting C_n to C. As indicated by the single-barbed arrows in Fig. 42, two simple closed contours K_1 and K_2 can be formed, each consisting of polygonal paths L_j or $-L_j$ and pieces of C and C_j and each described in such a direction that points enclosed by them lie to the left. The Cauchy-Goursat theorem can now be applied to f on K_1 and K_2, and the sum of the integrals over those contours is found to be zero. Since the integrals in opposite directions along each path L_j cancel, only the integral along B remains, and formula (2) follows.

To illustrate this theorem, we note that

$$\int_B \frac{dz}{z^2(z^2+9)} = 0$$

where B consists of the circle $|z| = 2$ described in the positive direction together with the circle $|z| = 1$ described in the negative direction (Fig. 43). The integrand is analytic except at the points $z = 0$ and $z = \pm 3i$, and these three points lie outside the annular region with boundary B.

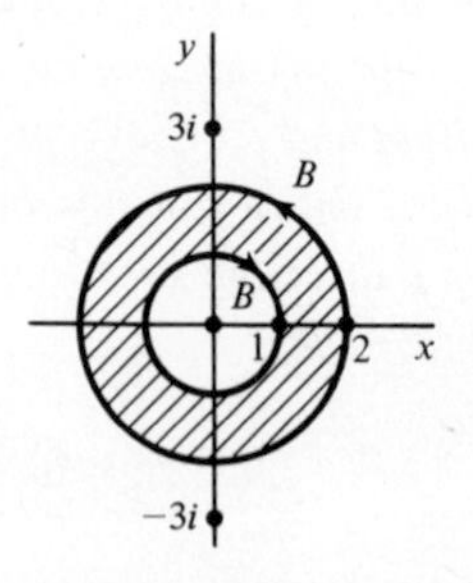

FIGURE 43

50. Indefinite Integrals

Let z_0 and z be any two points in a simply connected domain D throughout which f is analytic (Fig. 44). If C_1 and C_2 are two contours connecting z_0 to z and lying entirely within D, then C_1 and $-C_2$ together form a closed contour. Since the Cauchy-Goursat theorem holds for any closed contour in a simply connected domain, we find that

$$\int_{C_1} f(s)\,ds - \int_{C_2} f(s)\,ds = 0,$$

where s denotes points on C_1 and C_2. The integral from z_0 to z is thus independent of the contour taken as long as that contour lies within D. This integral then defines a function F on the simply connected domain D:

$$F(z) = \int_{z_0}^{z} f(s)\,ds. \tag{1}$$

We now show that the derivative of $F(z)$ exists and is equal to $f(z)$. Let $z + \Delta z$ be any point, distinct from z, lying in some neighborhood of z where that neighborhood is completely contained in D (Fig. 45). Then

$$F(z + \Delta z) - F(z) = \int_{z_0}^{z+\Delta z} f(s)\,ds - \int_{z_0}^{z} f(s)\,ds$$
$$= \int_{z}^{z+\Delta z} f(s)\,ds$$

where the path of integration from z to $z + \Delta z$ may be selected as a line segment. Since we can write (Exercise 13, Sec. 45)

$$f(z) = \frac{f(z)}{\Delta z}\int_{z}^{z+\Delta z} ds = \frac{1}{\Delta z}\int_{z}^{z+\Delta z} f(z)\,ds,$$

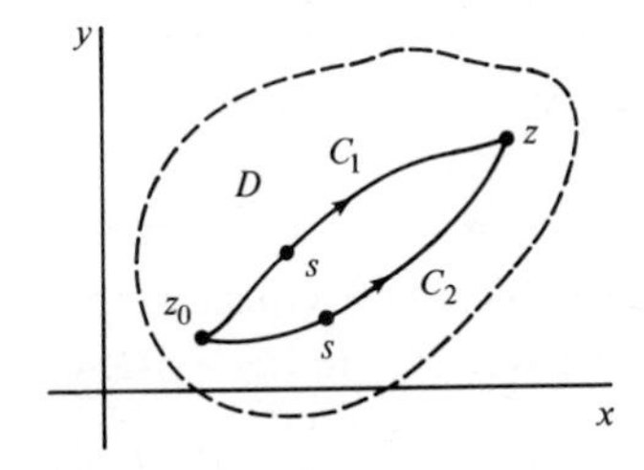

FIGURE 44

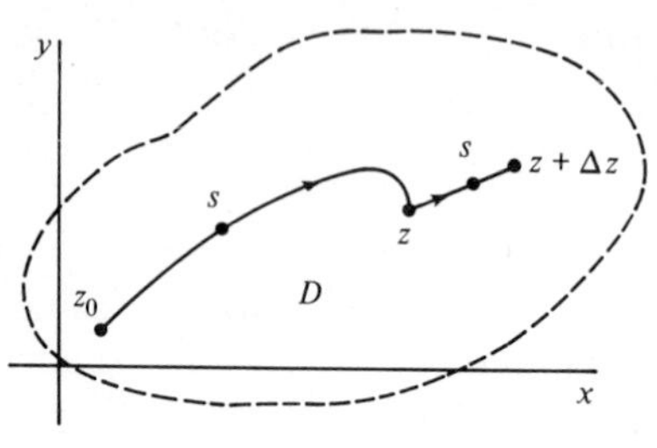

FIGURE 45

it follows that

$$\frac{F(z + \Delta z) - F(z)}{\Delta z} - f(z) = \frac{1}{\Delta z}\int_{z}^{z+\Delta z} [f(s) - f(z)]\, ds.$$

But f is continuous at the point z. Hence for each positive number ε, a positive number δ exists such that

$$|f(s) - f(z)| < \varepsilon$$

whenever $|s - z| < \delta$. Consequently, if the point $z + \Delta z$ is close enough to z so that $|\Delta z| < \delta$, then

$$\left|\frac{F(z + \Delta z) - F(z)}{\Delta z} - f(z)\right| < \frac{1}{|\Delta z|}\, \varepsilon |\Delta z| = \varepsilon;$$

that is,

$$\lim_{\Delta z \to 0} \frac{F(z + \Delta z) - F(z)}{\Delta z} = f(z).$$

Thus the derivative of integral (1) exists at each point z in D, and

$$F'(z) = f(z). \tag{2}$$

A definite integral of an analytic function is therefore an analytic function of its upper limit, provided the path of integration is confined to a simply connected domain throughout which the integrand is analytic.

We see from integral (1) that $F(z)$ is changed by an additive constant when the lower limit z_0 is replaced by a new constant. The function $F(z)$ is an *indefinite integral*, or antiderivative, of f and is written

$$F(z) = \int f(z)\, dz.$$

That is, $F(z)$ is an analytic function whose derivative is $f(z)$. In view of equation (1), *a definite integral can be evaluated as the change in the value of the indefinite integral*, as in elementary calculus:

$$\int_{\alpha}^{\beta} f(z)\,dz = \int_{z_0}^{\beta} f(z)\,dz - \int_{z_0}^{\alpha} f(z)\,dz = F(\beta) - F(\alpha) = F(z)\Big]_{\alpha}^{\beta}. \tag{3}$$

We continue to assume, of course, that the paths of integration are confined to a simply connected domain in which f is analytic.

It should be noted that if $G(z)$ is any analytic function other than $F(z)$ such that $G'(z) = f(z)$, then the derivative of the function $H(z) = F(z) - G(z)$ is zero. Hence, if $H(z) = u(x,y) + iv(x,y)$, it follows that

$$u_x(x,y) + iv_x(x,y) = 0;$$

so $u_x(x,y)$ and $v_x(x,y)$ are both zero throughout the domain in which the functions F and G are analytic. In view of the Cauchy-Riemann equations, the quantities $u_y(x,y)$ and $v_y(x,y)$ are also zero; $u(x,y)$ and $v(x,y)$ are therefore constant. Thus $H(z)$ is a constant, and it follows that the two indefinite integrals $F(z)$ and $G(z)$ differ at most by an additive complex constant. As a consequence, any indefinite integral of $f(z)$ can be used in place of $F(z)$ in formula (3).

An indefinite integral of the function $f(z) = z^2$, for example, is the entire function $F(z) = z^3/3$. Since the function z^2 is entire, we can write

$$\int_0^{1+i} z^2\,dz = \tfrac{1}{3}z^3\Big]_0^{1+i} = \tfrac{1}{3}(1+i)^3$$

for every contour between the points $z = 0$ and $z = 1 + i$.

To give another example, let us evaluate

$$\int_{-1}^{1} z^{1/2}\,dz \tag{4}$$

where

$$z^{1/2} = \sqrt{r}\exp\frac{i\theta}{2} \qquad (r > 0, 0 < \theta < 2\pi), \tag{5}$$

and the contour joining the two limits of integration lies above the real axis in the z plane. The function is not analytic at points on the ray $\theta = 0$, at $z = 1$ in particular. But another branch,

$$f(z) = \sqrt{r}\exp\frac{i\theta}{2} \qquad \left(r > 0, -\frac{\pi}{2} < \theta < \frac{3\pi}{2}\right),$$

of the multiple-valued function $z^{1/2}$ is analytic everywhere except on the ray $\theta = -\pi/2$. The values of $f(z)$ above the real axis coincide with those of our function (5), and so our integrand can be replaced by $f(z)$. Now an indefinite integral of $f(z)$ is the function

$$\tfrac{2}{3}z^{3/2} = \tfrac{2}{3}r^{3/2}\exp\frac{i3\theta}{2} \qquad \left(r > 0, -\frac{\pi}{2} < \theta < \frac{3\pi}{2}\right).$$

Thus

$$\int_{-1}^{1} z^{1/2}dz = \tfrac{2}{3}(e^0 - e^{i3\pi/2}) = \tfrac{2}{3}(1 + i).$$

Integral (4) over any contour below the real axis has another value. There we can replace the integrand by the branch

$$g(z) = \sqrt{r}\exp\frac{i\theta}{2} \qquad \left(r > 0, \frac{\pi}{2} < \theta < \frac{5\pi}{2}\right)$$

whose values coincide with those of the function (5) in the lower half plane. The analytic function

$$\tfrac{2}{3}z^{3/2} = \tfrac{2}{3}r^{3/2}\exp\frac{i3\theta}{2} \qquad \left(r > 0, \frac{\pi}{2} < \theta < \frac{5\pi}{2}\right)$$

is an indefinite integral of $g(z)$; thus

$$\int_{-1}^{1} z^{1/2}\, dz = \tfrac{2}{3}(e^{i3\pi} - e^{i3\pi/2}) = \tfrac{2}{3}(-1 + i).$$

The integral of the function (5) in the positive sense around a simple closed contour consisting of a path of the second type combined with one of the first therefore has the value

$$\tfrac{2}{3}(-1 + i) - \tfrac{2}{3}(1 + i) = -\tfrac{4}{3}.$$

EXERCISES

1. Determine the domain of analyticity of the function f and apply the Cauchy-Goursat theorem to show that

$$\int_C f(z)\, dz = 0$$

when the simple closed contour C is the circle $|z| = 1$ and when

(*a*) $f(z) = \dfrac{z^2}{z-3}$; (*b*) $f(z) = ze^{-z}$; (*c*) $f(z) = \dfrac{1}{z^2 + 2z + 2}$;

(*d*) $f(z) = \operatorname{sech} z$; (*e*) $f(z) = \tan z$; (*f*) $f(z) = \operatorname{Log}(z + 2)$.

2. Let B be the oriented boundary of the region bounded by the circle $|z| = 4$ and the square with sides along the lines $x = \pm 1$, $y = \pm 1$, where B is described so that the region lies to the left of B. State why

$$\int_B f(z)\,dz = 0$$

when

$$*(a)\ f(z) = \frac{1}{3z^2 + 1}; \qquad (b)\ f(z) = \frac{z + 2}{\sin(z/2)}; \qquad (c)\ f(z) = \frac{z}{1 - e^z}.$$

3. Let C be a simple closed contour interior to a simple closed contour C_0, where both C and C_0 are oriented in the positive direction. Show that if a function f is analytic in the closed region bounded by C and C_0, then

$$\int_C f(z)\,dz = \int_{C_0} f(z)\,dz.$$

4. Use results of Exercise 3 above and Exercise 16, Sec. 45, to show that

$$\int_C \frac{dz}{z - 2 - i} = 2\pi i, \qquad \int_C (z - 2 - i)^{n-1}\,dz = 0 \qquad (n = \pm 1, \pm 2, \ldots)$$

when C is the boundary of the rectangle $0 \leqq x \leqq 3$, $0 \leqq y \leqq 2$, described in the positive sense.

5. Use an indefinite integral to show that, for every contour C extending from a point α to a point β,

$$\int_C z^n\,dz = \frac{1}{n + 1}(\beta^{n+1} - \alpha^{n+1}) \qquad (n = 0, 1, 2, \ldots).$$

6. Evaluate each of these integrals where the path of integration is an arbitrary contour between the indicated limits:

$$(a)\ \int_i^{i/2} e^{\pi z}\,dz; \qquad (b)\ \int_0^{\pi + 2i} \cos\frac{z}{2}\,dz; \qquad (c)\ \int_1^3 (z - 2)^3\,dz.$$

Ans. (a) $(1 + i)/\pi$; (b) $e + 1/e$; (c) 0.

7. Show that if $z_1 \neq 0$, $z_2 \neq 0$, and $z_1 \neq z_2$, then

$$\int_{z_1}^{z_2} \frac{dz}{z^2} = \frac{1}{z_1} - \frac{1}{z_2}$$

whenever the path of integration is interior to a simply connected domain which does not contain the origin. Show how it follows that, for any simple closed contour C where the origin is either an interior point or an exterior point,

$$\int_C \frac{dz}{z^2} = 0.$$

8. Let z_0, z_1, and z_2 denote three distinct points of a simply connected domain D. Given that a function $f(z)$ and its derivative $f'(z)$ are both analytic throughout D except at z_0, generalize the result in Exercise 7 to show that, for each contour in D drawn from z_1 to z_2 but not passing through z_0,

$$\int_{z_1}^{z_2} f'(z)\,dz = f(z_2) - f(z_1); \qquad \text{thus} \qquad \int_C f'(z)\,dz = 0$$

when the simple closed contour C in D does not pass through z_0. Give examples of such functions and domains.

9. Use an indefinite integral to find the value of the integral

$$\int_{-2i}^{2i} \frac{dz}{z}$$

over every contour from $z = -2i$ to $z = 2i$ lying in the right half plane. Note that the principal branch Log z is an indefinite integral of $1/z$ that is analytic in the half plane $x \geqq 0$ except at the origin.

10. Do Exercise 9 for every contour that does not touch the nonnegative half of the real axis.

Ans. $-\pi i$.

11. Note that the single-valued function

$$f(z) = z^{1/2} = \sqrt{r} \exp \frac{i\theta}{2} \qquad \left(r > 0, \ -\frac{\pi}{2} \leqq \theta < \frac{3\pi}{2}\right),$$

where $f(0) = 0$, is continuous throughout the half plane $r \geqq 0$, $0 \leqq \theta \leqq \pi$. Let C denote the entire boundary of the half disk $0 \leqq r \leqq 1$, $0 \leqq \theta \leqq \pi$ where C is described in the positive direction. Show that

$$\int_C f(z)\,dz = 0$$

by computing the integrals of $f(z)$ over the semicircle and over the two radii on the x axis. Why does the Cauchy-Goursat theorem not apply here?

12. *Nested Intervals.* An infinite sequence of closed intervals $a_n \leqq x \leqq b_n$ ($n = 0, 1, 2, \ldots$) is determined according to some rule for selecting half of a given interval. The interval $a_1 \leqq x \leqq b_1$ is either the left-hand or right-hand half of the first interval $a_0 \leqq x \leqq b_0$, and the interval $a_2 \leqq x \leqq b_2$ is then one of the two halves of $a_1 \leqq x \leqq b_1$, etc. Prove that there is a point x_0 which belongs to every one of the closed intervals $a_n \leqq x \leqq b_n$.

Suggestion: Note that the left-hand end points a_n represent a bounded nondecreasing sequence of numbers since $a_0 \leqq a_n \leqq a_{n+1} < b_0$; hence they have a limit A as n tends to ∞. Show that the end points b_n also have a limit B; then show that $A = B = x_0$.

13. *Nested Squares.* A square σ_0: $a_0 \leqq x \leqq b_0$, $c_0 \leqq y \leqq d_0$, where $b_0 - a_0 = d_0 - c_0$, is divided into four equal squares by lines parallel to the coordinate axes. One of those four smaller squares σ_1: $a_1 \leqq x \leqq b_1$, $c_1 \leqq y \leqq d_1$, where $b_1 - a_1 =$

$d_1 - c_1$, is selected according to some rule. It in turn is divided into four equal squares one of which, called σ_2, is selected, etc. (Sec. 47). Prove that there is a point (x_0,y_0) which belongs to each of the closed regions of the infinite sequence $\sigma_0, \sigma_1, \sigma_2, \ldots$.

Suggestion: Apply the results of Exercise 12 to each of the sequences of closed intervals $a_n \leqq x \leqq b_n$ and $c_n \leqq y \leqq d_n$ $(n = 0,1,2,\ldots)$.

51. The Cauchy Integral Formula

Another fundamental result will now be established.

Theorem. *Let f be analytic everywhere within and on a simple closed contour C, taken in the positive sense. If z_0 is any point interior to C, then*

$$f(z_0) = \frac{1}{2\pi i}\int_C \frac{f(z)\,dz}{z - z_0}. \tag{1}$$

Formula (1) is called the *Cauchy integral formula.* It says that if a function f is to be analytic within and on a simple closed contour C, then the values of f inside C are completely determined by the values of f on C. Any change in the value of f at a point within C must, therefore, be accompanied by a change in its values on the boundary C.

To illustrate the use of formula (1) in evaluating integrals, let us show that

$$\int_C \frac{z\,dz}{(9 - z^2)(z + i)} = \frac{\pi}{5}$$

where C is the circle $|z| = 2$ taken in the positive sense. Since the function $f(z) = z/(9 - z^2)$ is analytic within and on C, we can apply the Cauchy integral formula to it when $z_0 = -i$. The desired result is $2\pi i f(-i)$, or $\pi/5$.

To prove the theorem, let C_0 be a circle

$$|z - z_0| = r_0$$

about z_0 where the radius r_0 is small enough so that C_0 is interior to C (Fig. 46). The function $f(z)/(z - z_0)$ is analytic at all points within and on C except at the point z_0. Hence, by the Cauchy-Goursat theorem for multiply connected domains, its integral around the oriented boundary of the region between C and C_0 is zero (Sec. 49). Thus

$$\int_C \frac{f(z)\,dz}{z - z_0} = \int_{C_0} \frac{f(z)\,dz}{z - z_0},$$

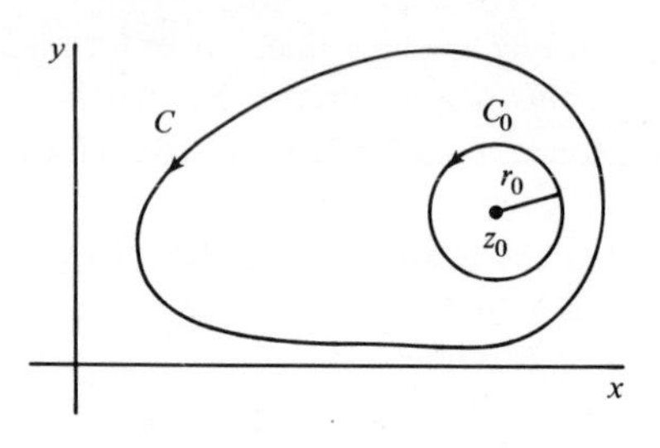

FIGURE 46

where both integrals are taken counterclockwise. Since the integrals of $f(z)/(z - z_0)$ around C and C_0 are then equal, we can write

$$\int_C \frac{f(z)\,dz}{z - z_0} = f(z_0)\int_{C_0} \frac{dz}{z - z_0} + \int_{C_0} \frac{f(z) - f(z_0)}{z - z_0}\,dz. \tag{2}$$

Observe that $z - z_0 = r_0 e^{i\theta}$ on C_0; and, according to Exercise 16, Sec. 45,

$$\int_{C_0} \frac{dz}{z - z_0} = 2\pi i. \tag{3}$$

We now show that the last integral in equation (2) has the value zero. Since f is continuous at the point z_0, there corresponds to each positive number ε a positive number δ such that

$$|f(z) - f(z_0)| < \varepsilon \qquad \text{whenever} \qquad |z - z_0| < \delta. \tag{4}$$

Let us then choose any positive number γ less than δ which is small enough so that the circle $|z - z_0| = \gamma$ is interior to C. Note that the first of inequalities (4) holds for any point z on that circle. Now the value of the last integral in equation (2) is independent of the choice of r_0 since this is true for the values of the other two integrals there. We thus take r_0 to be γ. Using property (9), Sec. 44, and noting that the length of C_0 is now $2\pi\gamma$, we find that

$$\left| \int_{C_0} \frac{f(z) - f(z_0)}{z - z_0}\,dz \right| < \frac{\varepsilon}{\gamma}\, 2\pi\gamma = 2\pi\varepsilon,$$

the absolute value of the integrand here being less than ε/γ. Consequently, the absolute value of the last integral in equation (2) can be made arbitrarily small, and the value of that integral must therefore be zero.

Equation (2) now reduces to

$$\int_C \frac{f(z)\,dz}{z - z_0} = f(z_0)2\pi i,$$

and the theorem is proved.

52. Derivatives of Analytic Functions

We are now ready to prove that if a function is analytic at a point, its derivatives of all orders exist at that point and are also analytic there.

We begin by assuming that f is analytic within and on a simple closed contour C, and we let z be any point within C. Denoting points on C by s and using the Cauchy integral formula,

$$f(z) = \frac{1}{2\pi i} \int_C \frac{f(s)\, ds}{s - z}, \tag{1}$$

we shall show that the derivative of f at z has the integral representation

$$f'(z) = \frac{1}{2\pi i} \int_C \frac{f(s)\, ds}{(s - z)^2}. \tag{2}$$

Note that expression (2) can be obtained formally by differentiating the integrand in formula (1) with respect to z.

To establish formula (2), we observe that, according to formula (1),

$$\begin{aligned} \frac{f(z + \Delta z) - f(z)}{\Delta z} &= \frac{1}{2\pi i} \int_C \left(\frac{1}{s - z - \Delta z} - \frac{1}{s - z} \right) \frac{f(s)}{\Delta z}\, ds \\ &= \frac{1}{2\pi i} \int_C \frac{f(s)\, ds}{(s - z - \Delta z)(s - z)}. \end{aligned}$$

We then use the fact that f is continuous on C to show that the last integral approaches the integral

$$\int_C \frac{f(s)\, ds}{(s - z)^2}$$

as Δz approaches zero. To accomplish this, we first write the modulus of their difference:

$$\left| \Delta z \int_C \frac{f(s)\, ds}{(s - z)^2 (s - z - \Delta z)} \right|.$$

We next let M denote the maximum value of $|f(s)|$ on C and we let L be the length of C. If d is the shortest distance from z to any point on C and if $|\Delta z| < d$, it follows that

$$\left| \Delta z \int_C \frac{f(s)\, ds}{(s - z)^2 (s - z - \Delta z)} \right| < \frac{|\Delta z|\, ML}{d^2 (d - |\Delta z|)},$$

where the last fraction approaches zero as Δz approaches zero. Consequently,

$$\lim_{\Delta z \to 0} \frac{f(z + \Delta z) - f(z)}{\Delta z} = \frac{1}{2\pi i} \int_C \frac{f(s)\,ds}{(s - z)^2};$$

and formula (2) is established.

If we apply the same techniques to formula (2), we find that

$$f''(z) = \frac{1}{\pi i} \int_C \frac{f(s)\,ds}{(s - z)^3}. \tag{3}$$

For it follows from formula (2) that

$$\begin{aligned} \frac{f'(z + \Delta z) - f'(z)}{\Delta z} &= \frac{1}{2\pi i} \int_C \left[\frac{1}{(s - z - \Delta z)^2} - \frac{1}{(s - z)^2} \right] \frac{f(s)\,ds}{\Delta z} \\ &= \frac{1}{2\pi i} \int_C \frac{2(s - z) - \Delta z}{(s - z - \Delta z)^2 (s - z)^2} f(s)\,ds. \end{aligned}$$

Again, since f is continuous on C, we can show that the limit of this last integral as Δz approaches zero is

$$2 \int_C \frac{f(s)\,ds}{(s - z)^3},$$

and formula (3) follows at once.

Formula (3) establishes the existence of the second derivative of the function f at each point z interior to C. In fact, it shows that if a function f is analytic at a point, then its derivative is also analytic at that point. For, if f is analytic at a point z, there must exist a circle about z such that f is analytic within and on the circle. Then according to formula (3), $f''(z)$ exists at each point interior to the circle, and the derivative of f is therefore analytic at z.

We can apply the same argument to the analytic function $f'(z)$ to conclude that its derivative $f''(z)$ is analytic, etc. The following fundamental result is thus obtained.

Theorem. *If a function f is analytic at a point, then its derivatives of all orders are also analytic functions at that point.*

Since $f'(z)$ is analytic and therefore continuous, and since

$$f'(z) = u_x(x,y) + iv_x(x,y) = v_y(x,y) - iu_y(x,y),$$

it follows that the first partial derivatives of u and v are continuous. Since $f''(z)$ is analytic and continuous and since

$$f''(z) = u_{xx}(x,y) + iv_{xx}(x,y) = v_{yx}(x,y) - iu_{yx}(x,y),$$

etc., it follows that the partial derivatives of u and v of all orders are continuous functions at any point where f is analytic. This result was anticipated in Sec. 20 for partial derivatives of the first and second order in the discussion of harmonic functions.

The arguments used in establishing formulas (2) and (3) can be applied successively to obtain an integral formula for the derivative of any given order. Indeed, mathematical induction yields the general formula

$$f^{(n)}(z) = \frac{n!}{2\pi i}\int_C \frac{f(s)\,ds}{(s-z)^{n+1}} \qquad (n = 1, 2, \ldots). \tag{4}$$

That is, the formula has been shown to hold for $n = 1$; and if we assume that it holds for any specific positive integer $n = k$, we can show by proceeding as before that it is valid when $n = k + 1$. The details of the proof are left to the reader, with the suggestion that in the algebraic simplifications the difference $s - z$ be retained throughout as a single term.

If we agree that $f^{(0)}(z_0)$ denotes $f(z_0)$ and that $0! = 1$, we may now write

$$f^{(n)}(z_0) = \frac{n!}{2\pi i}\int_C \frac{f(z)\,dz}{(z-z_0)^{n+1}} \qquad (n = 0, 1, 2, \ldots) \tag{5}$$

which becomes the Cauchy integral formula when $n = 0$ and is formula (4) in slightly different notation when $n = 1, 2, \ldots$.

For an application of formula (5), we note that when $f(z) = 1$,

$$\int_{C_0} \frac{dz}{z - z_0} = 2\pi i, \qquad \int_{C_0} \frac{dz}{(z-z_0)^{n+1}} = 0 \qquad (n = 1, 2, \ldots)$$

where C_0 is the circle of radius r_0 centered at z_0 taken in the positive sense. (See Exercise 16, Sec. 45.)

Formula (5), and the Cauchy integral formula in particular, can be extended to the case in which the simple closed contour C is replaced by *the oriented boundary B of a multiply connected domain* of the type described in the theorem of Sec. 49. This can be done when z_0 is a point in the domain and f is analytic in the region consisting of the domain and its boundary B.

53. Morera's Theorem

In Sec. 50 we proved that the derivative of the function

$$F(z) = \int_{z_0}^{z} f(s)\,ds \tag{1}$$

exists at each point of any simply connected domain D in which $f(z)$ is analytic; in fact,

$$F'(z) = f(z).$$

Although we assumed that f was analytic in D, in our proof we used only the continuity of f together with the condition that the integral of f around every simple closed contour interior to D was zero. Thus, knowing only that f has those two properties, we can conclude that the function F is analytic in D and that $F'(z) = f(z)$. It then follows that f is analytic in D since it is the derivative of an analytic function (Sec. 52). A theorem due to E. Morera (1856–1909) has therefore been established.

Theorem. *If a function f is continuous throughout a simply connected domain D and if, for each simple closed contour C lying in D,*

$$\int_C f(z)\,dz = 0, \tag{2}$$

then f is analytic throughout D.

Morera's theorem serves as a converse of the Cauchy-Goursat theorem.

We can extend Morera's theorem to an arbitrary domain D with condition (2) holding for each simple closed contour whose interior also lies in D. For, if z_0 is a point in D, there exists an ε neighborhood $|z - z_0| < \varepsilon$ contained in D, and Morera's theorem can then be applied to that ε neighborhood to show that f is analytic at z_0. The function f is thus analytic throughout D.

54. Maximum Moduli of Functions

Let f be a function which is analytic and *not constant* in an open disk $|z - z_0| < r_0$ centered at z_0. If C is any one of the circles $|z - z_0| = r$ where $0 < r < r_0$, then, according to the Cauchy integral formula,

$$f(z_0) = \frac{1}{2\pi i}\int_C \frac{f(z)\,dz}{z - z_0}. \tag{1}$$

The path C is taken in the positive sense, and a parametric representation for it is $z(\theta) = z_0 + re^{i\theta}$ $(0 \leqq \theta \leqq 2\pi)$. Expression (1) can therefore be written

$$f(z_0) = \frac{1}{2\pi}\int_0^{2\pi} f(z_0 + re^{i\theta})\,d\theta. \tag{2}$$

Formula (2) shows that the value of a function at the center of a circle within and on which the function is analytic is the arithmetic mean of its values on that circle.

From formula (2) we obtain the inequality

$$|f(z_0)| \leqq \frac{1}{2\pi}\int_0^{2\pi} |f(z_0 + re^{i\theta})|\, d\theta \qquad (0 \leqq r < r_0), \tag{3}$$

where the obvious special case $r = 0$ is included. On the other hand, if we assume that $|f(z)| \leqq |f(z_0)|$ for all z such that $|z - z_0| < r_0$, then

$$\frac{1}{2\pi}\int_0^{2\pi} |f(z_0 + re^{i\theta})|\, d\theta \leqq |f(z_0)| \qquad (0 \leqq r < r_0). \tag{4}$$

Combining inequalities (3) and (4), we find that

$$|f(z_0)| = \frac{1}{2\pi}\int_0^{2\pi} |f(z_0 + re^{i\theta})|\, d\theta,$$

or

$$\int_0^{2\pi} [|f(z_0)| - |f(z_0 + re^{i\theta})|]\, d\theta = 0 \qquad (0 \leqq r < r_0).$$

Since this last integrand is continuous and nonnegative, it follows that $|f(z_0)| - |f(z_0 + r_0 e^{i\theta})| = 0$; that is, $|f(z)| = |f(z_0)|$ for all z such that $|z - z_0| < r_0$. But, according to exercise 9(*c*), Sec. 20, this means that $f(z)$ must be constant in the domain $|z - z_0| < r_0$, and our original stipulation that $f(z)$ not be constant in that domain is violated. Our assumption that $|f(z)| \leqq |f(z_0)|$ for all z such that $|z - z_0| < r_0$ has therefore led us to a contradiction.

We have thus shown that *if a function f is analytic and not constant in a neighborhood of a point z_0, then there exists at least one point z in that neighborhood such that*

$$|f(z)| > |f(z_0)|. \tag{5}$$

An important consequence of this result is the following *maximum principle*.

Theorem. *If f is analytic and not constant in the interior of a region, then $|f(z)|$ has no maximum value in that interior.*

In order to complete the proof of the maximum principle, we need a result which is a direct corollary of the theorem in Sec. 106, Chap. 12; namely, if an analytic function is not constant over the interior of a region R, then it is not constant over any neighborhood of any point in the interior of R. Suppose now that $|f(z)|$ assumes a maximum value at a point z_0 in the interior of R.

It follows that $|f(z)| \leqq |f(z_0)|$ for each point z in a neighborhood of z_0, and this contradicts statement (5).

If a function f that is analytic at each point in the interior of a closed bounded region R is also continuous throughout R, the continuous function $|f(z)|$ has a maximum value in R (Sec. 13). That is, there exists a positive constant M such that $|f(z)| \leqq M$ for all points z in R, and equality holds for at least one such point. If f is a constant function, then $|f(z)| = M$ for all z in R. If, however, $f(z)$ is not constant, then, according to the maximum principle, $|f(z)| \neq M$ for any point z in the interior of R. Thus, *if a function f is continuous in a closed bounded region R and is analytic and not constant in the interior of R, then $|f(z)|$ assumes its maximum value on the boundary of R and never in the interior.*

Properties of minimum values of $|f(z)|$, as well as *maximum and minimum values of the harmonic function $u(x,y) = \text{Re}\,[f(z)]$*, are treated in the exercises.

When f is analytic within and on a circle $|z - z_0| = r_0$, taken in the positive sense and denoted by C_0, then

$$f^{(n)}(z_0) = \frac{n!}{2\pi i} \int_{C_0} \frac{f(z)\,dz}{(z - z_0)^{n+1}} \qquad (n = 1, 2, \ldots),$$

according to the integral representation for derivatives. If M is the maximum value of $|f(z)|$ on C_0, *Cauchy's inequality* follows:

$$|f^{(n)}(z_0)| \leqq \frac{n!\,M}{r_0^{\,n}} \qquad (n = 1, 2, \ldots). \tag{6}$$

When $n = 1$, we have the condition

$$|f'(z_0)| \leqq \frac{M}{r_0} \tag{7}$$

from which we can show that no entire function except a constant one is bounded for all z, a conclusion that can be stated as follows.

Liouville's Theorem. *If f is entire and bounded for all values of z in the complex plane, then $f(z)$ is constant.*

To prove this, observe that under the hypothesis a constant M exists such that $|f(z)| \leqq M$ for all z. Hence at each point z_0 inequality (7) is valid for any positive number r_0. Since r_0 can be made arbitrarily large and since $f'(z_0)$ is a fixed number, inequality (7) can hold only if $f'(z_0) = 0$. Thus, the derivative of $f(z)$ is zero for all values of z in the complex plane, and so $f(z)$ must be constant.

55. The Fundamental Theorem of Algebra

The following theorem is known as *the fundamental theorem of algebra.*

Theorem. *Any polynomial*

$$P(z) = a_0 + a_1 z + a_2 z^2 + \cdots + a_n z^n \qquad (a_n \neq 0),$$

where $n \geqq 1$, *has at least one zero. That is, there exists at least one point* z_0 *such that* $P(z_0) = 0$.

The proof of the theorem by purely algebraic methods is difficult, but it follows readily from Liouville's theorem of the preceding section. Let us suppose that $P(z)$ is not zero for any value of z. Then the function

$$f(z) = \frac{1}{P(z)}$$

is entire, and it is also bounded for all z. To see that it is bounded, note first that f is continuous and therefore bounded in any closed disk centered at the origin. Also, there exists a positive number R such that

$$|f(z)| = \frac{1}{|P(z)|} < \frac{2}{|a_n| R^n}$$

for all z *exterior* to the disk $|z| \leqq R$. (See Exercise 18 of this section.) Hence f is bounded for all values of z in the plane. It follows from Liouville's theorem that $f(z)$, and consequently $P(z)$, is constant. But $P(z)$ is not constant, and we have arrived at a contradiction.

In elementary algebra courses the fundamental theorem is often stated without proof. It then follows from the theorem that *any polynomial of degree* n, *where* $n \geqq 1$, *can be expressed as a product of linear factors*; that is,

$$P(z) = c(z - z_1)(z - z_2) \cdots (z - z_n)$$

where c and the z_k ($k = 1, 2, \ldots, n$) are complex constants. The fundamental theorem furnishes a number z_1 such that $P(z_1) = 0$. Then, by Exercise 19 of this section, the polynomial is divisible by $z - z_1$:

$$P(z) = (z - z_1)Q(z).$$

Here $Q(z)$ is a polynomial of degree $n - 1$. The proof can now be completed by mathematical induction.

It follows from this result that a polynomial of degree n ($n \geqq 1$) has no more than n *distinct* zeros.

EXERCISES

1. Let C be the circle $|z| = 3$ described in the positive sense. Show that if

$$g(z) = \int_C \frac{2s^2 - s - 2}{s - z}\, ds \qquad (|z| \neq 3),$$

then $g(2) = 8\pi i$. What is the value of $g(z)$ when $|z| > 3$?

2. Let C be a simple closed contour described in the positive sense and write

$$g(z) = \int_C \frac{s^3 + 2s}{(s - z)^3}\, ds.$$

Show why $g(z) = 6\pi i z$ when z is inside C and $g(z) = 0$ when z is outside C.

3. Let C denote the boundary of the square whose sides lie along the lines $x = \pm 2$ and $y = \pm 2$, where C is described in the positive sense. Evaluate each of these integrals:

$$(a)\ \int_C \frac{e^{-z}\, dz}{z - \pi i/2}; \qquad (b)\ \int_C \frac{\cos z}{z(z^2 + 8)}\, dz; \qquad (c)\ \int_C \frac{z\, dz}{2z + 1};$$

$$(d)\ \int_C \frac{\tan (z/2)}{(z - x_0)^2}\, dz \quad (-2 < x_0 < 2); \qquad (e)\ \int_C \frac{\cosh z}{z^4}\, dz.$$

Ans. (*a*) 2π; (*b*) $\pi i/4$; (*c*) $-\pi i/2$; (*d*) $i\pi \sec^2 (x_0/2)$; (*e*) 0.

4. Find the value of the integral of $g(z)$ around the simple closed contour $|z - i| = 2$ in the positive sense when

$$(a)\ g(z) = \frac{1}{z^2 + 4}; \qquad (b)\ g(z) = \frac{1}{(z^2 + 4)^2}.$$

Ans. (*a*) $\pi/2$; (*b*) $\pi/16$.

5. Show that when f is analytic within and on a simple closed contour C and z_0 is not on C, then

$$\int_C \frac{f'(z)\, dz}{z - z_0} = \int_C \frac{f(z)\, dz}{(z - z_0)^2}.$$

6. Let f denote a function that is *continuous* on a simple closed contour C. Following the procedure used in Sec. 52, prove that the function

$$g(z) = \frac{1}{2\pi i} \int_C \frac{f(s)\, ds}{s - z}$$

is *analytic* at each point z interior to C and that, for each such z,

$$g'(z) = \frac{1}{2\pi i} \int_C \frac{f(s)\, ds}{(s - z)^2}.$$

7. Let C be the unit circle $z = \exp(i\theta)$ described from $\theta = -\pi$ to $\theta = \pi$. First show that, for any real constant a,

$$\int_C \frac{e^{az}}{z}\, dz = 2\pi i;$$

then write the integral in terms of θ to derive the formula

$$\int_0^{\pi} e^{a \cos\theta} \cos(a \sin\theta)\, d\theta = \pi.$$

8. Let a function f be continuous in a closed bounded region R and let it be analytic and not constant throughout the interior of R. Assuming that $f(z) \neq 0$ anywhere in R, consider the function $1/f(z)$ to prove that $|f(z)|$ has a *minimum* value N somewhere in R and that $|f(z)| > N$ for each point z in the interior.

9. Give an example to show that in Exercise 8 the condition $f(z) \neq 0$ anywhere in R is necessary in order to prove the result of that exercise. That is, show that $|f(z)|$ can reach its minimum value at an interior point when that minimum value is zero.

10. Illustrate the maximum principle and Exercise 8 when $f(z) = (z+1)^2$ and the region R is the triangular region with vertices at the points $z = 0$, $z = 2$, and $z = i$. Do this by finding points in R where $|f(z)|$ has its maximum and minimum values.

Ans. $z = 2$, $z = 0$.

11. Let the function $f(z) = u(x,y) + iv(x,y)$ be continuous in a closed bounded region R and let it be analytic and not constant in the interior of R. Show that *the function $u(x,y)$ reaches its maximum value on the boundary of R and never in the interior*, where it is harmonic.

Suggestion: Apply the maximum principle to the function $\exp[f(z)]$.

12. Let $f(z) = u(x,y) + iv(x,y)$ be a function which is continuous in a closed bounded region R and analytic and not constant throughout the interior of R. Prove that *the function $u(x,y)$ reaches its minimum value on the boundary of R and never at an interior point.* (See Exercises 8 and 11.)

13. Illustrate Exercises 11 and 12 by taking the function $f(z) = e^z$ and taking R as the rectangular region $0 \leqq x \leqq 1$, $0 \leqq y \leqq \pi$ and finding those points in R where $u(x,y)$ reaches its maximum and minimum values.

Ans. $z = 1 + \pi i$, $z = 1$.

14. Let $f(z)$ be an entire function and let $u(x,y)$ be its real part. Show that if the harmonic function $u(x,y)$ has an upper bound u_0, $u(x,y) < u_0$ for all points in the xy plane, then $u(x,y)$ is constant.

15. Complete the derivation of formula (3), Sec. 52.

16. Carry out the induction argument used to establish formula (4), Sec. 52.

17. Let f be an entire function such that $|f(z)| \leqq A|z|$ for all z, where A is a fixed positive number. Show that either $f(z) = 0$ for all z or $f(z) = a_1 z$ $(a_1 \neq 0)$.

Suggestion: Use Cauchy's inequality (Sec. 54) to show that $f''(z)$ is zero everywhere in the complex plane.

18. Show that if

$$P(z) = a_0 + a_1 z + a_2 z^2 + \cdots + a_n z^n \qquad (a_n \neq 0)$$

is a polynomial of degree n where $n \geqq 1$, there exists a positive number R such that

$$|P(z)| > \frac{|a_n|\,|z|^n}{2}$$

for each value of z such that $|z| > R$.

Suggestion: Observe first that, for some positive number R, each of the numbers $|a_{n-1}/z|$, $|a_{n-2}/z^2|$, ..., $|a_1/z^{n-1}|$, $|a_0/z^n|$ can be made less than $|a_n|/(2n)$ when $|z| \geqq R$; hence

$$\left|\frac{a_{n-1}}{z} + \frac{a_{n-2}}{z^2} + \cdots + \frac{a_1}{z^{n-1}} + \frac{a_0}{z^n}\right| < \frac{|a_n|}{2}$$

when $|z| \geqq R$. Use this result and the inequality $|z_1 + z_2| \geqq ||z_1| - |z_2||$ to show that when $|z| \geqq R$,

$$\left|a_n + \left(\frac{a_{n-1}}{z} + \frac{a_{n-2}}{z^2} + \cdots + \frac{a_1}{z^{n-1}} + \frac{a_0}{z^n}\right)\right| > \frac{|a_n|}{2}.$$

The stated result is obtained by multiplying both sides of this inequality by $|z|^n$.

19. Let $P(z)$ be a polynomial of degree n ($n \geqq 1$); $P(z)$ is said to be *divisible* by $z - z_0$ if there exists a polynomial $Q(z)$, called the *quotient*, such that $P(z) = (z - z_0)Q(z)$. Show that (*a*) the polynomial $z^n - z_0{}^n$ ($n = 1, 2, \ldots$) is divisible by $z - z_0$; (*b*) the polynomial $P(z) - P(z_0)$ is divisible by $z - z_0$ and that the quotient is a polynomial of degree $n - 1$; (*c*) $P(z)$ is divisible by $z - z_0$ if and only if $P(z_0) = 0$.

6

SERIES

This chapter is devoted mainly to series representations of analytic functions. We present theorems which guarantee the existence of such representations, and we develop some facility in manipulating series.

56. Convergence of Sequences and Series

An infinite *sequence*

$$z_1, z_2, \ldots, z_n, \ldots$$

of complex numbers has a *limit* z if for each positive number ε there exists a positive integer n_0 such that

$$|z_n - z| < \varepsilon \qquad \text{whenever} \qquad n > n_0. \tag{1}$$

Geometrically, this means that for sufficiently large values of n the points z_n are arbitrarily close to the point z.

It is left as an exercise to show that a given sequence can have at most one limit; that is, a limit is unique. When the limit z exists, the sequence is said to *converge* to z, and we write

$$\lim_{n \to \infty} z_n = z.$$

If the sequence has no limit, it *diverges*.

Theorem 1. *Suppose that*

$$z_n = x_n + iy_n \qquad (n = 1, 2, \ldots)$$

and

$$z = x + iy.$$

Then

$$\lim_{n\to\infty} z_n = z \tag{2}$$

if and only if

$$\lim_{n\to\infty} x_n = x \quad \textit{and} \quad \lim_{n\to\infty} y_n = y. \tag{3}$$

To prove the theorem, we first assume that condition (2) holds and obtain conditions (3) from it. According to condition (2), for each positive number ε there exists a positive integer n_0 such that

$$|x_n - x + i(y_n - y)| < \varepsilon \quad \text{whenever} \quad n > n_0.$$

But

$$|x_n - x| \leqq |x_n - x + i(y_n - y)|$$

and

$$|y_n - y| \leqq |x_n - x + i(y_n - y)|.$$

Consequently,

$$|x_n - x| < \varepsilon \quad \text{and} \quad |y_n - y| < \varepsilon$$

whenever $n > n_0$; that is, conditions (3) are satisfied.

Conversely, if we start with conditions (3), we know that for each positive number ε there are positive integers n_1 and n_2 such that

$$|x_n - x| < \frac{\varepsilon}{2} \quad \text{whenever} \quad n > n_1$$

and

$$|y_n - y| < \frac{\varepsilon}{2} \quad \text{whenever} \quad n > n_2.$$

Hence

$$|x_n - x| < \frac{\varepsilon}{2} \quad \text{and} \quad |y_n - y| < \frac{\varepsilon}{2}$$

whenever $n > n_0$ where n_0 is the larger of the two integers n_1 and n_2. But

$$|x_n + iy_n - (x + iy)| \leqq |x_n - x| + |y_n - y|,$$

and so $|z_n - z| < \varepsilon$ whenever $n > n_0$. Condition (2) thus holds.

An infinite *series*

$$z_1 + z_2 + \cdots + z_n + \cdots$$

of complex numbers *converges* to a number S, called the *sum* of the series, if the sequence

$$S_N = \sum_{n=1}^{N} z_n \qquad (N = 1, 2, \ldots)$$

of *partial sums* converges to S; we then write

$$\sum_{n=1}^{\infty} z_n = S.$$

Note that since a sequence can have at most one limit, a series can have at most one sum. When a series does not converge, we say that it *diverges.*

Theorem 2. *Suppose that*

$$z_n = x_n + iy_n \qquad (n = 1, 2, \ldots)$$

and

$$S = X + iY.$$

Then

$$\sum_{n=1}^{\infty} z_n = S \tag{4}$$

if and only if

$$\sum_{n=1}^{\infty} x_n = X \quad \textit{and} \quad \sum_{n=1}^{\infty} y_n = Y. \tag{5}$$

To start the proof, let S_N denote the partial sum of the first N terms of the series in condition (4) and observe that

$$S_N = X_N + iY_N \tag{6}$$

where

$$X_N = \sum_{n=1}^{N} x_n, \qquad Y_N = \sum_{n=1}^{N} y_n.$$

Now condition (4) holds if and only if

$$\lim_{N \to \infty} S_N = S;$$

and, in view of relation (6) and Theorem 1, this condition holds if and only if

$$\lim_{N\to\infty} X_N = X \qquad \textit{and} \qquad \lim_{N\to\infty} Y_N = Y. \tag{7}$$

Condition (4) therefore implies conditions (7), and conversely. Since X_N and Y_N are the partial sums of series (5), the theorem is therefore true.

In establishing the fact that the sum of a series is a given number S, it is often convenient to define the *remainder* after N terms:

$$R_N = S - S_N.$$

Observe that $|S_N - S| = |R_N - 0|$ and so, by definition (1) of limit of a sequence, limit (7) exists when

$$\lim_{N\to\infty} R_N = 0, \tag{8}$$

and conversely. Hence *a series converges to a number S if and only if the sequence of remainders converges to* 0.

Especially important in the theory of complex variables are *power series.* They are series of the form

$$a_0 + \sum_{n=1}^{\infty} a_n(z - z_0)^n, \qquad or \qquad \sum_{n=0}^{\infty} a_n(z - z_0)^n,$$

where z_0, a_0, and the a_n are complex constants and z may be any number in a stated region. In such series involving a variable z we shall denote the sum, partial sums, and remainders by $S(z)$, $S_N(z)$, and $R_N(z)$, respectively.

EXERCISES

1. Show in two ways that the sequence

$$z_n = -2 + i\frac{(-1)^n}{n^2} \qquad (n = 1, 2, \ldots)$$

converges.

2. Let r_n denote the moduli and Θ_n the principal values of the arguments of the complex numbers z_n in Exercise 1. Show that the sequence r_n $(n = 1, 2, \ldots)$ converges but that the sequence Θ_n $(n = 1, 2, \ldots)$ does not.

3. By considering the remainder $R_N(z)$, show that

$$\sum_{n=1}^{\infty} z^n = \frac{z}{1-z}$$

where z is any complex number such that $|z| < 1$.

Suggestion: Use Exercise 14, Sec. 6, to show that $|R_N(z)| \leq |z|^{N+1}/(1 - |z|)$.

4. In the formula of Exercise 3 write $z = re^{i\theta}$, where $0 < r < 1$. Then show that

$$\sum_{n=1}^{\infty} r^n \cos n\theta = \frac{r\cos\theta - r^2}{1 - 2r\cos\theta + r^2},$$

$$\sum_{n=1}^{\infty} r^n \sin n\theta = \frac{r\sin\theta}{1 - 2r\cos\theta + r^2}.$$

5. Show that a limit of a sequence is unique.

6. Show that

$$\text{if} \quad \sum_{n=1}^{\infty} z_n = S, \qquad \text{then} \quad \sum_{n=1}^{\infty} \bar{z}_n = \bar{S}.$$

7. Let c be any complex number and show that

$$\text{if} \quad \sum_{n=1}^{\infty} z_n = S, \qquad \text{then} \quad \sum_{n=1}^{\infty} cz_n = cS.$$

8. Let a sequence z_n $(n = 1, 2, \ldots)$ converge to a number z. Show that there exists a positive number M such that the inequality $|z_n| < M$ holds for all n.

Suggestion: Note that there exists a positive integer n_0 such that

$$|z_n| \leqq |z| + |z_n - z| < |z| + 1$$

whenever $n > n_0$.

9. Show that if z is the limit of a sequence z_n $(n = 1, 2, \ldots)$ and if $|z_n| \leqq M$ for all n, then $|z| \leqq M$.

Suggestion: Suppose $|z| > M$ and observe that there exists a positive integer n_0 such that $|z - z_n| < |z| - M$ whenever $n > n_0$. Then, using the inequality $|z| - |z_n| \leqq |z - z_n|$, obtain the contradiction that $|z_n| > M$ for any n greater than n_0.

57. Taylor Series

We turn now to *Taylor's theorem* which is one of the most important results of the chapter.

Theorem. *Let f be analytic everywhere inside a circle C_0 with center at z_0 and radius r_0. Then at each point z inside C_0*

$$(1) \qquad f(z) = f(z_0) + f'(z_0)(z - z_0) + \frac{f''(z_0)}{2!}(z - z_0)^2 + \cdots + \frac{f^{(n)}(z_0)}{n!}(z - z_0)^n + \cdots;$$

that is, the power series here converges to $f(z)$ when $|z - z_0| < r_0$.

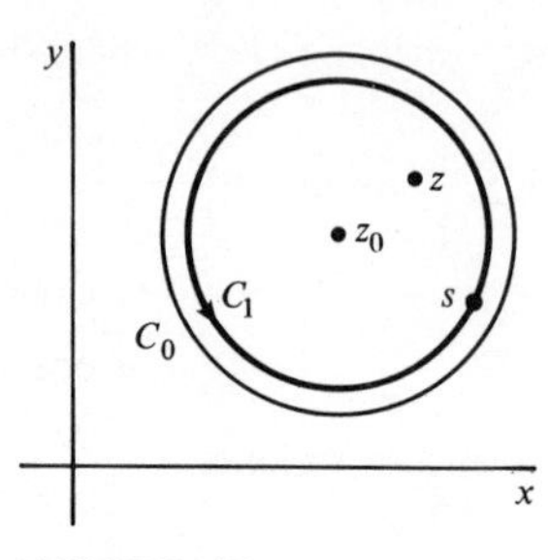

FIGURE 47

This is the expansion of $f(z)$ into a *Taylor series* about the point z_0. In the special case when all the terms in the expansion are real, we have the familiar Taylor series introduced in elementary calculus.

To prove the theorem, we let z be any fixed point inside the circle C_0 and write $|z - z_0| = r$; thus $r < r_0$. We also let s be any point lying on a circle C_1 centered at z_0 and with radius r_1 where $r < r_1 < r_0$; thus $|s - z_0| = r_1$ (Fig. 47). Since z is inside C_1 and f is analytic within and on that circle, it follows, according to the Cauchy integral formula, that

$$f(z) = \frac{1}{2\pi i} \int_{C_1} \frac{f(s)\,ds}{s - z} \tag{2}$$

where C_1 is taken in the positive sense.

Now

$$\frac{1}{s - z} = \frac{1}{(s - z_0) - (z - z_0)} = \frac{1}{s - z_0} \frac{1}{1 - (z - z_0)/(s - z_0)}.$$

Also, when c is any complex number other than unity,

$$\frac{1}{1 - c} = 1 + c + c^2 + \cdots + c^{N-1} + \frac{c^N}{1 - c}.$$

(See Exercise 14, Sec. 6.) Hence

$$\frac{1}{s - z} = \frac{1}{s - z_0} \times \left[1 + \frac{z - z_0}{s - z_0} + \cdots + \left(\frac{z - z_0}{s - z_0}\right)^{N-1} + \frac{1}{1 - (z - z_0)/(s - z_0)} \left(\frac{z - z_0}{s - z_0}\right)^N\right]$$

and, consequently,

$$\frac{f(s)}{s - z} = \frac{f(s)}{s - z_0} + \frac{f(s)}{(s - z_0)^2}(z - z_0) + \cdots + \frac{f(s)}{(s - z_0)^N}(z - z_0)^{N-1} + (z - z_0)^N \frac{f(s)}{(s - z)(s - z_0)^N}.$$

We next integrate each term counterclockwise around C_1 and divide through by $2\pi i$. In view of formula (2) and the integral formulas (Sec. 52)

$$\frac{1}{2\pi i}\int_{C_1}\frac{f(s)ds}{(s-z_0)^{n+1}}=\frac{1}{n!}f^{(n)}(z_0) \qquad (n=0,1,2,\ldots),$$

we can write the result as follows:

$$f(z)=f(z_0)+f'(z_0)(z-z_0)+\cdots+\frac{f^{(N-1)}(z_0)}{(N-1)!}(z-z_0)^{N-1}+R_N(z) \tag{3}$$

where

$$R_N(z)=\frac{(z-z_0)^N}{2\pi i}\int_{C_1}\frac{f(s)\,ds}{(s-z)(s-z_0)^N}. \tag{4}$$

Recalling that $|z-z_0|=r$ and $|s-z_0|=r_1$, we note that

$$|s-z|\geqq|s-z_0|-|z-z_0|=r_1-r.$$

Hence it follows from expression (4) that when M denotes the maximum value of $f(s)$ on C_1,

$$|R_N(z)|\leqq\frac{r^N}{2\pi}\frac{M2\pi r_1}{(r_1-r){r_1}^N}=\frac{Mr_1}{r_1-r}\left(\frac{r}{r_1}\right)^N.$$

But $r/r_1<1$, and therefore

$$\lim_{N\to\infty}R_N(z)=0.$$

Thus, for each point z interior to C_0, the limit as N tends to infinity of the sum of the first N terms on the right in equation (3) is $f(z)$. That is, if f is analytic inside a circle centered at z_0 with radius r_0, then $f(z)$ is represented by a Taylor series:

$$f(z)=f(z_0)+\sum_{n=1}^{\infty}\frac{f^{(n)}(z_0)}{n!}(z-z_0)^n \qquad \text{when } |z-z_0|<r_0. \tag{5}$$

If $z_0=0$, this reduces to a *Maclaurin series*:

$$f(z)=f(0)+\sum_{n=1}^{\infty}\frac{f^{(n)}(0)}{n!}z^n \qquad \text{when } |z|<r_0. \tag{6}$$

58. Observations and Examples

When it is known that f is analytic at all points within a circle centered at z_0, convergence of the Taylor series about z_0 to $f(z)$ for each point z within that circle is ensured; no test for the convergence of the series is required. In fact, according to Taylor's theorem, the series converges to $f(z)$ within the circle about z_0 whose radius is the distance from z_0 to the nearest point z_1 where f fails to be analytic. In Sec. 62 we shall find that this is actually the largest circle centered at z_0 such that the series converges to $f(z)$ for all z interior to it.

Our first example is a Maclaurin series expansion for $f(z) = e^z$. Here $f^{(n)}(z) = e^z$, and so $f^{(n)}(0) = 1$. Since e^z is analytic for every value of z, it follows that

$$e^z = 1 + \sum_{n=1}^{\infty} \frac{z^n}{n!} \qquad \text{when } |z| < \infty. \tag{1}$$

Note that when z is real, expansion (1) becomes

$$e^x = 1 + \sum_{n=1}^{\infty} \frac{x^n}{n!}$$

which is valid for each real number x.

In like manner, we find that

$$\sin z = \sum_{n=1}^{\infty} (-1)^{n+1} \frac{z^{2n-1}}{(2n-1)!} \qquad \text{when } |z| < \infty, \tag{2}$$

$$\cos z = 1 + \sum_{n=1}^{\infty} (-1)^n \frac{z^{2n}}{(2n)!} \qquad \text{when } |z| < \infty, \tag{3}$$

$$\sinh z = \sum_{n=1}^{\infty} \frac{z^{2n-1}}{(2n-1)!} \qquad \text{when } |z| < \infty, \tag{4}$$

$$\cosh z = 1 + \sum_{n=1}^{\infty} \frac{z^{2n}}{(2n)!} \qquad \text{when } |z| < \infty. \tag{5}$$

Another Maclaurin series that is readily obtained is

$$\frac{1}{1+z} = \sum_{n=0}^{\infty} (-1)^n z^n \qquad \text{when } |z| < 1. \tag{6}$$

By substituting Z^2 for z in this expansion, we note that

$$\frac{1}{1+Z^2} = \sum_{n=0}^{\infty} (-1)^n Z^{2n} \qquad \text{when } |Z| < 1$$

since $|Z^2| < 1$ when $|Z| < 1$. When we make the substitution $z = -c$, expansion (6) gives the sum of an infinite *geometric series* where c is the common ratio of adjacent terms; that is,

$$1 + c + c^2 + \cdots + c^n + \cdots = \frac{1}{1-c} \qquad \text{when } |c| < 1. \tag{7}$$

When $z \neq 0$, the derivatives of the function $f(z) = z^{-1}$ are

$$f^{(n)}(z) = (-1)^n n! z^{-n-1} \qquad (n = 1, 2, \ldots);$$

so $f^{(n)}(1) = (-1)^n n!$. Hence the expansion of this function into a Taylor series about the point $z = 1$ is

$$\frac{1}{z} = \sum_{n=0}^{\infty} (-1)^n (z-1)^n. \tag{8}$$

This expansion is valid when $|z - 1| < 1$ since the function is analytic at all points except $z = 0$.

For another example, let us expand the function

$$f(z) = \frac{1+2z}{z^2+z^3} = \frac{1}{z^2}\left(2 - \frac{1}{1+z}\right)$$

into a series involving both positive and negative powers of z. We cannot find a Maclaurin series for $f(z)$ itself since this function is not analytic at $z = 0$; but we have already found one for $1/(1+z)$. Thus, when $0 < |z| < 1$, it follows that

$$\begin{aligned}\frac{1+2z}{z^2+z^3} &= \frac{1}{z^2}(2 - 1 + z - z^2 + z^3 - \cdots)\\ &= \frac{1}{z^2} + \frac{1}{z} - 1 + z - z^2 + z^3 - \cdots.\end{aligned}$$

EXERCISES

1. Show that

$$e^z = e + e\sum_{n=1}^{\infty}\frac{(z-1)^n}{n!} \qquad \text{when } |z| < \infty.$$

2. Show that

(*a*) $\displaystyle\frac{1}{z^2} = 1 + \sum_{n=1}^{\infty}(n+1)(z+1)^n$ when $|z+1| < 1$;

(*b*) $\displaystyle\frac{1}{z^2} = \frac{1}{4} + \frac{1}{4}\sum_{n=1}^{\infty}(-1)^n(n+1)\left(\frac{z-2}{2}\right)^n$ when $|z-2| < 2$.

3. Expand $\cos z$ into a Taylor series about the point $z = \pi/2$.

4. Expand $\sinh z$ into a Taylor series about the point $z = \pi i$.

5. What is the largest circle within which the Maclaurin series for the function $\tanh z$ converges to $\tanh z$ for all z? Write the first two nonzero terms of the series.

6. Prove that when $0 < |z| < 4$,

$$\frac{1}{4z - z^2} = \sum_{n=0}^{\infty}\frac{z^{n-1}}{4^{n+1}}.$$

7. Make the substitution $z = Z - 1$ in Maclaurin series expansion (6), Sec. 58, to obtain a representation for $1/Z$ in powers of $Z - 1$ that is valid when $|Z - 1| < 1$. Your result should agree with Taylor series expansion (8) in the same section.
8. Substitute Z^{-1} for z in expansion (6), Sec. 58, as well as in its condition of validity, to obtain an expansion of the function $(1 + Z)^{-1}$ into negative powers of Z that is valid when $|Z| > 1$.

Ans. $(1 + Z)^{-1} = \sum_{n=0}^{\infty} (-1)^n Z^{-n-1}$.

9. Prove that when $z \neq 0$,

$$\frac{\sin(z^2)}{z^4} = \frac{1}{z^2} - \frac{z^2}{3!} + \frac{z^6}{5!} - \frac{z^{10}}{7!} + \cdots.$$

10. Represent the function

$$f(z) = \frac{z}{(z-1)(z-3)}$$

by a series involving positive and negative powers of $z - 1$ which converges to $f(z)$ when $0 < |z - 1| < 2$.

Ans. $f(z) = \dfrac{-1}{2(z-1)} - 3\sum_{n=1}^{\infty} \dfrac{(z-1)^{n-1}}{2^{n+1}}$.

59. Laurent Series

Let C_1 and C_2 be two concentric circles centered at a point z_0 and with radii r_1 and r_2, respectively, where $r_2 < r_1$ (Fig. 48). We now state *Laurent's theorem.*

Theorem. *If f is analytic on C_1 and C_2 and throughout the annular domain between those two circles, then at each point z in that domain $f(z)$ is represented by the expansion*

$$f(z) = \sum_{n=0}^{\infty} a_n (z - z_0)^n + \sum_{n=1}^{\infty} \frac{b_n}{(z - z_0)^n} \tag{1}$$

where

$$a_n = \frac{1}{2\pi i} \int_{C_1} \frac{f(s)\, ds}{(s - z_0)^{n+1}} \qquad (n = 0, 1, 2, \ldots), \tag{2}$$

$$b_n = \frac{1}{2\pi i} \int_{C_2} \frac{f(s)\, ds}{(s - z_0)^{-n+1}} \qquad (n = 1, 2, \ldots), \tag{3}$$

each path of integration being taken counterclockwise.

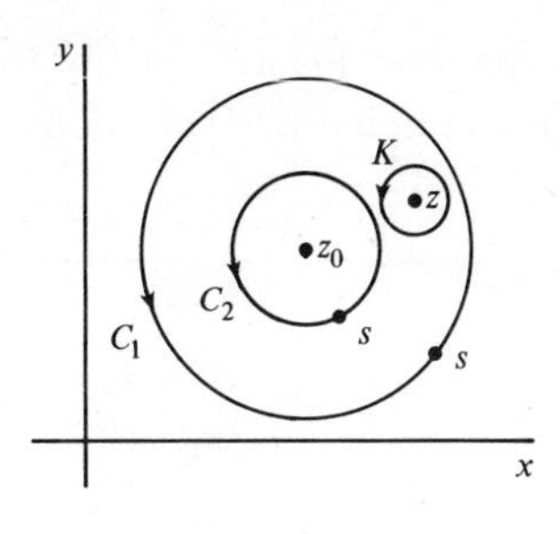

FIGURE 48

The series here is called a *Laurent series.*

If f is analytic at every point inside and on C_1 except at the point z_0 itself, the radius r_2 may be taken arbitrarily small. Expansion (1) is then valid when

$$0 < |z - z_0| < r_1.$$

If f is analytic at *all* points inside and on C_1, the function $f(z)/(z - z_0)^{-n+1}$ is analytic inside and on C_2 because $-n + 1 \leqq 0$. The integral in formula (3) therefore has the value zero, and expansion (1) reduces to a Taylor series.

Since the two functions $f(z)/(z - z_0)^{n+1}$ and $f(z)/(z - z_0)^{-n+1}$ are analytic throughout the annular region $r_2 \leqq |z - z_0| \leqq r_1$, any simple closed contour C around that annulus in the positive direction can be used as a path of integration in place of the circular paths C_1 and C_2. (See Exercise 3, Sec. 50.) Thus the Laurent series (1) can be written

$$f(z) = \sum_{n=-\infty}^{\infty} c_n(z - z_0)^n \qquad (r_2 < |z - z_0| < r_1) \tag{4}$$

where

$$c_n = \frac{1}{2\pi i}\int_C \frac{f(s)\,ds}{(s - z_0)^{n+1}} \qquad (n = 0, \pm 1, \pm 2, \ldots). \tag{5}$$

In particular cases, of course, some of the coefficients may be zero. The function

$$f(z) = \frac{1}{(z - 1)^2} \qquad (|z - 1| > 0),$$

for example, already has the form (4) where $z_0 = 1$. Here $c_{-2} = 1$ and all the other coefficients are zero; this is in agreement with formula (5) where the path C can be any positively oriented simple closed contour which encloses the point $z_0 = 1$.

The coefficients in expansion (4) are usually found by other means than by using formula (5). For example, the expansions

$$\frac{e^z}{z^2} = \frac{1}{z^2} + \frac{1}{z} + \frac{1}{2!} + \frac{z}{3!} + \frac{z^2}{4!} + \cdots \qquad (|z| > 0),$$

$$e^{1/z} = 1 + \sum_{n=1}^{\infty} \frac{1}{n!\,z^n} \qquad (|z| > 0)$$

follow from the Maclaurin series for e^z. We shall see (Sec. 63) that such representations are unique; so these must be the Laurent series when $z_0 = 0$.

To prove the theorem, we first observe that if z is a point in the annular domain, then

$$f(z) = \frac{1}{2\pi i} \int_{C_1} \frac{f(s)\,ds}{s - z} - \frac{1}{2\pi i} \int_{C_2} \frac{f(s)\,ds}{s - z}. \tag{6}$$

This equation is valid in view of the remarks at the end of Sec. 52 regarding the Cauchy integral formula when the path of integration is the oriented boundary of a multiply connected domain. To give the details in this instance, we construct a circle K, taken conterclockwise, about the point z such that K is completely contained in the annular domain (Fig. 48). It then follows from the extension of the Cauchy-Goursat theorem to functions analytic in a closed region whose interior is a multiply connected domain (Sec. 49) that

$$\int_{C_1} \frac{f(s)\,ds}{s - z} - \int_{C_2} \frac{f(s)\,ds}{s - z} - \int_{K} \frac{f(s)\,ds}{s - z} = 0.$$

According to the Cauchy integral formula, the value of the third integral here is $2\pi i f(z)$; hence equation (6) is valid.

In the first integral in equation (6), as in the proof of Taylor's theorem, we write

$$\frac{f(s)}{s - z} = \frac{f(s)}{s - z_0} + \frac{f(s)}{(s - z_0)^2}(z - z_0) + \cdots + \frac{f(s)}{(s - z_0)^N}(z - z_0)^{N-1} + (z - z_0)^N \frac{f(s)}{(s - z)(s - z_0)^N}. \tag{7}$$

As for the second integral, we note that

$$-\frac{1}{s - z} = \frac{1}{(z - z_0) - (s - z_0)} = \frac{1}{z - z_0} \frac{1}{1 - (s - z_0)/(z - z_0)}$$

and obtain the identity

$$
(8) \qquad -\frac{f(s)}{s-z} = f(s)\frac{1}{z-z_0} + \frac{f(s)}{(s-z_0)^{-1}}\frac{1}{(z-z_0)^2} + \cdots
$$

$$
+ \frac{f(s)}{(s-z_0)^{-N+1}}\frac{1}{(z-z_0)^N} + \frac{1}{(z-z_0)^N}\frac{(s-z_0)^N f(s)}{z-s}.
$$

It then follows from equation (6) that

$$
f(z) = a_0 + a_1(z-z_0) + \cdots + a_{N-1}(z-z_0)^{N-1}
$$

$$
+ R_N(z) + \frac{b_1}{z-z_0} + \frac{b_2}{(z-z_0)^2} + \cdots + \frac{b_N}{(z-z_0)^N} + Q_N(z)
$$

where a_n and b_n are the numbers given by formulas (2) and (3) and

$$
R_N(z) = \frac{(z-z_0)^N}{2\pi i}\int_{C_1}\frac{f(s)\,ds}{(s-z)(s-z_0)^N},
$$

$$
Q_N(z) = \frac{1}{2\pi i(z-z_0)^N}\int_{C_2}\frac{(s-z_0)^N f(s)}{z-s}\,ds.
$$

Let $r = |z - z_0|$; then $r_2 < r < r_1$. The proof that $R_N(z)$ approaches zero as N tends to infinity is the same as that used in the proof of Taylor's theorem. If M is the maximum value of $|f(s)|$ on C_2, then

$$
|Q_N(z)| \leq \frac{Mr_2}{r-r_2}\left(\frac{r_2}{r}\right)^N;
$$

hence $Q_N(z)$ also approaches zero as N tends to infinity. This completes the proof of Laurent's theorem.

60. Further Properties of Series

If a series

$$
(1) \qquad \sum_{n=1}^{\infty} z_n
$$

of complex numbers $z_n = x_n + iy_n$ converges, we know from Theorem 2, Sec. 56, that the two series of real numbers

$$
(2) \qquad \sum_{n=1}^{\infty} x_n \quad \text{and} \quad \sum_{n=1}^{\infty} y_n
$$

both converge. Now a necessary condition for the convergence of an infinite series of real numbers is that the nth term approach zero as n tends to infinity; so the numbers x_n and y_n must approach zero. Hence z_n approaches zero; that is, *a necessary condition for the convergence of series* (1) *is that*

$$
(3) \qquad \lim_{n\to\infty} z_n = 0.
$$

The terms of a convergent series of complex numbers are therefore *bounded.* That is, there exists a positive constant M such that $|z_n| < M$ for each positive integer n.

Suppose that series (1) is *absolutely convergent*; that is, the series

$$\sum_{n=1}^{\infty} |z_n| = \sum_{n=1}^{\infty} \sqrt{x_n^2 + y_n^2}$$

of real numbers converges. It follows from the comparison test for series of real numbers that the two series

$$\sum_{n=1}^{\infty} |x_n| \qquad \text{and} \qquad \sum_{n=1}^{\infty} |y_n|$$

both converge. Series (2) are then absolutely convergent, and they are therefore convergent because the absolute convergence of a series of real numbers implies the convergence of the series itself. But when series (2) converge, series (1) converges; thus, *absolute convergence of a series of complex numbers implies the convergence of the series itself.*

We now prove an important theorem on the convergence of power series. These and various results which follow apply to general power series of the form

$$\sum_{n=0}^{\infty} a_n(z - z_0)^n$$

but are given only for the special case when $z_0 = 0$. Proofs in the general case are essentially the same, and many of our results are generalized merely by substituting $z - z_0$ for z in certain formulas.

Theorem. *If a power series*

$$\sum_{n=0}^{\infty} a_n z^n \tag{4}$$

converges when $z = z_1$, *where* $z_1 \neq 0$, *it is absolutely convergent for every value of* z *such that* $|z| < |z_1|$.

Since the series whose terms are $a_n z_1^n$ converges, those terms are all bounded; that is,

$$|a_n z_1^n| < M \qquad (n = 0, 1, 2, \ldots)$$

for some positive constant M. Write

$$\frac{|z|}{|z_1|} = k \qquad \text{where } |z| < |z_1|;$$

then

$$|a_n z^n| = |a_n z_1{}^n| \left| \frac{z}{z_1} \right|^n < Mk^n.$$

Now the series whose terms are the positive real numbers Mk^n is a geometric series which is convergent because $k < 1$. We therefore conclude from the comparison test for series of real numbers that the series

$$\sum_{n=0}^{\infty} |a_n z^n|$$

converges, and the theorem is proved.

The set of all points inside some circle about the origin is therefore a region of convergence for power series (4). The greatest circle about the origin such that the series converges at each point inside is called *the circle of convergence* of the power series. *The series cannot converge at any point z_2 outside that circle*, according to the above theorem; for in that case it would converge everywhere inside the circle centered at the origin and passing through z_2. The first circle could not then be the circle of convergence.

If in series (4) we replace z by $z - z_0$, we have the series

$$\sum_{n=0}^{\infty} a_n (z - z_0)^n. \tag{5}$$

We see at once from the above discussion that if this series converges at z_1, then it is absolutely convergent at every point z interior to the circle centered at z_0 and passing through z_1; that is, it is absolutely convergent when

$$|z - z_0| < |z_1 - z_0|.$$

Likewise, if the series

$$\sum_{n=1}^{\infty} \frac{b_n}{(z - z_0)^n}$$

converges when $z = z_1$, then it is absolutely convergent at every point z exterior to the circle centered at z_0 and passing through z_1. The exterior of some circle about the point z_0 is therefore a region of convergence.

61. Uniform Convergence

Let C_1 denote a circle $|z| = r_1$ interior to which a power series about the point $z_0 = 0$ converges, and define in terms of this power series the function

$$S(z) = \sum_{n=0}^{\infty} a_n z^n \tag{1}$$

with domain of definition $|z| < r_1$. Then define on the same set the remainder function

$$R_N(z) = S(z) - \sum_{n=0}^{N-1} a_n z^n. \tag{2}$$

Since the power series converges for any fixed value of z where $|z| < r_1$, we know that the remainder $R_N(z)$ approaches zero for any such z as N tends to infinity; that is, for a given value of z where $|z| < r_1$, there is corresponding to each positive number ε a positive integer N_ε such that

$$|R_N(z)| < \varepsilon \qquad \text{whenever} \qquad N > N_\varepsilon. \tag{3}$$

If we take any point z such that $|z| \leqq |z_2|$ where $|z_2| < r_1$, condition (3) must, of course, hold; but we can also show that corresponding to each positive number ε, a single value of N_ε may be selected such that condition (3) holds regardless of what value of z is chosen in the closed disk $|z| \leqq |z_2|$. When the choice of N_ε may thus depend on the value of ε but not on the choice of z in a given region, we say that the convergence is *uniform* in that region.

To establish uniform convergence of the above power series in the region $|z| \leqq |z_2|$, we note first that for any two positive integers m and N, where $m > N$,

$$\left|\sum_{n=N}^{m} a_n z^n\right| \leqq \sum_{n=N}^{m} |a_n|\,|z|^n \leqq \sum_{n=N}^{m} |a_n|\,|z_2|^n = \sum_{n=N}^{m} |a_n z_2{}^n|. \tag{4}$$

The limit of this last sum as m tends to infinity is the remainder

$$Q_N = \lim_{m\to\infty} \sum_{n=N}^{m} |a_n z_2|^n \tag{5}$$

after N terms in the series of the absolute values of the terms in series (1) when $z = z_2$. We know from the theorem of the preceding section that series (1) is absolutely convergent when $z = z_2$. Observe now that since Q_N is a remainder of a convergent series, it approaches zero as N tends to infinity. That is, for each positive number ε, an integer N_ε exists such that $Q_N < \varepsilon$ whenever $N > N_\varepsilon$. Also, the terms in the series for which Q_N is a remainder are all nonnegative real numbers, and consequently

$$\sum_{n=N}^{m} |a_n z_2|^n \leqq Q_N.$$

In view of relation (4), then,

$$\left|\sum_{n=N}^{m} a_n z^n\right| \leqq Q_N \tag{6}$$

for every integer m greater than N. But, according to equation (2),

$$R_N(z) = \lim_{m\to\infty} \sum_{n=N}^{m} a_n z^n.$$

Hence

$$|R_N(z)| \leqq Q_N < \varepsilon \qquad \text{whenever} \qquad N > N_\varepsilon. \tag{7}$$

(See Exercise 9, Sec. 56.) Now N_ε is independent of z when $|z| \leqq |z_2|$, and the convergence is therefore uniform. We state our result in the following theorem.

Theorem. *The power series* (1) *is uniformly convergent for all points z within and on any circle that is interior to the circle of convergence.*

The partial sum

$$S_N(z) = \sum_{n=0}^{N-1} a_n z^n$$

of series (1) is a polynomial and is therefore continuous at the point z_2 which was arbitrarily selected inside the circle C_1. We now show that the sum $S(z)$ is also continuous at z_2; that is, for each positive number ε there is a positive number δ such that

$$|S(z) - S(z_2)| < \varepsilon \qquad \text{whenever} \qquad |z - z_2| < \delta. \tag{8}$$

To do this, we first note that the equation

$$S(z) = S_N(z) + R_N(z)$$

implies that

$$|S(z) - S(z_2)| = |S_N(z) - S_N(z_2) + R_N(z) - R_N(z_2)|,$$

or

$$|S(z) - S(z_2)| \leqq |S_N(z) - S_N(z_2)| + |R_N(z)| + |R_N(z_2)|. \tag{9}$$

But, in view of the uniform convergence established above, an integer M_ε exists such that

$$|R_N(z)| < \frac{\varepsilon}{3} \qquad \text{whenever} \qquad N > M_\varepsilon \tag{10}$$

where z is any point lying in some closed disk about the origin whose radius is larger than $|z_2|$ and less than the radius r_1 of the circle C_1. In particular, the first of inequalities (10) holds for all z lying in a neighborhood $|z - z_2| < \delta$ of z_2 that is small enough to be contained in the above closed disk.

Also, the polynomial $S_N(z)$ is continuous at z_2 for any value of N. In particular, when $N = M_\varepsilon + 1$, we can choose our δ so small that

$$|S_N(z) - S_N(z_2)| < \frac{\varepsilon}{3} \quad \text{whenever} \quad |z - z_2| < \delta.$$

Applying these observations to inequality (9) when $N = M_\varepsilon + 1$, we readily find that condition (8) holds.

We have thus shown that *a power series represents a continuous function of z at each point interior to its circle of convergence.*

By substituting $z - z_0$ or its reciprocal for z, we can extend at once the above results, with obvious modifications, to series of the types

$$\sum_{n=0}^{\infty} a_n(z - z_0)^n, \qquad \sum_{n=1}^{\infty} \frac{b_n}{(z - z_0)^n}.$$

If, for instance, the second series here is convergent in an annulus $r_1 \leqq |z - z_0| \leqq r_2$, it is uniformly convergent for all values of z in that annulus and its sum represents a continuous function of z there.

62. Integration and Differentiation of Power Series

We have just seen that a power series represents a continuous function S interior to the circle of convergence. We shall prove in this section that S is actually analytic within that circle.

Theorem 1. *Let C denote any contour interior to the circle of convergence of the power series*

$$S(z) = \sum_{n=0}^{\infty} a_n z^n \tag{1}$$

and let g be any function that is continuous on C. The series formed by multiplying each term of the power series by g(z) can be integrated term by term over C; that is

$$\int_C g(z)S(z)\,dz = \sum_{n=0}^{\infty} a_n \int_C g(z)z^n\,dz. \tag{2}$$

Since the sum $S(z)$ of the power series is a continuous function, the integral of the product

$$g(z)S(z) = \sum_{n=0}^{N-1} a_n g(z)z^n + g(z)R_N(z),$$

where $R_N(z)$ is the remainder of the given series after N terms, exists. The terms of the finite sum here are also continuous on the contour C, and so their integrals over C exist. Consequently, the integral of the quantity $g(z)R_N(z)$ must exist, and

$$\int_C g(z)S(z)\,dz = \sum_{n=0}^{N-1} a_n \int_C g(z)z^n\,dz + \int_C g(z)R_N(z)\,dz. \tag{3}$$

Let M be the maximum value of $|g(z)|$ on C, and let L denote the length of C. In view of the uniform convergence of the given power series (Sec. 61), we know that for every positive number ε there exists an integer N_ε such that for all points z on C

$$|R_N(z)| < \varepsilon \qquad \text{whenever} \qquad N > N_\varepsilon.$$

Since ε and N_ε are independent of z, we can write

$$\left|\int_C g(z)R_N(z)\,dz\right| < M\varepsilon L \qquad \text{whenever} \qquad N > N_\varepsilon.$$

It follows, therefore, from equation (3) that

$$\int_C g(z)S(z)\,dz = \lim_{N\to\infty} \sum_{n=0}^{N-1} a_n \int_C g(z)z^n\,dz.$$

This is the same as equation (2), and Theorem 1 is proved.

When $g(z) = 1$ for each value of z on any *simple closed contour* C interior to the circle of convergence of the given power series, then

$$\int_C g(z)z^n\,dz = \int_C z^n\,dz = 0 \qquad (n = 0, 1, 2, \ldots).$$

It thus follows from equation (2) that

$$\int_C S(z)\,dz = 0$$

for every simple closed contour interior to the circle of convergence; and, according to Morera's theorem (Sec. 53), the function S is analytic inside that circle. This result can be stated as follows.

Theorem 2. *A power series represents a function that is analytic at every point interior to the circle of convergence of that series.*

Theorem 2 is often helpful in establishing the analyticity of functions or in evaluating limits. To illustrate, let us show that the function

$$f(z) = \begin{cases} \dfrac{\sin z}{z} & (z \neq 0) \\ 1 & (z = 0) \end{cases}$$

is entire. Since the Maclaurin series for the sine function converges to sin z for every z, the series obtained by dividing each term of that series by z,

$$1 - \frac{z^2}{3!} + \frac{z^4}{5!} - \cdots = 1 + \sum_{n=1}^{\infty} (-1)^n \frac{z^{2n}}{(2n+1)!}, \tag{4}$$

converges to $f(z)$ if $z \neq 0$. But series (4) clearly converges to $f(0)$ when $z = 0$. Hence $f(z)$ is represented by the convergent power series (4) for all z, and f is therefore an entire function. Note that since f is continuous at $z = 0$ and $(\sin z)/z = f(z)$ when $z \neq 0$,

$$\lim_{z \to 0} \frac{\sin z}{z} = \lim_{z \to 0} f(z) = f(0) = 1. \tag{5}$$

This is a result known beforehand because the limit here is the definition of the derivative of sin z at $z = 0$.

We observed in Sec. 58 that a Taylor series for a function f about a point z_0 converges to $f(z)$ at each point z interior to the circle centered at z_0 and passing through the nearest point z_1 where f fails to be analytic. In view of Theorem 2, we now know that *there is no larger circle about* z_0 *such that at each point* z *interior to it the Taylor series converges to* $f(z)$. For if there were such a circle, f would be analytic at z_1; but f is not analytic at z_1.

It should be noted, however, that even though there is no larger circle about z_0 within which the Taylor series *converges to* $f(z)$ at each point z, the Taylor series itself can *converge* everywhere within such a circle. For example, $|z| = 1$ is the largest circle about the origin within which the Maclaurin series for the function $f(z) = e^z(z - 1)/(z - 1)$ converges to $f(z)$ for all z. Yet the series actually converges everywhere in the plane.

We now present a companion to Theorem 1.

Theorem 3. *The power series* (1) *can be differentiated term by term; that is, at each point* z *interior to the circle of convergence,*

$$S'(z) = \sum_{n=1}^{\infty} n a_n z^{n-1}. \tag{6}$$

To establish this result, let z denote any point interior to the circle of convergence and let C be some simple closed contour surrounding z and interior to that circle. Also, define the function

$$g(s) = \frac{1}{2\pi i} \frac{1}{(s - z)^2} \tag{7}$$

at each point s on C. Since the function

$$S(z) = \sum_{n=0}^{\infty} a_n z^n$$

is analytic inside and on C, we can write

$$\int_C g(s)S(s)\,ds = \frac{1}{2\pi i}\int_C \frac{S(s)}{(s-z)^2}\,ds = S'(z)$$

where the integral representation for the derivative (Sec. 52) has been used. Furthermore,

$$\int_C g(s)s^n\,ds = \frac{1}{2\pi i}\int_C \frac{s^n}{(s-z)^2}\,ds = \frac{d}{dz}z^n \qquad (n = 0, 1, 2, \ldots).$$

Hence, according to equation (2) where the dummy variable z is replaced by s and $g(s)$ is defined by equation (7),

$$S'(z) = \sum_{n=0}^{\infty} a_n \frac{d}{dz} z^n = \sum_{n=1}^{\infty} n a_n z^{n-1}.$$

This completes the proof.

The results here are readily extended to series involving positive or negative powers of $z - z_0$.

EXERCISES

1. By differentiating the Maclaurin series for $1/(1-z)$, obtain the representations

$$\frac{1}{(1-z)^2} = \sum_{n=1}^{\infty} nz^{n-1}, \qquad \frac{2}{(1-z)^3} = \sum_{n=2}^{\infty} n(n-1)z^{n-2} \qquad (|z| < 1).$$

2. Expand the function $1/z$ into powers of $z - 1$; then obtain by differentiation the expansion of $1/z^2$ in powers of $z - 1$. Give regions of validity.
3. Integrate the Maclaurin series for $1/(1+s)$ along a contour interior to the circle of convergence from $s = 0$ to $s = z$ to obtain the representation

$$\text{Log}\,(z+1) = \sum_{n=1}^{\infty} (-1)^{n+1} \frac{z^n}{n} \qquad (|z| < 1).$$

4. Prove that if $f(z) = (e^{cz} - 1)/z$ when $z \neq 0$ and $f(0) = c$, then f is entire.
5. Expand $\sinh z$ into powers of $z - \pi i$ to prove that

$$\lim_{z \to \pi i} \frac{\sinh z}{z - \pi i} = -1.$$

6. Prove that if $f(z) = z^{-1}\,\text{Log}\,(z+1)$ when $z \neq 0$ and $f(0) = 1$, then f is analytic throughout the domain $|z| < 1$.

7. Write $f(z) = (z^2 - \pi^2/4)^{-1} \cos z$ when $z^2 \neq \pi^2/4$ and $f(\pm\pi/2) = -1/\pi$; then prove that f is an entire function.

8. Suppose that a function f is analytic at z_0 and $f(z_0) = 0$. Use series to show that

$$\lim_{z \to z_0} \frac{f(z)}{z - z_0} = f'(z_0).$$

Also note that this follows directly from the definition of $f'(z_0)$.

9. Suppose that f and g are analytic at z_0 and $f(z_0) = g(z_0) = 0$ while $g'(z_0) \neq 0$. Prove that

$$\lim_{z \to z_0} \frac{f(z)}{g(z)} = \frac{f'(z_0)}{g'(z_0)}.$$

10. Prove that if f is analytic at z_0 and $f(z_0) = f'(z_0) = \cdots = f^{(m)}(z_0) = 0$, the function

$$g(z) = \begin{cases} \dfrac{f(z)}{(z - z_0)^{m+1}} & (z \neq z_0) \\[2ex] \dfrac{f^{(m+1)}(z_0)}{(m+1)!} & (z = z_0) \end{cases}$$

is analytic at z_0.

63. Uniqueness of Representations

The series in equation (6) of the preceding section is a power series which converges to $S'(z)$ everywhere within the circle of convergence C_0 of the series

$$S(z) = \sum_{n=0}^{\infty} a_n z^n. \tag{1}$$

Consequently, that series for $S'(z)$ can be differentiated term by term; that is,

$$S''(z) = \sum_{n=2}^{\infty} n(n-1) a_n z^{n-2}$$

for all z within C_0. Indeed, the derivative of $S(z)$ of any order can be found by successively differentiating its series representation term by term. Moreover,

$$S(0) = a_0, \qquad S'(0) = a_1, \qquad S''(0) = 2!\,a_2, \ldots,$$

and the coefficients a_n are those in the Maclaurin series expansion for $S(z)$:

$$a_n = \frac{S^{(n)}(0)}{n!}.$$

The generalization to series involving positive powers of $z - z_0$ is immediate. We thus have the following theorem on the uniqueness of representations of functions in power series.

Theorem 1. *If the series*

$$\sum_{n=0}^{\infty} a_n(z - z_0)^n \tag{2}$$

converges to $f(z)$ at all points interior to some circle $|z - z_0| = r_0$, it is the Taylor series expansion of $f(z)$ in powers of $z - z_0$.

To illustrate, let us substitute z^2 for z in the Maclaurin series for $\sin z$:

$$\sin(z^2) = \sum_{n=1}^{\infty} (-1)^{n+1} \frac{z^{4n-2}}{(2n-1)!} \qquad (|z| < \infty). \tag{3}$$

This series must be identical to the series that would be found by expanding the function $\sin(z^2)$ directly into a Maclaurin series.

It follows from Theorem 1 that if series (2) converges to zero at every point in some neighborhood of z_0, then each of the coefficients a_n must be zero.

Theorem 2. *If the series*

$$\sum_{n=-\infty}^{\infty} c_n(z - z_0)^n = \sum_{n=0}^{\infty} a_n(z - z_0)^n + \sum_{n=1}^{\infty} \frac{b_n}{(z - z_0)^n} \tag{4}$$

converges to $f(z)$ at all points in some annular domain about z_0, then it is the Laurent series expansion for $f(z)$ in powers of $z - z_0$ for that domain.

We prove this theorem with the aid of Theorem 1 of the preceding section extended to series involving positive and negative powers of $z - z_0$. We write

$$\int_C g(z)f(z)\,dz = \sum_{n=-\infty}^{\infty} c_n \int_C g(z)(z - z_0)^n\,dz \tag{5}$$

where

$$g(z) = \frac{1}{2\pi i(z - z_0)^{m+1}} \qquad (m = 0, \pm 1, \pm 2, \ldots)$$

and C is a circle around the given annulus, centered at z_0 and taken in the positive sense. Observe now that since

$$\frac{1}{2\pi i}\int_C \frac{dz}{(z - z_0)^{m-n+1}} = \begin{cases} 0 & (m \neq n) \\ 1 & (m = n) \end{cases}$$

(see Exercise 16, Sec. 45), equation (5) reduces to

$$\frac{1}{2\pi i}\int_C \frac{f(z)\,dz}{(z-z_0)^{m+1}} = c_m$$

which is an expression for the coefficients in the Laurent series for $f(z)$ in the annulus.

64. Multiplication and Division

Suppose that each of the power series

$$\sum_{n=0}^{\infty} a_n z^n \quad \text{and} \quad \sum_{n=0}^{\infty} b_n z^n \tag{1}$$

converges within some circle $|z| = r_0$. The sums $f(z)$ and $g(z)$ are then analytic functions in the disk $|z| < r_0$, and the product of those sums has a Maclaurin series expansion which is valid there:

$$f(z)g(z) = \sum_{n=0}^{\infty} c_n z^n \qquad (|z| < r_0). \tag{2}$$

The coefficients c_n are given by the formulas

$$\begin{aligned}
c_0 &= f(0)g(0) = a_0 b_0,\\
c_1 &= f(0)g'(0) + f'(0)g(0) = a_0 b_1 + a_1 b_0,\\
c_2 &= \frac{1}{2!}[f(0)g''(0) + 2f'(0)g'(0) + f''(0)g(0)]\\
&= a_0 b_2 + a_1 b_1 + a_2 b_0,
\end{aligned}$$

etc. Here we have used the fact that the two series (1) are the Maclaurin series for $f(z)$ and $g(z)$, respectively. With the aid of the formula for the nth derivative of the product of two functions, we can see that

$$\begin{aligned}
f(z)g(z) &= a_0 b_0 + (a_0 b_1 + a_1 b_0)z + (a_0 b_2 + a_1 b_1 + a_2 b_0)z^2 + \cdots\\
&\quad + \left(\sum_{k=0}^{n} a_k b_{n-k}\right) z^n + \cdots \qquad (|z| < r_0).
\end{aligned} \tag{3}$$

Series (3) is the same as the series obtained by multiplying the two series (1) together term by term and collecting the resulting terms in like powers of z; it is called the *Cauchy product* of the two given series. We can now state the following theorem.

Theorem. *The Cauchy product of the two power series* (1) *converges to the product of their sums at all points interior to both their circles of convergence.*

Continuing to let $f(z)$ and $g(z)$ denote the sums of series (1), suppose now that $g(z) \neq 0$ in some neighborhood of the origin. The quotient $h(z) = f(z)/g(z)$ is analytic in that neighborhood and thus has a Maclaurin series expansion

$$h(z) = \sum_{n=0}^{\infty} d_n z^n \tag{4}$$

where $d_0 = h(0)$, $d_1 = h'(0)$, $d_2 = h''(0)/2!$, etc. The first few of these coefficients can be found in terms of the coefficients a_n and b_n in series (1) by differentiating the quotient $f(z)/g(z)$ successively. The results are the same as those found by carrying out the division of the first of series (1) by the second. This method identifies the first few terms of the quotient of the given power series with those of the Maclaurin series for $f(z)/g(z)$. This is generally the result that is needed, although it can be shown that the series obtained by the two methods are identical.

The addition of two power series term by term is always valid within their common region of convergence. This follows from the definition of the sum of the series. Multiplication by a constant is a special case of the above theorem on the multiplication of two series; consequently, two power series can be subtracted term by term.

65. Examples

Consider first the function

$$f(z) = \frac{-1}{(z-1)(z-2)} = \frac{1}{z-1} - \frac{1}{z-2} \tag{1}$$

which is analytic everywhere except at the two points $z = 1$ and $z = 2$.

EXAMPLE 1. Obtain the Maclaurin series for $f(z)$, which represents $f(z)$ in the open disk $|z| < 1$.

Observe that $|z/2| < 1$ at any point z in that disk. Consequently, knowing the sum of a geometric series (Sec. 58), we can write

$$f(z) = \frac{1}{2}\frac{1}{1 - z/2} - \frac{1}{1-z} = \sum_{n=0}^{\infty} \left[\frac{1}{2}\left(\frac{z}{2}\right)^n - z^n\right] \qquad (|z| < 1).$$

Since this series in powers of z converges to $f(z)$ when $|z| < 1$, it is the Maclaurin series for $f(z)$. That is, the coefficient of z^n in the expansion

$$f(z) = \sum_{n=0}^{\infty} (2^{-n-1} - 1)z^n \qquad (|z| < 1) \tag{2}$$

must have the value $f^{(n)}(0)/n!$; hence $f^{(n)}(0) = n!(2^{-n-1} - 1)$.

EXAMPLE 2. Write the Laurent series that represents $f(z)$ throughout the annulus $1 < |z| < 2$.

In that annular domain $|1/z| < 1$ and $|z/2| < 1$. Consequently,

$$f(z) = \frac{1}{z}\frac{1}{1 - 1/z} + \frac{1}{2}\frac{1}{1 - z/2} = \sum_{n=0}^{\infty} \frac{1}{z^{n+1}} + \sum_{n=0}^{\infty} \frac{z^n}{2^{n+1}} \qquad (1 < |z| < 2). \tag{3}$$

Now there is only one such representation for $f(z)$ in that annulus, and so expansion (3) is the Laurent series for $f(z)$ there. Since the coefficient of z^{-1} has the value $c_{-1} = 1$, our formula (5), Sec. 59, for the coefficients c_n shows that if C is any simple closed contour around the annulus in the positive direction, then

$$\int_C f(z)\,dz = 2\pi i.$$

EXAMPLE 3. Obtain the Laurent series for $f(z)$ in the domain $|z| > 2$.

In that domain $|1/z| < 1$ and $|2/z| < 1$; therefore

$$f(z) = \frac{1}{z}\left(\frac{1}{1 - 1/z} - \frac{1}{1 - 2/z}\right) = \sum_{n=0}^{\infty} \frac{1 - 2^n}{z^{n+1}} \qquad (|z| > 2). \tag{4}$$

This is the required Laurent series. Here the coefficient of z^{-1} is zero; hence the integral of $f(z)$ around each simple closed contour about the origin and exterior to the circle $|z| = 2$ has the value zero.

EXAMPLE 4. Find the first few terms of the Laurent series for the function

$$h(z) = \frac{1}{z^2 \sinh z} = \frac{1}{z^3}\frac{1}{1 + z^2/3! + z^4/5! + \cdots}$$

in the domain $0 < |z| < \pi$.

The denominator of the last fraction consists of a power series that converges to $z^{-1} \sinh z$ when $z \neq 0$ and to unity when $z = 0$. Thus the sum of that series is not zero anywhere in the domain $|z| < \pi$, and the power series representation of the fraction is found by division to be

$$\frac{1}{1 + z^2/3! + z^4/5! + \cdots} = 1 - \frac{1}{3!}z^2 + \left[\frac{1}{(3!)^2} - \frac{1}{5!}\right]z^4 + \cdots \qquad (|z| < \pi).$$

Hence the first few terms of the Laurent series for $h(z)$ in the domain specified are readily obtained:

$$\frac{1}{z^2 \sinh z} = \frac{1}{z^3} - \frac{1}{6}\frac{1}{z} + \frac{7}{360}z + \cdots \qquad (0 < |z| < \pi). \tag{5}$$

66. Zeros of Analytic Functions

If a function f is analytic at z_0, there is a circle about z_0 interior to which f is represented by a Taylor series

$$f(z) = a_0 + \sum_{n=1}^{\infty} a_n(z - z_0)^n \qquad (|z - z_0| < r_0), \tag{1}$$

where $a_0 = f(z_0)$ and $a_n = f^{(n)}(z_0)/n!$. If z_0 is a zero of f, then $a_0 = 0$. If, in addition,

$$f'(z_0) = f''(z_0) = \cdots = f^{(m-1)}(z_0) = 0 \tag{2}$$

but $f^{(m)}(z_0) \neq 0$, then z_0 is called a *zero of order m* and

$$f(z) = (z - z_0)^m \sum_{n=0}^{\infty} a_{m+n}(z - z_0)^n \qquad (a_m \neq 0, |z - z_0| < r_0). \tag{3}$$

Let $g(z)$ denote the sum of the series in equation (3):

$$g(z) = \sum_{n=0}^{\infty} a_{m+n}(z - z_0)^n \qquad (|z - z_0| < r_0). \tag{4}$$

Note that $g(z_0) = a_m \neq 0$. Since series (4) converges, g is continuous at z_0. Consequently, for each positive number ε there exists a positive number δ such that

$$|g(z) - a_m| < \varepsilon \qquad \text{whenever} \qquad |z - z_0| < \delta.$$

If $\varepsilon = |a_m|/2$ and δ_0 is a corresponding value of δ, then

$$|g(z) - a_m| < \frac{|a_m|}{2} \qquad \text{whenever} \qquad |z - z_0| < \delta. \tag{5}$$

It follows that $g(z) \neq 0$ at any point in the neighborhood $|z - z_0| < \delta_0$ because the first of inequalities (5) becomes $|a_m| < |a_m|/2$ if $g(z) = 0$ at any such z.

We have therefore established the following theorem.

Theorem. *Let a function f be analytic at a point z_0 which is a zero of f. There is a neighborhood of z_0 throughout which f has no other zeros, unless f is identically zero. That is, the zeros of an analytic function are isolated.*

EXERCISES

1. Write $g(z) = \sin(z^2)$. Use Maclaurin series (3), Sec. 63, for $g(z)$ to show that $g^{(2n-1)}(0) = 0$ and $g^{(4n)}(0) = 0$ $(n = 1, 2, \ldots)$.
2. Use expansion (5), Sec. 65, to show that if C is the circle $|z| = 1$, taken counterclockwise, then

$$\int_C \frac{dz}{z^2 \sinh z} = -\frac{\pi i}{3}.$$

3. Obtain the Maclaurin series representation

$$z \cosh (z^2) = z + \sum_{n=1}^{\infty} \frac{1}{(2n)!} z^{4n+1} \qquad (|z| < \infty).$$

4. Represent the function $(z+1)/(z-1)$ by (*a*) its Maclaurin series and give the region of validity for that representation; (*b*) its Laurent series for the domain $|z| > 1$.

Ans. (*a*) $-1 - 2\sum_{n=1}^{\infty} z^n \ (|z| < 1)$; (*b*) $1 + 2\sum_{n=1}^{\infty} z^{-n} \ (|z| > 1)$.

5. Obtain the expansion of the function $(z-1)/z^2$ into (*a*) its Taylor series in powers of $z - 1$ and give the region of validity; (*b*) its Laurent series for the domain $|z-1| > 1$.

Ans. (*a*) $\sum_{n=1}^{\infty} (-1)^{n+1} n(z-1)^n \ (|z-1| < 1)$; (*b*) $\sum_{n=1}^{\infty} (-1)^{n+1} n (z-1)^{-n}$ $(|z-1| > 1)$.

6. Obtain the Laurent series expansion

$$\frac{\sinh z}{z^2} = \frac{1}{z} + \sum_{n=1}^{\infty} \frac{1}{(2n+1)!} z^{2n-1} \qquad (|z| > 0).$$

7. Give two Laurent series expansions in powers of z for the function

$$f(z) = \frac{1}{z^2(1-z)}$$

and specify the regions in which those expansions are valid.

Ans. $\sum_{n=0}^{\infty} z^{n-2} \quad (0 < |z| < 1); -\sum_{n=0}^{\infty} z^{-n-3} \quad (|z| > 1)$.

8. Write the two Laurent series in powers of z that represent the function $z^{-1}(1+z^2)^{-1}$ in certain domains and specify those domains.

9. Obtain the first four nonzero terms of the Laurent series expansion

$$\frac{e^z}{z(z^2+1)} = \frac{1}{z} + 1 - \frac{1}{2} z - \frac{5}{6} z^2 + \cdots \qquad (0 < |z| > 1).$$

10. Obtain the first few nonzero terms of the Laurent series expansions

(*a*) $$\csc z = \frac{1}{z} + \frac{1}{3!} z - \left[\frac{1}{5!} - \frac{1}{(3!)^2}\right] z^3 + \cdots \qquad (0 < |z| < \pi);$$

(*b*) $$\frac{1}{e^z - 1} = \frac{1}{z} - \frac{1}{2} + \frac{1}{12} z - \frac{1}{720} z^3 + \cdots \qquad (0 < |z| < 2\pi).$$

11. Write the Laurent series expansion of the function $1/(z-k)$ for the domain $|z| > |k|$, where k is real and $-1 < k < 1$. Then write $z = e^{i\theta}$ to obtain the formulas

$$\sum_{n=1}^{\infty} k^n \cos n\theta = \frac{k \cos \theta - k^2}{1 - 2k \cos \theta + k^2},$$

$$\sum_{n=1}^{\infty} k^n \sin n\theta = \frac{k \sin \theta}{1 - 2k \cos \theta + k^2}.$$

Compare Exercise 4, Sec. 56.

12. Let $F(r,\theta)$ denote a function of $z = r \exp (i\theta)$ which is analytic in some annular domain about the origin that includes the circle $r = 1$. Take that circle as the curve C in the formula for the coefficients c_n in a Laurent series expansion of $F(r,\theta)$ in powers of z, and show that

$$F(1,\theta) = \frac{1}{2\pi} \int_{-\pi}^{\pi} F(1,\phi)\, d\phi + \frac{1}{\pi} \sum_{n=1}^{\infty} \int_{-\pi}^{\pi} F(1,\phi) \cos [n(\theta - \phi)]\, d\phi.$$

This is one form of the Fourier series expansion of the complex-valued function $F(1,\theta)$ of the real variable θ on the unit circle centered at the origin. Let $u(\theta)$ and $v(\theta)$ denote the real and imaginary parts of $F(1,\theta)$ and show that the above expansion is valid when the function F is replaced everywhere by u or everywhere by v. The restrictions on the real-valued functions u and v here are more severe than they need be in order that those functions be represented by Fourier series.[1]

[1] For other sufficient conditions, see, for instance, R. V. Churchill, "Fourier Series and Boundary Value Problems," 2d ed., pp. 90 and 110–111, 1963.

7

RESIDUES AND POLES

The Cauchy-Goursat theorem of Chap. 5 assures us that if a function is analytic everywhere inside and on a simple closed contour C, then the integral of the function around that contour is zero. If, however, the function fails to be analytic at a finite number of points inside C, there is, as we shall see in this chapter, a specific number, called a residue, which each of those points contributes to the value of the integral.

We develop here the theory of residues and illustrate that theory by using it to evaluate certain types of real definite integrals occurring in applied mathematics.

67. Residues

Recall (Sec. 19) that a point z_0 is called a singular point of a function f if f fails to be analytic at z_0 but is analytic at some point in every neighborhood of z_0. A singular point z_0 is said to be *isolated* if, in addition, there is some neighborhood of z_0 throughout which f is analytic except at the point itself.

The function $1/z$ furnishes a simple example. It is analytic everywhere except at $z = 0$; hence the origin is an isolated singular point of that function. The function

$$\frac{z+1}{z^3(z^2+1)}$$

has the three isolated singular points $z = 0$ and $z = \pm i$.

Observe, however, that while the origin is a singular point of $\operatorname{Log} z$, it is not an isolated singular point since every neighborhood of the origin contains points on the negative real axis and $\operatorname{Log} z$ fails to be analytic at each of those points. The function

$$\frac{1}{\sin(\pi/z)}$$

has the singular points $z = 1/n$ $(n = \pm 1, \pm 2, \ldots)$ and $z = 0$, all lying on the segment of the real axis between $z = -1$ and $z = 1$. Each singular point except $z = 0$ is isolated. The singular point $z = 0$ is not isolated because every neighborhood of the origin contains other singular points of the function.

When z_0 is an isolated singular point of f, there is a positive number r_1 such that the function is analytic at each point z for which $0 < |z - z_0| < r_1$. In that domain the function is represented by the Laurent series

$$f(z) = \sum_{n=0}^{\infty} a_n(z - z_0)^n + \frac{b_1}{z - z_0} + \frac{b_2}{(z - z_0)^2} + \cdots, \tag{1}$$

where the coefficients are given by formulas (2) and (3), Sec. 59. In particular,

$$b_1 = \frac{1}{2\pi i} \int_C f(z)\, dz \tag{2}$$

where C is any simple closed contour around z_0, described in the positive sense, such that f is analytic on C and interior to it except at the point z_0 itself. The complex number b_1, which is the coefficient of $1/(z - z_0)$ in expansion (1), is called the *residue* of f at the isolated singular point z_0.

Formula (2) gives us a powerful method for evaluating certain integrals around simple closed contours. Consider, for example, the integral

$$\int_C \frac{e^{-z}}{(z-1)^2}\, dz \tag{3}$$

where C is the circle $|z| = 2$ described in the positive sense. The integrand

$$f(z) = \frac{e^{-z}}{(z-1)^2}$$

is analytic on C and interior to it except at the isolated singular point $z = 1$. Thus, according to formula (2), the value of integral (3) is $2\pi i$ times the residue of f at $z = 1$. To determine this residue we use the Taylor series for e^{-z} about $z = 1$ to write the Laurent expansion

$$\frac{e^{-z}}{(z-1)^2} = \frac{e^{-1}}{(z-1)^2} - \frac{e^{-1}}{z-1} + e^{-1}\sum_{n=2}^{\infty}(-1)^n\frac{(z-1)^{n-2}}{n!} \qquad (|z-1| > 0). \tag{4}$$

From this we find that the residue of f at $z = 1$ is $-e^{-1}$. Consequently,

$$\int_C \frac{e^{-z}}{(z-1)^2}\,dz = -\frac{2\pi i}{e}. \tag{5}$$

For another example, let us show that with this same contour

$$\int_C \exp\left(\frac{1}{z^2}\right) dz = 0. \tag{6}$$

Since $1/z^2$ is analytic everywhere except at the origin, so is the integrand. The isolated singular point $z = 0$ is interior to C; and with the aid of the Maclaurin series for the exponential function, we can write the Laurent expansion

$$\exp\left(\frac{1}{z^2}\right) = 1 + \frac{1}{z^2} + \frac{1}{2!}\frac{1}{z^4} + \frac{1}{3!}\frac{1}{z^6} + \cdots \qquad (|z| > 0).$$

The residue of the integrand at its isolated singular point $z = 0$ is therefore zero ($b_1 = 0$), and the value of integral (6) is established.

68. The Residue Theorem

If a function f has only a finite number of singular points interior to some simple closed contour C, then they must be isolated. The following theorem is a precise statement of the fact that the value of the integral of f around C is $2\pi i$ times the sum of the residues associated with those singular points.

Theorem. *Let C be a simple closed contour within and on which a function f is analytic except for a finite number of singular points $z_1, z_2, \ldots, z_n$ interior to C. If $B_1, B_2, \ldots, B_n$ denote the residues of f at those points, then*

$$\int_C f(z)\,dz = 2\pi i(B_1 + B_2 + \cdots + B_n) \tag{1}$$

where C is described in the positive sense.

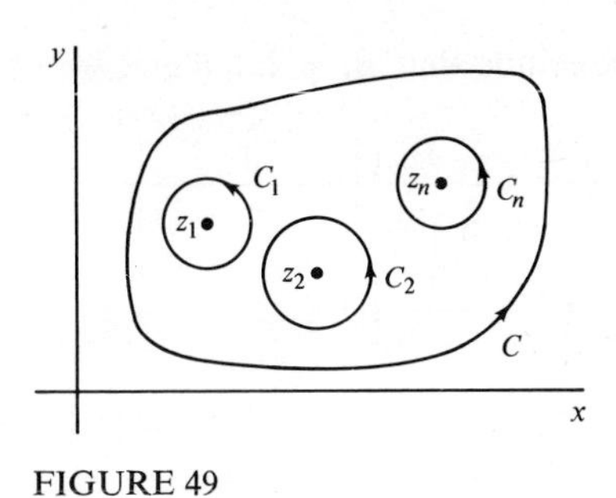

FIGURE 49

To prove the theorem, let the points z_j be centers of positively oriented circles C_j which are interior to C and are so small that no two of the circles have points in common (Fig. 49). The circles C_j together with the simple closed contour C form the boundary of a region throughout which f is analytic and whose interior is a multiply connected domain. Hence, according to the extension of the Cauchy-Goursat theorem to such regions (Sec. 49),

$$\int_C f(z)\,dz - \int_{C_1} f(z)\,dz - \int_{C_2} f(z)\,dz - \cdots - \int_{C_n} f(z)\,dz = 0.$$

This equation reduces to formula (1) because

$$B_j = \frac{1}{2\pi i}\int_{C_j} f(z)\,dz \qquad (j = 1, 2, \ldots, n);$$

and the theorem is proved.

To illustrate the theorem, let us evaluate the integral

$$\int_C \frac{5z-2}{z(z-1)}\,dz \tag{2}$$

where C is the circle $|z| = 2$, described counterclockwise. The integrand has two singularities, $z = 0$ and $z = 1$, both interior to C. We can find the residues B_1 at $z = 0$ and B_2 at $z = 1$ with the aid of the Maclaurin series

$$\frac{1}{1+z} = 1 - z + z^2 - \cdots \qquad (|z| < 1).$$

We first write the Laurent expansion

$$\frac{5z-2}{z(z-1)} = \left(5 - \frac{2}{z}\right)\left(\frac{-1}{1-z}\right) = \left(-5 + \frac{2}{z}\right)(1 + z + z^2 + \cdots)$$

$$= \frac{2}{z} - 3 - 3z - 3z^2 - \cdots \qquad (0 < |z| < 1)$$

of the integrand and conclude that $B_1 = 2$. Next, we observe that

$$\frac{5z-2}{z(z-1)} = \left(5 + \frac{3}{z-1}\right)\left[\frac{1}{1+(z-1)}\right]$$
$$= \left(5 + \frac{3}{z-1}\right)[1 - (z-1) + (z-1)^2 - \cdots]$$

when $0 < |z-1| < 1$. The coefficient of $1/(z-1)$ in the Laurent expansion valid for $0 < |z-1| < 1$ is therefore 3. It follows that $B_2 = 3$, and

$$\int_C \frac{5z-2}{z(z-1)}\,dz = 2\pi i(B_1 + B_2) = 10\pi i.$$

In this example it is, of course, simpler to write the integrand as the sum of its partial fractions. Then

$$\int_C \frac{5z-2}{z(z-1)}\,dz = \int_C \frac{2}{z}\,dz + \int_C \frac{3}{z-1}\,dz = 4\pi i + 6\pi i = 10\pi i.$$

69. The Principal Part of a Function

We have seen that if a function f has an isolated singular point z_0, the function is represented by a Laurent series

$$f(z) = \sum_{n=0}^{\infty} a_n(z-z_0)^n + \sum_{n=1}^{\infty} \frac{b_n}{(z-z_0)^n} \tag{1}$$

in a domain $0 < |z - z_0| < r_1$ centered at z_0. The portion of the series involving negative powers of $z - z_0$ is called the *principal part* of f at z_0. We now use the principal part to distinguish between three types of isolated singular points. The behavior of the function near the isolated singular point is fundamentally different in each case.

If the principal part of f at z_0 contains at least one nonzero term but the number of such terms is finite, there exists a positive integer m such that $b_m \neq 0$ and $b_{m+1} = b_{m+2} = \cdots = 0$. That is, expansion (1) takes the form

$$f(z) = \sum_{n=0}^{\infty} a_n(z-z_0)^n + \frac{b_1}{z-z_0} + \frac{b_2}{(z-z_0)^2} + \cdots + \frac{b_m}{(z-z_0)^m} \tag{2}$$
$$(0 < |z - z_0| < r_1).$$

In this case the isolated singular point z_0 is called a *pole of order m*. A pole of order $m = 1$ is called a *simple pole*.

The function

$$\frac{z^2 - 2z + 3}{z - 2} = 2 + (z - 2) + \frac{3}{z - 2} \qquad (|z - 2| > 0),$$

for example, has a simple pole at $z = 2$. Its residue there is 3. The function

$$\frac{\sinh z}{z^4} = \frac{1}{z^3} + \frac{1}{3!}\frac{1}{z} + \frac{1}{5!} z + \frac{1}{7!} z^3 + \cdots \qquad (|z| > 0)$$

has a pole of order 3 at $z = 0$, with a residue there of 1/6.

As will be shown in the following section, $f(z)$ always tends to infinity as z approaches a pole.

When the principal part of f at z_0 has an infinite number of nonzero terms, the point is called an *essential singular point*. An example is the function

$$\exp\left(\frac{1}{z}\right) = 1 + \sum_{n=1}^{\infty} \frac{1}{n!}\frac{1}{z^n} \qquad (|z| > 0) \tag{3}$$

which has an essential singular point at $z = 0$. Its residue there is unity.

An important result concerning the behavior of a function near an essential singular point is due to Picard. It states that in any neighborhood of an essential singular point the function assumes every finite value, with one possible exception, an infinite number of times. We will not prove Picard's theorem but will prove a closely related result later in Sec. 112.[1]

To illustrate Picard's theorem, let us show that the function exp $(1/z)$ given in equation (3) assumes the value -1 an infinite number of times in any neighborhood of the origin. To do this, recall from Sec. 22 that $\exp z = -1$ when $z = (1 + 2n)\pi i$ $(n = 0, \pm 1, \pm 2, \ldots)$. This means that $\exp(1/z) = -1$ at the points

$$z = \frac{1}{(1 + 2n)\pi i} \qquad (n = 0, \pm 1, \pm 2, \ldots),$$

and an infinite number of these points lie in any neighborhood of the origin. Note that $|\exp(1/z)| \neq 0$ for any value of z, and zero is then the exceptional value which is not assumed by the function.

When all the coefficients b_n in the principal part of f at an isolated singular point z_0 are zero, the point z_0 is called a *removable singular point* of f. In this case the Laurent series (1) contains only nonnegative powers of $z - z_0$, and the series is in fact a power series. If we define f to be a_0 at z_0, the function becomes analytic at z_0. (See Theorem 2, Sec. 62.) Thus, a function f with a removable

[1] For a proof of Picard's theorem, see Sec. 51 in vol. III of the books by Markushevich cited in Appendix 1.

singular point can be made analytic at that point by assigning a suitable value to the function there.

Consider, for instance, the function

$$f(z) = \frac{e^z - 1}{z} = 1 + \frac{z}{2!} + \frac{z^2}{3!} + \cdots \qquad (|z| > 0).$$

If we write $f(0) = 1$, the function becomes entire.

70. Poles

Assume that f has a pole of order m at z_0. Let us define a new function ϕ by means of the equation

$$\phi(z) = (z - z_0)^m f(z).$$

In view of equation (2) of the preceding section,

$$\phi(z) = b_m + b_{m-1}(z - z_0) + b_{m-2}(z - z_0)^2 + \cdots + b_1(z - z_0)^{m-1} + \sum_{n=0}^{\infty} a_n(z - z_0)^{m+n} \qquad (0 < |z - z_0| < r_1) \tag{1}$$

where $b_m \neq 0$; hence the point z_0 is a removable singular point of the function ϕ. Let us write

$$\phi(z_0) = b_m$$

in order to make the function ϕ analytic at z_0. Recall now that analyticity at a point implies continuity there; thus our definition of $\phi(z_0)$ can be written

$$\phi(z_0) = \lim_{z \to z_0} (z - z_0)^m f(z) = b_m. \tag{2}$$

Since this limit exists and $b_m \neq 0$, it follows that $f(z)$ *tends to infinity as* z *approaches a pole* z_0. (See Exercise 11, Sec. 71.)

The function ϕ can, moreover, be used to determine the residue of f at the pole z_0. This residue is the coefficient b_1 in the Laurent series (2) of the preceding section. Since expression (1) is now the Taylor series for ϕ about z_0, the number b_1 is given by the formula

$$b_1 = \frac{\phi^{(m-1)}(z_0)}{(m-1)!}. \tag{3}$$

When $m = 1$, this expression for the residue of f at the simple pole z_0 can be written, according to equation (2), as

$$b_1 = \phi(z_0) = \lim_{z \to z_0} (z - z_0) f(z). \tag{4}$$

Suppose now that we are simply given some function f such that the product

$$(z - z_0)^m f(z)$$

can be defined at z_0 so that it is analytic there. As before, m is a positive integer. Let $\phi(z)$ denote the above product. Then, in some open disk about z_0,

$$\phi(z) = (z - z_0)^m f(z) = \phi(z_0) + \phi'(z_0)(z - z_0) + \cdots + \frac{\phi^{(m)}(z_0)}{m!}(z - z_0)^m + \cdots.$$

Consequently, at each point except z_0 in that disk

$$f(z) = \frac{\phi(z_0)}{(z - z_0)^m} + \frac{\phi'(z_0)}{(z - z_0)^{m-1}} + \cdots + \frac{\phi^{(m-1)}(z_0)}{(m-1)!}\frac{1}{z - z_0}$$
$$+ \sum_{n=m}^{\infty} \frac{\phi^{(n)}(z_0)}{n!}(z - z_0)^{n-m}.$$

If $\phi(z_0) \neq 0$, it follows that f has a pole of order m at z_0 with the residue there given by equation (3) or (4). We state this test for poles as follows.

Theorem. *Given a function f, suppose that for some positive integer m the function*

$$\phi(z) = (z - z_0)^m f(z)$$

can be defined at z_0 so that it is analytic there and $\phi(z_0) \neq 0$. Then f has a pole of order m at z_0. Its residue there is given by formula (3) *if $m > 1$ and by formula* (4) *if $m = 1$.*

Note that the conditions in the theorem are satisfied whenever $f(z)$ has the form

$$f(z) = \frac{\phi(z)}{(z - z_0)^m} \qquad (m = 1, 2, \ldots)$$

where the function ϕ is analytic at z_0 and $\phi(z_0) \neq 0$.

As an illustration, the function $f(z) = e^{-2z}/z^3$ has a pole of order 3 at $z = 0$. In this case $\phi(z) = e^{-2z}$. Thus, according to formula (3), the residue of f at $z = 0$ is $\phi^{(2)}(0)/2!$, or 2.

To illustrate formula (4), we note that the function $f(z) = (z + 1)/(z^2 + 9)$ has a simple pole at $z = 3i$ and the residue there is

$$\lim_{z \to 3i} (z - 3i)\frac{z + 1}{z^2 + 9} = \lim_{z \to 3i} \frac{z + 1}{z + 3i} = \frac{3 - i}{6}.$$

The point $z = -3i$ is also a simple pole, with residue $(3 + i)/6$.

71. Quotients of Analytic Functions

The basic method for computing the residue of a function at an isolated singular point z_0 is that of appealing directly to the appropriate Laurent series and noting the coefficient of $1/(z - z_0)$. When z_0 is an essential singular point, we offer no other method; but for residues at poles, formulas (3) and (4) of the preceding section may be used to advantage when the function ϕ is simple enough.

Another method for finding the residue of a function f at a pole z_0 is available if f can be written as a quotient

$$f(z) = \frac{p(z)}{q(z)} \tag{1}$$

where *p and q are both analytic at z_0 and $p(z_0) \neq 0$.* We first note that z_0 is an isolated singular point of f if and only if $q(z_0) = 0$. For if $q(z_0) = 0$, then $q(z) \neq 0$ at any other point in some neighborhood of z_0; this is because the zeros of an analytic function which is not identically zero are isolated (Sec. 66). It follows that f is analytic everywhere in that neighborhood of z_0 except at the point z_0 itself, and z_0 is therefore an isolated singular point of f. Conversely, if z_0 is an isolated singular point of f, then $q(z_0) = 0$. For, when $q(z_0) \neq 0$, it follows by continuity that $q(z) \neq 0$ throughout some neighborhood of z_0. (See Exercise 11, Sec. 15.) It then follows that f is analytic at z_0, and this contradicts the fact that z_0 is an isolated singular point.

The function f given in equation (1) *has a simple pole at z_0 if, in addition to the conditions given there, $q(z_0) = 0$ and $q'(z_0) \neq 0$. The residue of f at the simple pole z_0 is given by the formula*

$$b_1 = \frac{p(z_0)}{q'(z_0)}. \tag{2}$$

To prove this statement, we expand each of the analytic functions p and q into a Taylor series valid in a disk $|z - z_0| < r_1$, and we write

$$(z - z_0)f(z) = \frac{p(z_0) + p'(z_0)(z - z_0) + \cdots}{q'(z_0) + q''(z_0)(z - z_0)/2! + \cdots} \qquad (0 < |z - z_0| < r_1). \tag{3}$$

The quotient of the two series here represents a function ϕ that is analytic at z_0; and since $\phi(z_0) = p(z_0)/q'(z_0) \neq 0$, the proof is easily completed by means of the theorem in the preceding section.

In like manner, we can show that *if, in addition to the conditions on p and q given with equation* (1),

$$q(z_0) = q'(z_0) = \cdots = q^{(m-1)}(z_0) = 0$$

and $q^{(m)}(z_0) \neq 0$, then the function f has a pole of order m at z_0. When $m = 2$, the residue of f at the second order pole z_0 is given by the formula

$$b_1 = 2\frac{p'(z_0)}{q''(z_0)} - \frac{2}{3}\frac{p(z_0)q'''(z_0)}{[q''(z_0)]^2}, \tag{4}$$

as we can see by computing $\phi'(z_0)$ where

$$\phi(z) = \frac{p(z_0) + p'(z_0)(z - z_0) + \cdots}{q''(z_0)/2! + q'''(z_0)(z - z_0)/3! + \cdots} \qquad (|z - z_0| < r_1).$$

When $m > 2$, the corresponding formulas for the residues are lengthy.

To illustrate formula (2), we consider the function

$$\cot z = \frac{\cos z}{\sin z}$$

which has the isolated singular points $z = n\pi$ $(n = 0, \pm 1, \pm 2, \ldots)$. Writing $p(z) = \cos z$ and $q(z) = \sin z$, we find that each of these points is a simple pole with residue

$$b_1 = \frac{p(n\pi)}{q'(n\pi)} = \frac{\cos n\pi}{\cos n\pi} = 1.$$

For another example, we compute the residue of the function

$$f(z) = \frac{1}{z(e^z - 1)}$$

at the origin. Here $p(z) = 1$, $q(z) = z(e^z - 1)$, $q(0) = q'(0) = 0$, $q''(0) = 2$, and $q'''(0) = 3$. Thus the origin is a pole of the second order and, according to formula (4), the residue of f there is $-1/2$.

EXERCISES

1. In each case write the principal part of the function at its isolated singular point. Determine if that point is a pole, an essential singular point, or a removable singular point of the given function.

(*a*) $ze^{1/z}$; (*b*) $\dfrac{z^2}{1+z}$; (*c*) $\dfrac{\sin z}{z}$; (*d*) $\dfrac{\cos z}{z}$.

2. Show that all the singular points of each of the following functions are poles. Determine the order m of each pole and the corresponding residue B.

(*a*) $\dfrac{z+1}{z^2 - 2z}$; (*b*) $\tanh z$; (*c*) $\dfrac{1 - \exp(2z)}{z^4}$;

(*d*) $\dfrac{\exp(2z)}{(z-1)^2}$; (*e*) $\dfrac{z}{\cos z}$; (*f*) $\dfrac{\exp z}{z^2 + \pi^2}$.

Ans. (*a*) $m = 1$, $B = -\frac{1}{2}, \frac{3}{2}$; (*b*) $m = 1$, $B = 1$; (*c*) $m = 3$, $B = -\frac{4}{3}$.

3. Find the residue at $z = 0$ of the function

$$(a)\ \csc^2 z; \qquad (b)\ z^{-3}\csc(z^2); \qquad (c)\ z\cos\frac{1}{z}.$$

Ans. (*a*) 0; (*b*) 1/6; (*c*) $-1/2$.

4. Find the value of the integral

$$\int_C \frac{3z^3 + 2}{(z-1)(z^2+9)}\,dz$$

taken counterclockwise around the circle (*a*) $|z - 2| = 2$; (*b*) $|z| = 4$.

Ans. (*a*) πi; (*b*) $6\pi i$.

5. Find the value of the integral

$$\int_C \frac{dz}{z^3(z+4)}$$

taken counterclockwise around the circle (*a*) $|z| = 2$; (*b*) $|z + 2| = 3$.

Ans. (*a*) $\pi i/32$; (*b*) 0.

6. Let C be the circle $|z| = 2$ described in the positive sense and evaluate the integral

$$(a)\ \int_C \tan z\,dz; \qquad (b)\ \int_C \frac{dz}{\sinh 2z}; \qquad (c)\ \int_C \frac{\cosh \pi z\,dz}{z(z^2+1)}.$$

Ans. (*a*) $-4\pi i$; (*b*) $-\pi i$.

7. Evaluate the integral of f in the positive sense around the unit circle about the origin when $f(z)$ is

$$(a)\ z^{-2}e^{-z}; \qquad (b)\ z^{-1}\csc z; \qquad (c)\ z^{-2}\csc z; \qquad (d)\ z\exp\frac{1}{z}.$$

Ans. (*a*) $-2\pi i$; (*b*) 0; (*d*) πi.

8. Evaluate integral (2), Sec. 68, by finding the coefficient of $1/z$ in the Laurent expansion of the integrand

$$\frac{5z-2}{z(z-1)} = \frac{5z-2}{z^2}\,\frac{1}{1-(1/z)}$$

into powers of z where the region of validity is $|z| > 1$. Note, however, that the coefficient sought is *not* the residue of the integrand at $z = 1$.

9. Find the residue at $z = 1$ of the branch of the multiple-valued function

$$f(z) = \frac{\sqrt{z}}{1-z}$$

obtained by restricting $\arg z$ such that $(2n-1)\pi < \arg z < (2n+1)\pi$, where n is an integer.

Ans. $(-1)^{n+1}$.

10. Let f be a function which is analytic at the point z_0. Show that z_0 is a removable singular point of the function

$$g(z) = \frac{f(z)}{z - z_0}$$

when $f(z_0) = 0$. Show that when $f(z_0) \neq 0$, the point z_0 is a simple pole of g, with residue $f(z_0)$.

11. With the aid of expression (2), Sec. 70, for the number b_m, show that

$$\lim_{z \to z_0} f(z) = \infty$$

when z_0 is a pole of f.

Suggestion: Observe that there exists a positive number δ such that

$$|b_m - (z - z_0)^m f(z)| < \tfrac{1}{2}|b_m| \qquad \text{whenever} \qquad 0 < |z - z_0| < \delta;$$

then use the inequality $|z_1| - |z_2| \leqq |z_1 - z_2|$.

12. Prove that if a function $f(z)$ is analytic at z_0 and if z_0 is a zero of order m of $f(z)$, then the function $1/f(z)$ has a pole of order m at z_0.

13. Let a function $f(z)$ be analytic throughout a simply connected domain D and let z_0 be the only zero of $f(z)$ in D. Prove that if C is a positively oriented simple closed contour in D that encloses z_0, then

$$\frac{1}{2\pi i} \int_C \frac{f'(z)}{f(z)}\, dz = m$$

where the positive integer m is the order of the zero. The quotient $f'(z)/f(z)$ is the derivative of $\log f(z)$ and is known as the *logarithmic derivative* of $f(z)$.

14. Using the result of Exercise 13, prove the following property of the logarithmic derivative defined there. Let D be a simply connected domain throughout which a function f is analytic and $f'(z) \neq 0$. Let C denote a simple closed contour in D, described in the positive sense, such that $f(z) \neq 0$ at any point on C. Then, if f has N zeros interior to C, that number is given by the formula

$$N = \frac{1}{2\pi i} \int_C \frac{f'(z)}{f(z)}\, dz.$$

72. Evaluation of Improper Real Integrals

An important application of the theory of residues is the evaluation of certain types of real definite integrals. The examples treated here and in the remainder of the chapter illustrate this application of our theory.

Recall from elementary calculus that an improper integral of the form

$$\int_{-\infty}^{\infty} f(x)\, dx, \tag{1}$$

where f is continuous for all x, is said to *converge* and have the value

$$\lim_{R\to\infty}\int_{-R}^{0} f(x)\,dx + \lim_{R\to\infty}\int_{0}^{R} f(x)\,dx \tag{2}$$

when the individual limits exist. Another number associated with integral (1) is also useful. Namely, the *Cauchy principal value* of integral (1) is defined by the equation

$$\text{P.V.}\int_{-\infty}^{\infty} f(x)\,dx = \lim_{R\to\infty}\int_{-R}^{R} f(x)\,dx, \tag{3}$$

provided the limit on the right exists.

If integral (1) converges, the value obtained is also the Cauchy principal value. On the other hand, when $f(x) = x$, for example, the Cauchy principal value of integral (1) is 0, whereas that integral does not converge according to definition (2). But suppose that f is an even function; that is, $f(-x) = f(x)$ for all real numbers x. Then if the Cauchy principal value of integral (1) exists, integral (1) actually converges. For when f is even,

$$\int_{-R}^{0} f(x)\,dx = \int_{0}^{R} f(x)\,dx = \frac{1}{2}\int_{-R}^{R} f(x)\,dx;$$

and the existence of the limit in expression (3) thus implies the existence of both the limits in expression (2).

Suppose now that the integrand $f(x)$ in integral (1) can be written $f(x) = p(x)/q(x)$ where $p(x)$ and $q(x)$ are real polynomials with no factors in common and $q(x)$ has no real zeros. If the degree of $q(x)$ is at least two greater than the degree of $p(x)$, the integral converges. The value to which that integral converges can often be found quite easily by determining its Cauchy principal value by means of the theory of residues.

To illustrate the method, let us evaluate the convergent integral

$$\int_{0}^{\infty} \frac{2x^2-1}{x^4+5x^2+4}\,dx = \frac{1}{2}\int_{-\infty}^{\infty} \frac{2x^2-1}{x^4+5x^2+4}\,dx. \tag{4}$$

Note that the integral on the right represents an integration of the function

$$f(z) = \frac{2z^2-1}{z^4+5z^2+4} = \frac{2z^2-1}{(z^2+1)(z^2+4)}$$

along the entire real axis. This function has simple poles at the points $z = \pm i$, $\pm 2i$ and is analytic everywhere else.

When $R > 2$, the singular points of f in the upper half plane lie in the interior of the semicircular region bounded by the segment $-R \leqq x \leqq R$, $y = 0$

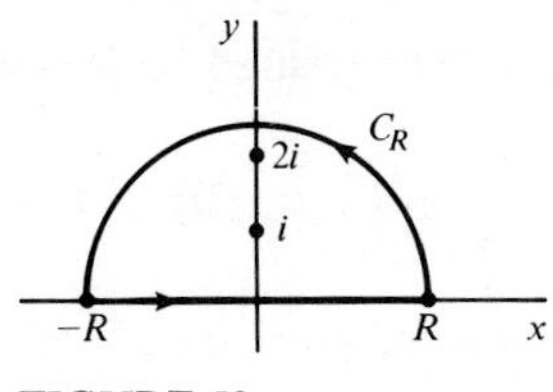

FIGURE 50

of the x axis and the upper half C_R of the circle $|z| = R$ (Fig. 50). Integrating f counterclockwise around the boundary of this semicircular region, we see that

$$\int_{-R}^{R} f(x)\,dx + \int_{C_R} f(z)\,dz = 2\pi i(B_1 + B_2) \tag{5}$$

where B_1 is the residue of f at the point $z = i$ and B_2 is the residue of f at the point $z = 2i$. According to formula (4), Sec. 70,

$$B_1 = \lim_{z\to i} (z - i)f(z) = \frac{i}{2}$$

and

$$B_2 = \lim_{z\to 2i} (z - 2i)f(z) = -\frac{3i}{4};$$

therefore, equation (5) can be written

$$\int_{-R}^{R} f(x)\,dx = \frac{\pi}{2} - \int_{C_R} f(z)\,dz. \tag{6}$$

This last equation is valid for all values of R greater than 2.

We now show that the integral on the right in equation (6) approaches 0 as R tends to ∞. To do this, we observe that

$$|z^4 + 5z^2 + 4| = |z^2 + 1||z^2 + 4| \geqq (|z|^2 - 1)(|z|^2 - 4).$$

Hence, when z is a point on C_R,

$$|z^4 + 5z^2 + 4| \geqq (R^2 - 1)(R^2 - 4).$$

Also, on C_R,

$$|2z^2 - 1| \leqq 2|z|^2 + 1 = 2R^2 + 1.$$

Consequently,

$$\left|\int_{C_R} \frac{2z^2 - 1}{z^4 + 5z^2 + 4}\,dz\right| \leqq \frac{2R^2 + 1}{(R^2 - 1)(R^2 - 4)}\,\pi R,$$

where πR is the length of C_R. The desired limit is now evident; that is,

$$\lim_{R\to\infty}\int_{C_R} f(z)\,dz = 0.$$

It thus follows from equation (6) that

$$\lim_{R\to\infty}\int_{-R}^{R}\frac{2x^2-1}{x^4+5x^2+4}\,dx = \frac{\pi}{2},$$

or

$$\text{P.V.}\int_{-\infty}^{\infty}\frac{2x^2-1}{x^4+5x^2+4}\,dx = \frac{\pi}{2};$$

and, since the integral here actually converges, we arrive at the result

$$\int_{0}^{\infty}\frac{2x^2-1}{x^4+5x^2+4}\,dx = \frac{\pi}{4}.$$

73. Improper Integrals Involving Trigonometric Functions

Residue theory can be useful in evaluating convergent improper integrals of the form

$$\int_{-\infty}^{\infty}\frac{p(x)}{q(x)}\cos x\,dx \quad\text{and}\quad \int_{-\infty}^{\infty}\frac{p(x)}{q(x)}\sin x\,dx \tag{1}$$

where $p(x)$ and $q(x)$ are real polynomials and $q(x)$ has no real zeros. The method of the previous section cannot be applied directly here since $|\cos z|$ and $|\sin z|$ increase like $\sinh y$, or e^y, as y tends to infinity (Sec. 24). We note, however, that integrals (1) are the real and imaginary parts of the integral

$$\int_{-\infty}^{\infty}\frac{p(x)}{q(x)}e^{ix}\,dx$$

and that e^{iz} has modulus e^{-y} which is bounded in the upper half plane.

To illustrate our modification of the earlier method, let us now show that

$$\int_{-\infty}^{\infty}\frac{\cos x}{(x^2+1)^2}\,dx = \frac{\pi}{e}. \tag{2}$$

This integral is the real part of the integral

$$\int_{-\infty}^{\infty}\frac{e^{ix}}{(x^2+1)^2}\,dx$$

which represents an integration of the function

$$f(z) = \frac{e^{iz}}{(z^2+1)^2}$$

along the real axis.

The function f is analytic except for poles of order 2 at the points $z = \pm i$. The pole $z = i$ lies in the interior of the semicircular region whose boundary is the segment $-R \leqq x \leqq R$, $y = 0$ of the real axis and the upper half C_R of the circle $|z| = R$, where $R > 1$. Integrating f counterclockwise around this boundary, we find that

$$\int_{-R}^{R} \frac{e^{ix}}{(x^2+1)^2}\,dx = 2\pi i B_1 - \int_{C_R} \frac{e^{iz}}{(z^2+1)^2}\,dz \tag{3}$$

where B_1 is the residue of f at the pole $z = i$. To compute this residue, write

$$\phi(z) = (z-i)^2 f(z) = \frac{e^{iz}}{(z+i)^2}.$$

Then, according to formula (3), Sec. 70,

$$B_1 = \phi'(i) = -\frac{i}{2e}. \tag{4}$$

To show that the second integral in equation (3) approaches 0 as R tends to infinity, we note that when z is on C_R,

$$|z^2+1|^2 \geqq (R^2-1)^2.$$

Then, since

$$|e^{iz}| = |e^{-y}| \leqq 1$$

when $y \geqq 0$,

$$\left| \int_{C_R} \frac{e^{iz}}{(z^2+1)^2}\,dz \right| \leqq \frac{\pi R}{(R^2-1)^2}.$$

From this inequality and equations (3) and (4) it follows that

$$\lim_{R\to\infty} \int_{-R}^{R} \frac{e^{ix}}{(x^2+1)^2}\,dx = \frac{\pi}{e}. \tag{5}$$

That is,

$$\lim_{R\to\infty} \int_{-R}^{R} \frac{\cos x}{(x^2+1)^2}\,dx = \frac{\pi}{e},$$

where the real parts on each side of equation (5) have been equated.

Thus, the Cauchy principal value of integral (2) exists and is equal to π/e. Moreover, since integral (2) has an even integrand, we can conclude that it actually converges to π/e.

EXERCISES

Establish the following integration formulas with the aid of residues:

1. $\displaystyle\int_0^\infty \frac{dx}{x^2+1} = \frac{\pi}{2}.$

2. $\displaystyle\int_0^\infty \frac{dx}{x^4+1} = \frac{\pi}{2\sqrt{2}}.$

3. $\displaystyle\int_0^\infty \frac{x^2\,dx}{(x^2+1)(x^2+4)} = \frac{\pi}{6}.$

4. $\displaystyle\int_0^\infty \frac{dx}{(x^2+1)^2} = \frac{\pi}{4}.$

5. $\displaystyle\int_0^\infty \frac{x^2\,dx}{(x^2+9)(x^2+4)^2} = \frac{\pi}{200}.$

6. $\displaystyle\int_0^\infty \frac{x^2\,dx}{x^6+1} = \frac{\pi}{6}.$

7. $\displaystyle\int_0^\infty \frac{\cos ax}{x^2+1}\,dx = \frac{\pi}{2}e^{-a}$ $\qquad (a \geqq 0).$

8. $\displaystyle\int_0^\infty \frac{\cos ax}{(x^2+b^2)^2}\,dx = \frac{\pi}{4b^3}(1+ab)e^{-ab}$ $\qquad (a > 0, b > 0).$

9. $\displaystyle\int_{-\infty}^\infty \frac{\cos x\,dx}{(x^2+a^2)(x^2+b^2)} = \frac{\pi}{a^2-b^2}\left(\frac{e^{-b}}{b} - \frac{e^{-a}}{a}\right)$ $\qquad (a > b > 0).$

10. $\displaystyle\int_{-\infty}^\infty \frac{x \sin ax}{x^4+4}\,dx = \frac{\pi}{2}e^{-a}\sin a$ $\qquad (a > 0).$

Determine the Cauchy principal value of each of the following convergent integrals:

11. $\displaystyle\int_{-\infty}^\infty \frac{dx}{x^2+2x+2}.$

12. $\displaystyle\int_{-\infty}^\infty \frac{x\,dx}{(x^2+1)(x^2+2x+2)}.$ $\qquad$ *Ans.* $-\pi/5$.

13. $\displaystyle\int_{-\infty}^\infty \frac{x^2\,dx}{(x^2+1)^2}.$

14. $\displaystyle\int_{-\infty}^\infty \frac{x \sin x\,dx}{(x^2+1)(x^2+4)}.$

15. $\displaystyle\int_{-\infty}^\infty \frac{\sin x\,dx}{x^2+4x+5}.$ $\qquad$ *Ans.* $-(\pi/e)\sin 2$.

16. $\displaystyle\int_{-\infty}^\infty \frac{\cos x\,dx}{(x+a)^2+b^2}$ $\qquad (b > 0).$

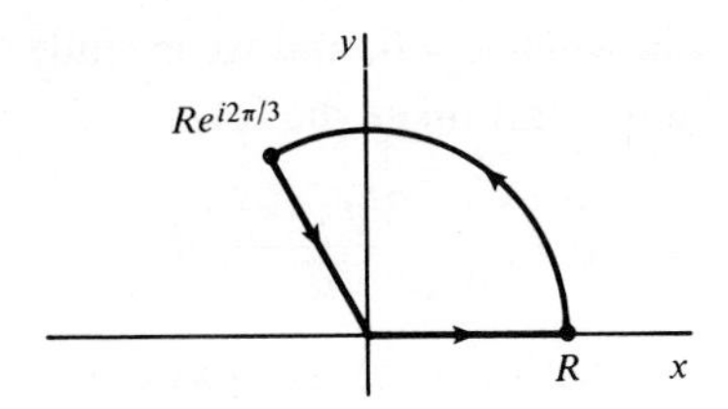

FIGURE 51

17. Use residues and the contour shown in Fig. 51 to establish the integration formula

$$\int_0^\infty \frac{dx}{x^3+1} = \frac{2\pi}{3\sqrt{3}}.$$

74. Definite Integrals of Trigonometric Functions

The method of residues is also useful in evaluating certain definite integrals of the type

$$\int_0^{2\pi} F(\sin\theta, \cos\theta)\, d\theta. \tag{1}$$

The fact that θ varies from 0 to 2π suggests that we consider θ as an argument of a point z on the unit circle C centered at the origin; hence we write $z = e^{i\theta}$, $0 \leqq \theta \leqq 2\pi$. When we make this substitution using the equations

$$\sin\theta = \frac{z - z^{-1}}{2i}, \qquad \cos\theta = \frac{z + z^{-1}}{2}, \qquad d\theta = \frac{dz}{iz}, \tag{2}$$

integral (1) becomes the contour integral

$$\int_C F\left(\frac{z - z^{-1}}{2i}, \frac{z + z^{-1}}{2}\right) \frac{dz}{iz} \tag{3}$$

of a function of z around the circle C described in the positive sense. Integral (1) is, of course, simply a parametric form of integral (3), in accordance with expression (2), Sec. 44. When the integrand of integral (3) is a rational function of z, we can evaluate that integral by means of the residue theorem once the zeros of the polynomial in the denominator have been located, provided none lie on C.

For an illustration, let us show that

$$\int_0^{2\pi} \frac{d\theta}{1 + a\sin\theta} = \frac{2\pi}{\sqrt{1 - a^2}} \qquad (-1 < a < 1). \tag{4}$$

This formula is clearly valid when $a = 0$, and we exclude that case in our proof. With substitutions (2), the integral takes the form

$$\int_C \frac{2/a}{z^2 + (2i/a)z - 1}\, dz \tag{5}$$

where C is the circle $|z| = 1$ with counterclockwise orientation. The denominator of the integrand here has the zeros

$$z_1 = \left(\frac{-1 + \sqrt{1 - a^2}}{a}\right)i, \qquad z_2 = \left(\frac{-1 - \sqrt{1 - a^2}}{a}\right)i.$$

The integrand can therefore be expressed as the function

$$f(z) = \frac{2/a}{(z - z_1)(z - z_2)}.$$

Note that, because $-1 < a < 1$,

$$|z_2| = \frac{1 + \sqrt{1 - a^2}}{|a|} > 1.$$

Also, since $|z_1 z_2| = 1$, it follows that $|z_1| < 1$. Hence there are no singular points on C, and the only one interior to it is the simple pole z_1. The corresponding residue is

$$B_1 = \lim_{z \to z_1} (z - z_1) f(z) = \frac{2/a}{z_1 - z_2} = \frac{1}{i\sqrt{1 - a^2}}.$$

Consequently,

$$\int_C \frac{2/a}{z^2 + (2i/a)z - 1}\, dz = 2\pi i B_1 = \frac{2\pi}{\sqrt{1 - a^2}};$$

and integration formula (4) follows.

75. Integration Around a Branch Point

For our final illustration of the use of the residue theorem in evaluating real integrals, we now consider an example involving branch points and branch cuts.

Let x^{-a}, where $x > 0$ and $0 < a < 1$, denote the principal value of the indicated power of x; that is, x^{-a} is the positive real number $\exp(-a \operatorname{Log} x)$. We shall evaluate the improper real integral

$$\int_0^\infty \frac{x^{-a}}{x + 1}\, dx \qquad (0 < a < 1) \tag{1}$$

which is important in the study of the gamma function.[1] The integral exists when $0 < a < 1$ because the integrand behaves like x^{-a} near $x = 0$ and like x^{-a-1} as x tends to infinity.

To evaluate integral (1), we consider the two line integrals

$$\int_{C_1} f_1(z)\,dz, \qquad \int_{C_2} f_2(z)\,dz$$

where

$$f_1(z) = \frac{z^{-a}}{z+1} \qquad \left(|z| > 0,\ -\frac{\pi}{2} < \arg z < \frac{3\pi}{2}\right),$$

$$f_2(z) = \frac{z^{-a}}{z+1} \qquad \left(|z| > 0,\ \frac{\pi}{2} < \arg z < \frac{5\pi}{2}\right)$$

and C_1 and C_2 are the simple closed contours shown in Fig. 52. In that figure $\rho < 1 < R$, and the angle ϕ is chosen so that $\pi/2 < \phi < \pi$.

Observe that the function f_1 is analytic within and on C_1; hence

$$\int_{C_1} f_1(z)\,dz = 0. \tag{2}$$

Moreover, the function f_2 is analytic within and on C_2 except for the simple pole at the point $z = -1$ which is interior to C_2. Now in the definition of f_2,

$$z^{-a} = \exp\left[-a(\text{Log}\,|z| + i\arg z)\right] \qquad \text{where} \qquad \frac{\pi}{2} < \arg z < \frac{5\pi}{2},$$

and the residue of f_2 at $z = -1$ is

$$\lim_{z\to -1}\,(z+1)f_2(z) = \lim_{z\to -1} z^{-a} = \exp(-a\pi i).$$

[1] See, for example, p. 4 of the book by Lebedev cited in Appendix 1.

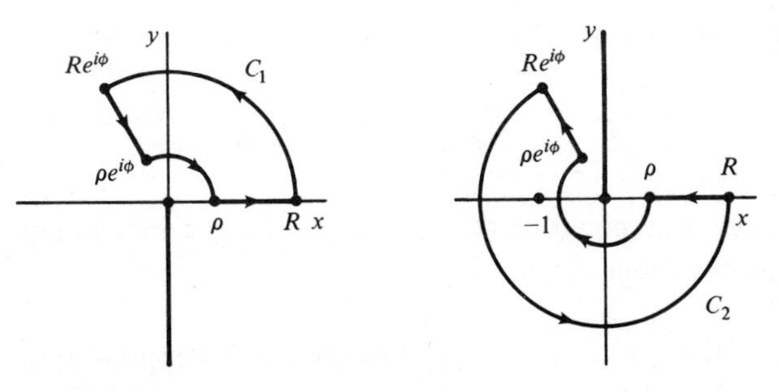

FIGURE 52

Thus

$$\int_{C_2} f_2(z)\,dz = 2\pi i \exp(-a\pi i). \tag{3}$$

Since $f_1(z) = f_2(z)$ on the ray $\arg z = \phi$, it is also true that

$$\begin{aligned}\int_{C_1} f_1(z)\,dz + \int_{C_2} f_2(z)\,dz &= \int_\rho^R f_1(x)\,dx - \int_\rho^R f_2(x)\,dx \\ &+ \int_{\Gamma_1} f_1(z)\,dz + \int_{\Gamma_2} f_2(z)\,dz + \int_{\gamma_1} f_1(z)\,dz + \int_{\gamma_2} f_2(z)\,dz\end{aligned} \tag{4}$$

where Γ_k is the large circular arc and γ_k is the small circular arc of the simple closed contour C_k $(k = 1, 2)$ shown in Fig. 52.

When z is on Γ_k $(k = 1, 2)$,

$$|f_k(z)| = \left|\frac{z^{-a}}{z+1}\right| \leqq \frac{R^{-a}}{R-1};$$

and since the arc Γ_k is a portion of a circle whose circumference is $2\pi R$,

$$\left|\int_{\Gamma_k} f_k(z)\,dz\right| \leqq \frac{R^{-a}}{R-1}\,2\pi R.$$

Hence

$$\lim_{R\to\infty} \int_{\Gamma_k} f_k(z)\,dz = 0 \qquad (k = 1, 2). \tag{5}$$

When z is on γ_k $(k = 1, 2)$,

$$|f_k(z)| = \left|\frac{z^{-a}}{z+1}\right| \leqq \frac{\rho^{-a}}{1-\rho}.$$

Consequently,

$$\left|\int_{\gamma_k} f_k(z)\,dz\right| \leqq \frac{\rho^{-a}}{1-\rho}\,2\pi\rho,$$

and

$$\lim_{\rho\to 0} \int_{\gamma_k} f_k(z)\,dz = 0 \qquad (k = 1, 2). \tag{6}$$

It follows from equation (4) and the results obtained in equations (5) and (6) as well as equations (2) and (3) that

$$\lim_{\substack{R\to\infty \\ \rho\to 0}} \left(\int_\rho^R f_1(x)\,dx - \int_\rho^R f_2(x)\,dx\right) = 2\pi i \exp(-a\pi i).$$

Since

$$\int_\rho^R f_1(x)\,dx - \int_\rho^R f_2(x)\,dx = \int_\rho^R \frac{1}{x+1}\,[e^{-a\,\mathrm{Log}\,x} - e^{-a(\mathrm{Log}\,x + 2\pi i)}]\,dx$$

$$= \int_\rho^R \frac{x^{-a}}{x+1}(1 - e^{-2\pi a i})\,dx,$$

we thus arrive at the result

$$\lim_{\substack{R\to\infty\\ \rho\to 0}} \int_\rho^R \frac{x^{-a}}{x+1}\,dx = \frac{2\pi i \exp(-a\pi i)}{1 - \exp(-2a\pi i)}.$$

That is,

$$\int_0^\infty \frac{x^{-a}}{x+1}\,dx = \frac{\pi}{\sin a\pi} \qquad (0 < a < 1).$$

EXERCISES

Use residues to establish the following integration formulas:

1. (*a*) $\displaystyle\int_0^{2\pi} \frac{d\theta}{5 + 4\sin\theta} = \frac{2\pi}{3}$; (*b*) $\displaystyle\int_{-\pi}^{\pi} \frac{d\theta}{1 + \sin^2\theta} = \pi\sqrt{2}$.

2. $\displaystyle\int_0^{2\pi} \frac{d\theta}{1 + a\cos\theta} = \frac{2\pi}{\sqrt{1 - a^2}} \qquad (-1 < a < 1).$

3. $\displaystyle\int_0^{2\pi} \frac{\cos^2 3\theta\,d\theta}{5 - 4\cos 2\theta} = \frac{3\pi}{8}.$

4. $\displaystyle\int_0^{\pi} \frac{\cos 2\theta\,d\theta}{1 - 2a\cos\theta + a^2} = \frac{\pi a^2}{1 - a^2} \qquad (-1 < a < 1).$

5. $\displaystyle\int_0^{\pi} \frac{d\theta}{(a + \cos\theta)^2} = \frac{\pi a}{(a^2 - 1)^{3/2}} \qquad (a > 1).$

6. $\displaystyle\int_0^{\pi} \sin^{2n}\theta\,d\theta = \frac{(2n)!}{2^{2n}(n!)^2}\,\pi \qquad (n = 1, 2, \ldots).$

7. Derive the integration formula

$$\int_0^\infty \exp(-x^2)\cos(2bx)\,dx = \frac{\sqrt{\pi}}{2}\exp(-b^2) \qquad (b > 0)$$

by integrating the function $\exp(-z^2)$ around the rectangular path shown in Fig. 53 and then letting a tend to infinity. Use the fact that

$$\int_0^\infty \exp(-x^2)\,dx = \frac{\sqrt{\pi}}{2}.$$

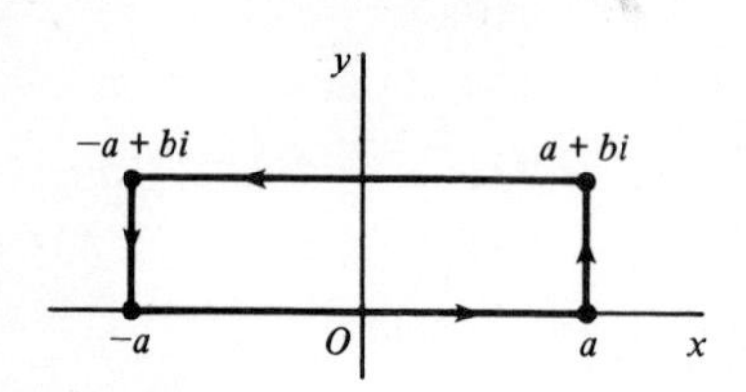

FIGURE 53

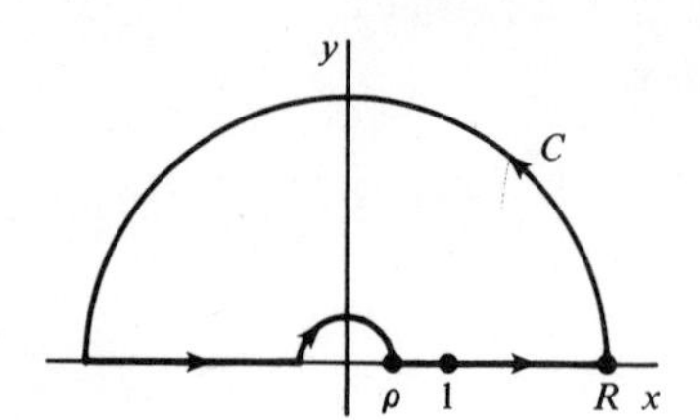

FIGURE 54

8. Derive the formulas

$$(a)\ \int_0^\infty \frac{\text{Log}\, x}{x^2+1}\,dx = 0; \qquad (b)\ \int_0^\infty \frac{\text{Log}\, x}{(x^2+1)^2}\,dx = -\frac{\pi}{4}.$$

Suggestion: The simple closed contour shown in Fig. 54 can be used here, together with the results of Exercises 1 and 4, Sec. 73.

9. The *beta function* is this function of two real variables:

$$B(p,q) = \int_0^1 t^{p-1}(1-t)^{q-1}\,dt \qquad (p>0, q>0).$$

Make the substitution $t = 1/(x+1)$ and use the result obtained in Sec. 75 to show that

$$B(p, 1-p) = \frac{\pi}{\sin p\pi} \qquad (0<p<1).$$

10. With the aid of the contours shown in Fig. 52, derive the following integration formulas:

$$(a)\ \int_0^\infty \frac{x^{-1/2}}{x^2+1}\,dx = \frac{\pi}{\sqrt{2}} \qquad \text{where} \qquad x^{-1/2} = \exp\left(-\tfrac{1}{2}\,\text{Log}\, x\right);$$

$$(b)\ \int_0^\infty \frac{x^a}{(x^2+1)^2}\,dx = \frac{\pi}{4}\,\frac{1-a}{\cos(a\pi/2)} \qquad \text{where} \qquad -1<a<3,\ x^a = \exp(a\,\text{Log}\, x).$$

11. Derive *Jordan's lemma*:

$$\int_0^{\pi/2} e^{-R\sin\theta}\,d\theta < \frac{\pi}{2R} \qquad (R>0).$$

Suggestion: Note first that $\sin\theta \geq 2\theta/\pi$ when $0 \leq \theta \leq \pi/2$ by considering the graph of the sine function. Then write $\exp(-R\sin\theta) \leq \exp(-2R\theta/\pi)$.

12. The Fresnel integrals

$$\int_0^\infty \cos(x^2)\,dx = \int_0^\infty \sin(x^2)\,dx = \frac{\sqrt{\pi}}{2\sqrt{2}}$$

are important in diffraction theory. Given that

$$\int_0^\infty \exp(-x^2)\,dx = \frac{\sqrt{\pi}}{2},$$

evaluate these integrals by integrating $\exp(iz^2)$ around the boundary of the sector $0 \leqq r \leqq R$, $0 \leqq \theta \leqq \pi/4$ and then letting R tend to ∞. Use Jordan's lemma (Exercise 11) to show that the integral along the circular arc $r = R$, $0 \leqq \theta \leqq \pi/4$ approaches zero as R tends to infinity.

13. Let the point $z = x_0$ on the x axis be a simple pole of a function f and let B_0 be the residue of f at that pole. Let γ denote the upper half of the circle $|z - x_0| = \rho$ (Fig. 55), described in the clockwise sense, where ρ is sufficiently small so that f is analytic within and on that circle except for the pole at x_0. Note that

$$f(z) = \frac{B_0}{z - x_0} + g(z) \qquad (0 < |z - x_0| < \rho)$$

where g is analytic, and hence continuous, throughout the neighborhood $|z - x_0| < \rho$ and show that

$$\lim_{\rho \to 0} \int_\gamma f(z)\,dz = -B_0 \pi i.$$

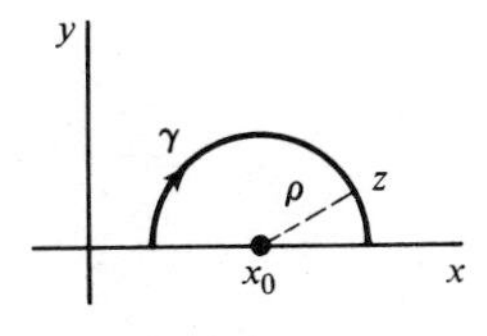

FIGURE 55

14. The formula

$$\int_0^\infty \frac{\sin x}{x}\,dx = \frac{\pi}{2}$$

is important in the theory of Fourier series.[1] Derive this formula by integrating e^{iz}/z around the simple closed contour C shown in Fig. 54 and then letting R and ρ tend to infinity and zero, respectively. Use Jordan's lemma (Exercise 11) to show that the integral along the semicircle $z = Re^{i\theta}$, $0 \leqq \theta \leqq \pi$ approaches zero as R tends to infinity. Also, use Exercise 13 to show that the integral along the smaller semicircle in Fig. 54 approaches $-\pi i$ as ρ tends to zero.

[1] See R. V. Churchill, "Fourier Series and Boundary Value Problems," 2d ed., pp. 85–86, 1963.

15. Derive the integration formula

$$\int_0^\infty \frac{\sin^2 x}{x^2}\,dx = \frac{\pi}{2}.$$

Suggestion: Note that $2\sin^2 x = \operatorname{Re}(1 - e^{i2x})$ and integrate the function $(1 - e^{i2z})/z^2$ around the contour shown in Fig. 54.

16. The integral of the function

$$f(x) = \frac{1}{x(x^2 - 4x + 5)}$$

over an interval that includes the origin does not exist. Show that the *principal value* of the integral of that function along the entire x axis,

$$\text{P.V.} \int_{-\infty}^{\infty} f(x)\,dx = \lim_{\rho\to 0}\left[\int_{-\infty}^{-\rho} f(x)\,dx + \int_{\rho}^{\infty} f(x)\,dx\right] \qquad (\rho > 0),$$

does exist by finding that value with the aid of the contour in Fig. 54 and the result found in Exercise 13.

Ans. $2\pi/5$.

8

CONFORMAL MAPPING

In this chapter we introduce the concept of a conformal mapping and then obtain results concerning the behavior of functions which are harmonic in the interior of a region and differentiable on its boundary under a change of variables determined by such a mapping. Applications of these results will follow in the next chapter.

76. Basic Properties

Let us examine the changes in direction of curves through a point z_0 under a transformation $w = f(z)$ when the function f is analytic at that point and $f'(z_0) \neq 0$.

Suppose C is a smooth arc passing through z_0. If $z(t) = x(t) + iy(t)$, $a \leqq t \leqq b$, is a parametric representation of C, then $w(t) = f[z(t)]$, $a \leqq t \leqq b$, is a parametric representation of the image Γ of C under the transformation $w = f(z)$. According to the chain rule given in Exercise 7, Sec. 43,

$$w'(t) = f'[z(t)]z'(t). \tag{1}$$

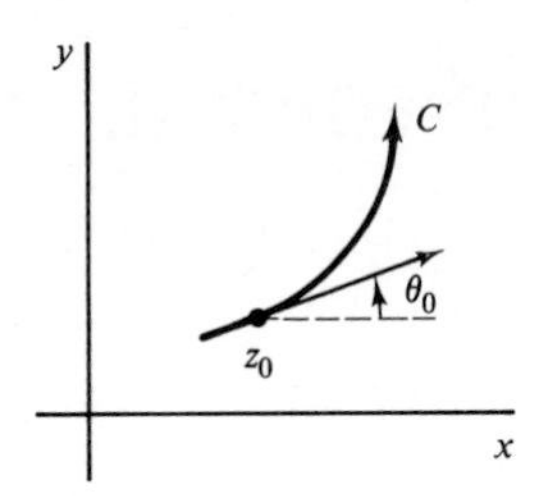

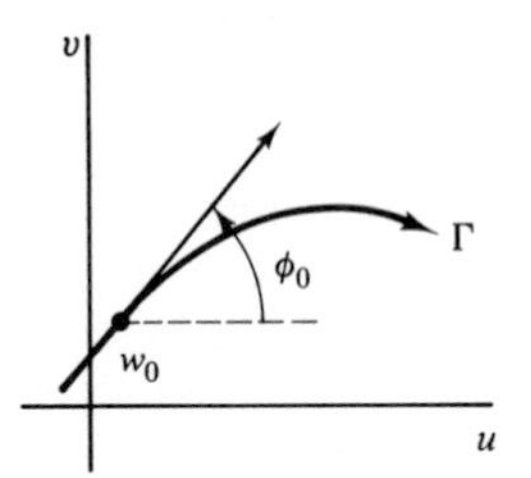

FIGURE 56. $\phi_0 = \psi_0 + \theta_0$.

Thus, when the arc C lies in a domain which contains the point z_0 and throughout which f is analytic and $f'(z) \neq 0$, the image curve Γ is also a smooth arc. Moreover, from equation (1) we obtain the relation

$$\arg w'(t) = \arg f'[z(t)] + \arg z'(t). \tag{2}$$

The angle of inclination of a directed line tangent to C at the point $z_0 = z(t_0)$, $a < t_0 < b$, is any value θ_0 of $\arg z'(t_0)$ (Sec. 43). If ψ_0 is a value of $\arg f'(z_0)$, then, according to equation (2), the quantity

$$\phi_0 = \psi_0 + \theta_0$$

is a value of $\arg w'(t_0)$ and is therefore the angle of inclination of a directed line tangent to Γ at the point $w_0 = f(z_0)$ (Fig. 56). Thus, *when a function f is analytic at a point z_0 and $f'(z_0) \neq 0$, a directed line tangent to a smooth arc C at z_0 is rotated through the angle*

$$\psi_0 = \arg f'(z_0) \tag{3}$$

by the transformation $w = f(z)$.

Let C_1 and C_2 be two smooth arcs passing through z_0 and let θ_1 and θ_2 be angles of inclination of directed lines tangent to C_1 and C_2, respectively, at z_0. Then, according to the preceding paragraph, the quantities

$$\phi_1 = \psi_0 + \theta_1 \qquad \text{and} \qquad \phi_2 = \psi_0 + \theta_2$$

are angles of inclination of directed lines tangent to the image curves Γ_1 and Γ_2, respectively, at the point $w_0 = f(z_0)$. Thus, $\phi_2 - \phi_1 = \theta_2 - \theta_1$; that is, the angle $\phi_2 - \phi_1$ from Γ_1 to Γ_2 is the same in *magnitude* and *sense* as the angle $\theta_2 - \theta_1$ from C_1 to C_2. Those angles are denoted by α in Fig. 57.

A mapping that preserves the magnitude and sense of the angle between any two smooth arcs passing through a specific point is said to be *conformal* at that point. We state our result as follows.

Theorem. *At each point z where a function f is analytic and $f'(z) \neq 0$ the mapping $w = f(z)$ is conformal.*

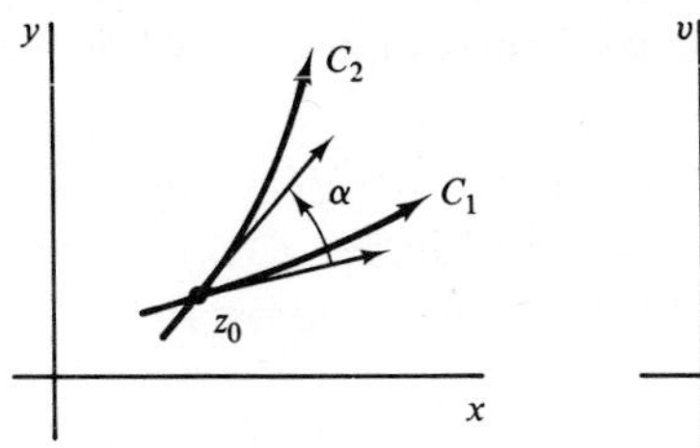

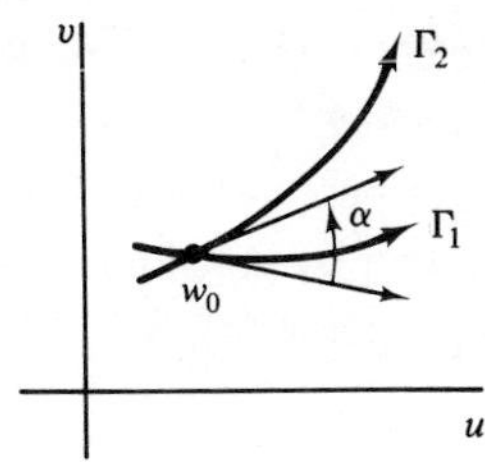

FIGURE 57

Henceforth the term *conformal mapping*, or conformal transformation, will denote a mapping by means of an analytic function defined on a domain where the derivative is never zero.

A mapping that preserves the magnitude of each angle but not necessarily the sense is called an *isogonal* mapping. The transformation $w = \bar{z}$, a reflection in the real axis, is isogonal but not conformal. If this is followed by a conformal transformation, the resulting transformation $w = f(\bar{z})$ is also isogonal but not conformal.

Suppose f is not a constant function and is analytic at a point z_0. If $f'(z_0) = 0$, z_0 is called a *critical point*. The point $z = 0$ is, for example, a critical point of the transformation

$$w = z^2.$$

The ray $\theta = c$ from the point $z = 0$ is mapped onto the ray $\phi = 2c$ from the point $w = 0$. Thus, the angle between any two rays drawn from the critical point $z = 0$ is doubled by the transformation.

More generally, it can be shown that if z_0 is a critical point of the transformation $w = f(z)$, then there is a positive integer m such that the angle between any two smooth arcs passing through z_0 is multiplied by m under that transformation. The integer m is the smallest positive integer such that $f^{(m)}(z_0) \neq 0$. Details are left to the exercises.

77. Further Properties and Examples

If the images of two curves under a conformal mapping are orthogonal, then those curves must be orthogonal. In particular, if the transformation

$$u + iv = f(x + iy)$$

is conformal at a point (x_0,y_0) and if $u_0 + iv_0 = f(x_0 + iy_0)$, the level curves $u(x,y) = u_0$ and $v(x,y) = v_0$ are mapped into the orthogonal lines $u = u_0$ and

$v = v_0$, respectively. Hence those level curves must be orthogonal. (Compare Exercise 13, Sec. 20.)

Another property of a transformation $w = f(z)$ which is conformal at a point z_0 is obtained by considering the modulus of $f'(z_0)$. By the definition of derivative and property (8), Sec. 12, of limits, we know that

$$|f'(z_0)| = \lim_{z \to z_0} \frac{|f(z) - f(z_0)|}{|z - z_0|}.$$

Evidently then, the length $|z - z_0|$ of a small line segment with one end at z_0 is increased or decreased by approximately the factor $|f'(z_0)|$ under the transformation $w = f(z)$ since $|f(z) - f(z_0)|$ is the corresponding length in the w plane. Moreover, the image of a small region in a neighborhood of the point z_0 conforms to the original region in the sense that it has approximately the same shape. Both the *angle of rotation* ψ_0, given in equation (3), Sec. 76, and the *scale factor* $|f'(z_0)|$ of a conformal transformation change in general from point to point, and a large region may be transformed into a region that bears no resemblance to the original one.

A transformation $w = f(z)$ which is conformal at a point z_0 has a *local inverse* there. That is, if $w_0 = f(z_0)$, there exist rectangular domains R and S centered at z_0 and w_0, respectively, such that to each point w in S there corresponds a unique point z in R with the property that $w = f(z)$. The inverse transformation, denoted by $z = g(w)$, is analytic at w_0 and its derivative there is given by the formula

$$g'(w_0) = \frac{1}{f'(z_0)}. \tag{1}$$

The existence of such an inverse function is a direct consequence of a result in the calculus of real variables.[1] We state that result here and leave the details of its application for the exercises. Let the functions

$$u = u(x,y), \qquad v = v(x,y) \tag{2}$$

be continuous in a neighborhood of a point (x_0,y_0) in the xy plane, and suppose that they possess continuous first partial derivatives throughout that neighborhood. These functions represent a transformation into the uv plane, and we assume further that the Jacobian

$$J(x,y) = \begin{vmatrix} u_x(x,y) & u_y(x,y) \\ v_x(x,y) & v_y(x,y) \end{vmatrix} = u_x(x,y)v_y(x,y) - v_x(x,y)u_y(x,y)$$

[1] See, for instance, A. E. Taylor and W. R. Mann, "Advanced Calculus," 2d ed., pp. 251–252, 1972.

of the transformation is not zero at (x_0,y_0). It follows that if $u_0 = u(x_0,y_0)$ and $v_0 = v(x_0,y_0)$, there are rectangular domains R and S centered at (x_0,y_0) and (u_0,v_0), respectively, such that to each point (u,v) in S there corresponds a unique point (x,y) in R with the property that $u = u(x,y)$ and $v = v(x,y)$. This defines inverse functions

$$x = x(u,v), \qquad y = y(u,v) \tag{3}$$

on S which are continuous with continuous first partial derivatives. Those partial derivatives, moreover, satisfy the conditions

$$\begin{aligned} x_u(u,v) &= \frac{1}{J(x,y)} v_y(x,y), & x_v(u,v) &= -\frac{1}{J(x,y)} u_y(x,y), \\ y_u(u,v) &= -\frac{1}{J(x,y)} v_x(x,y), & y_v(u,v) &= \frac{1}{J(x,y)} u_x(x,y), \end{aligned} \tag{4}$$

where the points (x,y) and (u,v) are related by equations (2) and (3).

Note that even though a conformal transformation is one to one in a neighborhood of each of the points in its domain of definition, it need not be one to one in that domain of definition. An example is the function $w = z^2$ which is conformal in the domain $1 < |z| < 2$ but is not one to one there.

Each of the elementary functions studied in Chap. 4 is analytic in some domain. Hence each of the transformations defined by those functions is conformal at any point in the domain of analyticity which is not a critical point (Sec. 76). As an illustration, the transformation

$$w = z^2 = x^2 - y^2 + i2xy$$

is conformal at the point $z = 1 + i$, where the lines $y = x$ and $x = 1$ intersect. The line $y = x$ is transformed into the half line $u = 0$, $v \geqq 0$ and the line $x = 1$ is transformed into the curve whose parametric equations are

$$u = 1 - y^2, \qquad v = 2y;$$

this is the parabola $v^2 = -4(u - 1)$ (Fig. 58). If the direction of increasing y is taken as the positive sense on the two lines in the z plane, the angle from the first to the second is $\pi/4$. When $y > 0$ and y increases along the line $y = x$, v increases along the line $u = 0$, since $v = 2y^2$, and so the positive sense of the first image is upward when $y > 0$. This is also true for the parabola, as we see from the second parametric equation $v = 2y$. It is readily verified that the angle from the first image to the second at the point $w = 2i$, which is the image of the point $z = 1 + i$, is $\pi/4$ as required by the conformality of the mapping.

Note that the angle of rotation of the transformation $w = z^2$ at the point $z = 1 + i$ is a value of arg $[2(1 + i)]$, or $\pi/4$. The scale factor at that point is $2\sqrt{2}$.

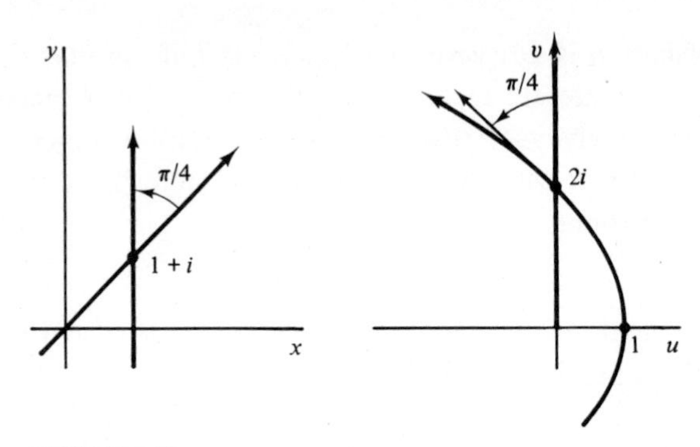

FIGURE 58

EXERCISES

1. Determine the angle of rotation at the point $z = 2 + i$ when the transformation is $w = z^2$. Illustrate the angle of rotation for some particular curve. Show that the scale factor of the transformation at that point is $2\sqrt{5}$.
2. What angle of rotation is produced by the transformation $w = 1/z$
   (*a*) at the point $z = 1$; (*b*) at the point $z = i$?

   Ans. (*a*) π; (*b*) 0.
3. Show that under the transformation $w = 1/z$ the images of the lines $y = x - 1$ and $y = 0$ are the circle $u^2 + v^2 - u - v = 0$ and the line $v = 0$, respectively. Sketch these curves, determine corresponding directions along them, and verify the conformality of the mapping at the point $z = 1$.
4. Show that the angle of rotation at the nonzero point $z_0 = r_0 \exp(i\theta_0)$ under the transformation $w = z^n$, where n is a positive integer, is $(n - 1)\theta_0$. Determine the scale factor of the transformation at that point.

   Ans. $nr_0^{\,n-1}$.
5. Show that the transformation $w = \exp z$ is everywhere conformal. Note that the mapping of directed line segments shown in Figs. 7 and 8 of Appendix 2 agrees with this.
6. Show that the transformation $w = \sin z$ is conformal at all points except $z = (2n - 1)\pi/2$ $(n = 0, \pm 1, \pm 2, \ldots)$. Note that this is in agreement with the mapping of directed line segments shown in Figs. 9, 10, and 11 of Appendix 2.
7. Let a transformation $w = f(z)$ be conformal at a point z_0. Write $f(z) = u(x,y) + iv(x,y)$ and use the results from the calculus of real variables stated in Sec. 77 to show that at z_0 the function f has a local inverse g which is analytic at $w_0 = f(z_0)$.

   Suggestion: First express the transformation $w = f(z)$ in terms of equations (2), Sec. 77. In order to show that the Jacobian is not zero at (x_0, y_0), use the Cauchy-Riemann equations to show that its value there is $|f'(z_0)|^2$. Then define g in terms of the functions (3), Sec. 77, and use conditions (4) there to show that the first partial derivatives of those functions satisfy the Cauchy-Riemann equations at (u_0,v_0).

8. Show that if $z = g(w)$ is the local inverse of the conformal transformation $w = f(z)$ at the point z_0, then

$$g'(w_0) = \frac{1}{f'(z_0)}$$

where $w_0 = f(z_0)$. Note that the existence of $g'(w_0)$ is verified in Exercise 7 and that the above formula reveals that g is actually conformal at w_0.

Suggestion: Write $g[f(z)] = z$ and then apply the chain rule for differentiating composite functions.

9. Determine the local inverse of the transformation $w = \exp z$ at the point (*a*) $z_0 = 0$; (*b*) $z_0 = 2\pi i$. Verify formula (1), Sec. 77, for the derivative of the local inverse, obtained in Exercise 8.

Ans. (*a*) $\operatorname{Log} w$; (*b*) $\operatorname{Log} w + 2\pi i$.

10. Let z_0 be a critical point of the function f and let m be the smallest positive integer such that $f^{(m)}(z_0) \neq 0$. Let Γ be the image of the smooth arc C under the transformation $w = f(z)$, as shown in Fig. 56. Show that the angles of inclination now satisfy the relation

$$\phi_0 = m\theta_0 + \arg [f^{(m)}(z_0)].$$

Then show that if α denotes the angle between two smooth arcs C_1 and C_2, as shown in Fig. 57, it follows that the corresponding angle between the images Γ_1 and Γ_2 is $\beta = m\alpha$.

Suggestion: From the Taylor series for f at z_0 obtain the relation

$$\arg [f(z) - f(z_0)] = m \arg (z - z_0) + \arg \left[\frac{f^{(m)}(z_0)}{m!} + \frac{f^{(m+1)}(z_0)}{(m+1)!}(z - z_0) + \cdots\right].$$

Then use the fact that the angle of inclination of the arc C at z_0 and that of its image Γ at $f(z_0)$ are limits of $\arg (z - z_0)$ and $\arg [f(z) - f(z_0)]$, respectively, as z approaches z_0 along the curve C.

78. Harmonic Conjugates

We saw in Sec. 20 that if a function

$$f(z) = u(x,y) + iv(x,y)$$

is analytic in a domain D, then the real-valued function $v(x,y)$ is a harmonic conjugate of the real-valued function $u(x,y)$. That is, the functions $u(x,y)$ and $v(x,y)$ are harmonic in D and their first partial derivatives satisfy the Cauchy-Riemann equations

$$u_x(x,y) = v_y(x,y), \qquad u_y(x,y) = -v_x(x,y) \tag{1}$$

throughout D.

We now show that if $u(x,y)$ is any given harmonic function defined on a *simply connected domain* D, then a harmonic conjugate always exists. To show this, we first consider the line integral

$$\int_{(x_0,y_0)}^{(x,y)} -u_t(r,t)\,dr + u_r(r,t)\,dt \tag{2}$$

where the path of integration is any contour lying in D and joining the fixed point (x_0,y_0) to a point (x,y) which is allowed to vary. We use r and t as our variables of integration to distinguish them from the variables appearing in the upper limit. Integral (2) is suggested by the fact that if v is a harmonic conjugate of u, then

$$dv = v_x\,dx + v_y\,dy = -u_y\,dx + u_x\,dy.$$

Since u is harmonic throughout D, it satisfies Laplace's equation,

$$u_{xx}(x,y) + u_{yy}(x,y) = 0,$$

from which it follows that the partial derivative of $-u_y(x,y)$ with respect to y is equal to the partial derivative of $u_x(x,y)$ with respect to x. That is, the integrand in integral (2) is an exact differential.[1] Thus, integral (2) is independent of path and it defines a real-valued function

$$v(x,y) = \int_{(x_0,y_0)}^{(x,y)} -u_t(r,t)\,dr + u_r(r,t)\,dt \tag{3}$$

of its upper limit.

It remains to show that $v(x,y)$ is a harmonic conjugate of $u(x,y)$. According to formulas in the calculus of real variables for the derivatives of line integrals with variable upper limits of integration, we have

$$v_x(x,y) = -u_y(x,y), \qquad v_y(x,y) = u_x(x,y) \tag{4}$$

which are the Cauchy-Riemann equations (1). Since the first partial derivatives of u are continuous, it is evident from equations (4) that those derivatives of v are also continuous. It follows (Sec. 17) that $u(x,y) + iv(x,y)$ is an analytic function in D and, therefore, v is a harmonic conjugate of u.

The function v defined by formula (3) is, of course, not the only harmonic conjugate of u. For, the function $v(x,y) + c$ where c is an arbitrary real constant is also a harmonic conjugate of u.

[1] For properties of exact differentials used here, see, for example, A. E. Taylor and W. R. Mann, "Advanced Calculus," 2d ed., pp. 495–504, 1972.

To illustrate the above, consider the function $u(x,y) = xy$ which is harmonic throughout the entire xy plane. According to equation (3), the function

$$v(x,y) = \int_{(0,0)}^{(x,y)} -r\,dr + t\,dt$$

is a harmonic conjugate of $u(x,y)$. The integral here is readily evaluated by inspection; it can also be evaluated by first integrating along the horizontal path from the point (0,0) to the point $(x,0)$ and then along the vertical path from $(x,0)$ to the point (x,y). The result is

$$v(x,y) = -\tfrac{1}{2}x^2 + \tfrac{1}{2}y^2,$$

and the corresponding analytic function is then

$$f(z) = xy - \frac{i}{2}(x^2 - y^2) = -\frac{i}{2}z^2.$$

79. Transformations of Harmonic Functions

The problem of finding a function that is harmonic in a specified domain and satisfies prescribed conditions on the boundary of that domain is prominent in applied mathematics. If the values of the function are prescribed along the boundary, the problem is known as a boundary value problem of the *first kind*, or a *Dirichlet problem*. If values of the normal derivative of the function are prescribed on the boundary, the boundary value problem is one of the *second kind*, or a *Neumann problem*. Modifications and combinations of those types of boundary conditions also arise.

Every analytic function furnishes a pair of harmonic functions. Since the function $-ie^{iz}$, for example, is entire, its components

$$H(x,y) = e^{-y}\sin x, \qquad G(x,y) = -e^{-y}\cos x \tag{1}$$

are everywhere harmonic. The function H satisfies the conditions

$$H_{xx}(x,y) + H_{yy}(x,y) = 0, \tag{2}$$

$$H(0,y) = 0, \qquad H(\pi,y) = 0, \tag{3}$$

$$H(x,0) = \sin x, \qquad \lim_{y\to\infty} H(x,y) = 0 \tag{4}$$

which make up a Dirichlet problem for the strip $0 < x < \pi$, $y > 0$. Of course, the same function satisfies other boundary conditions for this and other domains; for instance, its normal derivative $H_x(x,y)$ is zero on the line $x = \pi/2$.

Sometimes the solution of a given problem can be discovered by identifying it as the real or imaginary part of an analytic function. But the success of this procedure depends on the simplicity of the problem and on our familiarity with the real and imaginary parts of a variety of analytic functions. An important additional aid in solving such problems will now be given.

Theorem. *Suppose that the analytic function*

$$f(z) = u(x,y) + iv(x,y)$$

maps a domain D_z *in the z plane onto a domain* D_w *in the w plane. If* $h(u,v)$ *is a harmonic function defined on* D_w*, then the function*

$$H(x,y) = h[u(x,y), v(x,y)]$$

is harmonic in D_z.

Our proof of the theorem is for the case when D_w is a simply connected domain; this is the situation most frequently encountered in the applications. Recall that, according to the preceding section, there is corresponding to a given harmonic function $h(u,v)$ a harmonic conjugate $g(u,v)$. Hence the function $\Phi(w) = h(u,v) + ig(u,v)$ is analytic in D_w. Since the function $f(z)$ is analytic in D_z, the composite function $\Phi[f(z)]$ is also analytic in D_z. Therefore the real part $h[u(x,y),v(x,y)]$ of this composition is harmonic in D_z.

The proof of the theorem for the general case when D_w is not necessarily simply connected can be accomplished by means of the chain rule for partial derivatives. The details are left for the exercises.

As an illustration of the theorem, the function $h(u,v) = e^{-v} \sin u$ is harmonic in the domain D_w consisting of all points in the upper half plane $v > 0$. Under the transformation

$$w = z^2,$$

we have $u = x^2 - y^2$ and $v = 2xy$; moreover, the domain D_z in the z plane consisting of the points in the first quadrant $x > 0$, $y > 0$ is mapped onto the domain D_w. Hence the function

$$H(x,y) = e^{-2xy} \sin (x^2 - y^2)$$

is harmonic in D_z.

For another illustration, consider the function $h(u,v) = v$, which is harmonic in the strip $-\pi/2 < v < \pi/2$, and observe that the transformation $w = \text{Log}\, z$ maps the right half plane $x > 0$ onto that strip. Writing

$$\text{Log}\, z = \text{Log}\sqrt{x^2 + y^2} + i \arctan \frac{y}{x},$$

where $-\pi/2 < \arctan t < \pi/2$, we find that the function

$$H(x,y) = \arctan \frac{y}{x}$$

is harmonic in the half plane $x > 0$.

80. Transformations of Boundary Conditions

The conditions that a function or its normal derivative have prescribed values along the boundary of a domain in which it is harmonic are the most common although not the only important types of boundary conditions. In this section we show that certain of these conditions remain unaltered under the change of variables associated with a conformal transformation. These results are applied in the next chapter to obtain solutions to boundary value problems. The technique used there is to transform a given boundary problem in the *xy* plane into a simpler one in the *uv* plane and then to use the theorems of this and the preceding section to write the solution of the original problem in terms of the solution obtained for the latter.

Let the analytic function

$$f(z) = u(x,y) + iv(x,y) \tag{1}$$

map an arc C in the z plane onto an arc Γ in the w plane, and let a function $h(u,v)$ be defined on Γ. Now write the function

$$H(x,y) = h[u(x,y), v(x,y)]$$

and let c denote any real number. It follows that if $h(u,v) = c$ on Γ, then $H(x,y) = c$ on C.

Suppose that, in addition, $f(z)$ is conformal on C and $h(u,v)$ is differentiable on Γ. If the normal derivative dh/dn of $h(u,v)$ along Γ is zero, then the normal derivative of $H(x,y)$ along C is also zero. To see this, we recall from the calculus of real variables that the gradient of $h(u,v)$ is a vector in whose direction the directional derviative of h is maximum.[1] It can be written in terms of the first partial derivatives of h with respect to u and v as

$$\text{grad } h(u,v) = h_u(u,v) + ih_v(u,v).$$

The magnitude of grad $h(u,v)$ is the value of the maximum directional derivative, and the component of grad $h(u,v)$ in any direction is the value of the directional derivative of h in that direction. The gradient vector at a point is, moreover, orthogonal to a level curve $h(u,v) = c$ through that point.

[1] For properties of gradients used here, see for example, A. E. Taylor and W. R. Mann, "Advanced Calculus," 2d ed., pp. 295–298, 1972.

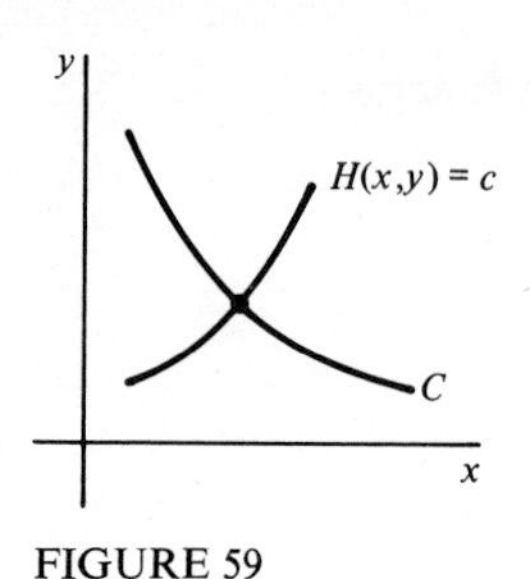

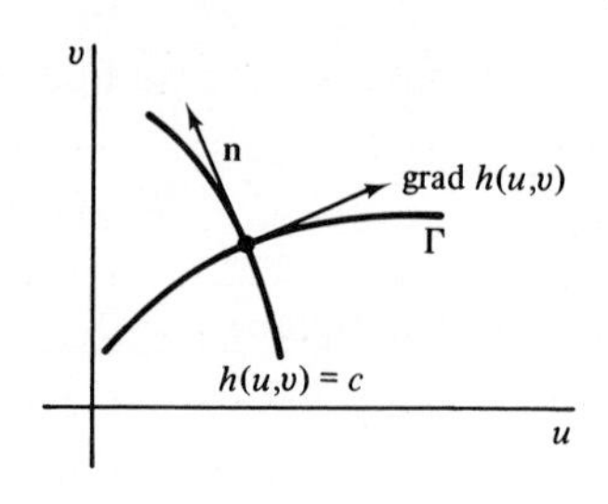

FIGURE 59

Consider now any point on Γ. Since the quantity dh/dn at that point is the component of the gradient vector there in a direction normal to Γ and since $dh/dn = 0$, the gradient vector must be tangent to Γ (Fig. 59). But the gradient vector is orthogonal to the level curve $h(u,v) = c$ through the point, and so Γ is orthogonal to that level curve. Since the transformation $w = f(z)$ is conformal at the point where the curves C and $H(x,y) = c$ intersect, it follows that these curves must be orthogonal. Consequently, the component of the gradient of $H(x,y)$ in the direction of a vector normal to C is zero. That is, the normal derivative of $H(x,y)$ at each point of C is zero.

In the above discussion we have assumed that grad $h(u,v) \neq 0$. If grad $h(u,v) = 0$, it follows from Exercise 9(*a*) of this section that grad $H(x,y) = 0$; hence dh/dn and the corresponding normal derivative of H are both zero.

We now summarize these results in a form which will be useful in the applications.

Theorem. *Suppose that the analytic function*

$$f(z) = u(x,y) + iv(x,y)$$

maps an arc C in the z plane onto an arc Γ in the w plane. Let $f(z)$ be conformal on C and let a function $h(u,v)$ be differentiable on Γ. If the function $h(u,v)$ satisfies either of the conditions

$$h = c \qquad or \qquad \frac{dh}{dn} = 0$$

along Γ, then the function

$$H(x,y) = h[u(x,y), v(x,y)]$$

satisfies the corresponding condition along C.

A boundary condition that is not of one of the two types mentioned in the theorem may be transformed into a condition that is substantially different from the original one. New boundary conditions for the transformed problem may be obtained for a particular transformation in any case. It is interesting to note that under a conformal transformation the ratio of a directional derivative of H along an arc C in the z plane to the directional derivative of h along the image curve Γ at the corresponding point in the w plane is $|f'(z)|$; usually this ratio is not constant along a given arc. (See Exercises 5 and 9 of this section.)

EXERCISES

1. Use formula (3), Sec. 78, to find a harmonic conjugate of the harmonic function $u(x,y) = x^3 - 3xy^2$. Write the resulting analytic function in terms of the complex variable z.

2. Let $u(x,y)$ be harmonic in a simply connected domain D. Show that the partial derivatives of u of all orders are continuous throughout D.

3. Under the transformation $w = e^z$ the image of the line segment $x = 0$, $0 \leqq y \leqq \pi$ is the semicircle $u^2 + v^2 = 1$, $v \geqq 0$. Also, the function

$$h(u,v) = 2 - u + \frac{u}{u^2 + v^2}$$

is harmonic, and therefore differentiable, everywhere in the w plane except for the origin; and it assumes the value 2 on the semicircle. Write $H(x,y) = h[u(x,y), v(x,y)]$ according to the indicated change of variables and show directly that $H = 2$ along the line segment, thus illustrating the theorem in Sec. 80.

4. Under the transformation $w = z^2$ the u axis is the image of the positive axes in the z plane together with the origin there. Consider now the harmonic function

$$h(u,v) = e^{-u} \cos v$$

and note that its normal derivative along the u axis is zero; that is, $h_v(u,0) = 0$. Show directly that the normal derivative of the function $H(x,y)$, defined in the theorem of Sec. 80, is zero along the positive axes in the z plane. Observe that the transformation $w = z^2$ is not conformal at the origin.

5. Replace the function $h(u,v)$ in Exercise 4 by the harmonic function

$$h(u,v) = 2v + e^{-u} \cos v.$$

Then show that $h_v(u,0) = 2$ but that $H_y(x,0) = 4x$ along the positive x axis and $H_x(0,y) = 4y$ along the positive y axis. Thus a condition of the type $dh/dn = c$ is not necessarily transformed into a condition of the type $dH/dn = c$.

6. Show that if a function $H(x,y)$ is a solution to a Neumann problem (Sec. 79), then $H(x,y) + c$, where c is any real constant, is also a solution to that problem.

7. Let the analytic function $f(z) = u(x,y) + iv(x,y)$ map a domain D_z in the z plane onto a domain D_w in the w plane. Show that if $h(u,v)$ is a harmonic function defined on D_w and if $H(x,y) = h[u(x,y),v(x,y)]$, then

$$H_{xx}(x,y) + H_{yy}(x,y) = [h_{uu}(u,v) + h_{vv}(u,v)]|f'(z)|^2.$$

Conclude that the function $H(x,y)$ is harmonic in D_z.

8. Let p be a function of the variables u and v which satisfies *Poisson's equation*

$$p_{uu}(u,v) + p_{vv}(u,v) = \Phi(u,v)$$

in a domain D_w of the w plane, where Φ is a prescribed function. Show that if $f(z) = u(x,y) + iv(x,y)$ is an analytic function which maps a domain D_z onto D_w, the function

$$P(x,y) = p[u(x,y),v(x,y)]$$

satisfies the Poisson equation

$$P_{xx}(x,y) + P_{yy}(x,y) = \Phi[u(x,y),v(x,y)]|f'(z)|^2.$$

(See Exercise 7.)

9. Let the function $f(z) = u(x,y) + iv(x,y)$ be an analytic function which defines a conformal mapping of a domain D_z in the z plane onto a domain D_w in the w plane. Let $h(u,v)$ be a harmonic function defined on D_w, and write $H(x,y) = h[u(x,y),v(x,y)]$. (*a*) Show that under the indicated change of variables $|\text{grad } H(x,y)| = |\text{grad } h(u,v)|\,|f'(z)|$. (*b*) Why is the angle at a point in D_z between an arc C and the vector grad $H(x,y)$ equal to the angle at the corresponding point in D_w between the image Γ of C and grad $h(u,v)$? (*c*) Using the results of parts (*a*) and (*b*), show that if σ is distance along C and τ is distance along Γ, the directional derivative is transformed as follows:

$$\frac{dH}{d\sigma} = \frac{dh}{d\tau}|f'(z)|.$$

9

APPLICATIONS OF CONFORMAL MAPPING

We shall now use conformal mapping to solve a number of physical problems involving Laplace's equation in two independent variables. Problems in heat conduction, electrostatic potential, and fluid flow will be treated. Since these problems are intended to illustrate methods, they will be kept on a fairly elementary level.

81. Steady Temperatures

In the theory of heat conduction the *flux* across a surface within a solid body at a point on that surface is the quantity of heat flowing in a specified direction normal to the surface per unit time per unit area at the point. Flux is therefore measured in such units as calories per second per square centimeter. It is denoted here by Φ and it varies with the normal derivative of the temperature T at the point on the surface:

$$\Phi = -K\frac{dT}{dn} \qquad (K > 0). \tag{1}$$

The constant K is known as the *thermal conductivity* of the material of the solid, which is assumed to be homogeneous.

The points in the solid are assigned cartesian coordinates in three-dimensional space, and we restrict our attention to those cases when the temperature T varies with only the x and y coordinates. Since T does not vary with the coordinate along the axis perpendicular to the xy plane, the flow of heat is then two-dimensional and parallel to that plane. We assume, moreover, that the flow is in a steady state; that is, T does not vary with time.

It is assumed that no thermal energy is created or destroyed within the solid. That is, no heat sources or sinks are present there. Also, the temperature function $T(x,y)$ and all its partial derivatives of the first and second order are continuous at each point interior to the solid. This statement and formula (1) for the flux of heat are postulates for the mathematical theory of heat conduction, postulates that also apply at points within a solid containing a continuous distribution of sources or sinks.

Consider now an interior element of the solid, the element having the shape of a rectangular prism of unit length perpendicular to the xy plane with base Δx by Δy in that plane (Fig. 60). The time rate of flow of heat toward the right across the left-hand face is $-KT_x(x,y)\,\Delta y$, and toward the right across the right-hand face it is $-KT_x(x+\Delta x,y)\,\Delta y$. Subtracting the first rate from the second, we obtain the rate of loss of heat from the element through those two faces. This resultant rate can be written

$$-K\left[\frac{T_x(x+\Delta x,y)-T_x(x,y)}{\Delta x}\right]\Delta x\,\Delta y,$$

or

$$-KT_{xx}(x,y)\,\Delta x\,\Delta y \tag{2}$$

if Δx is very small. All the expressions here are, of course, approximations whose accuracy increases as Δx and Δy are made smaller.

In like manner, the resultant rate of heat loss through the lower and upper faces of the element is found to be

$$-KT_{yy}(x,y)\,\Delta x\,\Delta y. \tag{3}$$

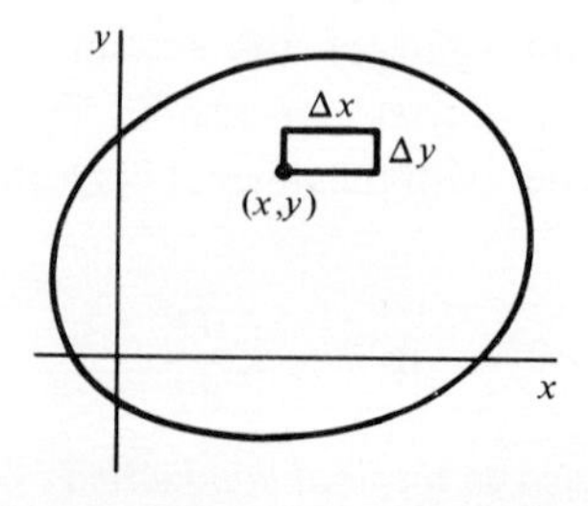

FIGURE 60

Heat enters or leaves the element only through these four faces, and the temperatures within the element are steady. Hence the sum of expressions (2) and (3) is zero; that is,

$$T_{xx}(x,y) + T_{yy}(x,y) = 0. \tag{4}$$

The temperature function thus satisfies Laplace's equation at each interior point of the solid.

In view of equation (4) and the continuity of the temperature function and its partial derivatives, *T is a harmonic function of x and y* in the domain representing the interior of the solid body.

The surfaces $T(x,y) = c$, where c is any real constant, are the *isotherms* within the solid. They can also be considered as curves in the xy plane; for $T(x,y)$ can be interpreted as the temperature in a thin sheet of material in that plane with the faces of the sheet thermally insulated. The isotherms are the level curves of the function T.

The gradient of T is perpendicular to the isotherm at each point, and the maximum flux at a point is in the direction of the gradient there. If $T(x,y)$ denotes temperatures in a thin sheet and if S is a harmonic conjugate of the function T, then a curve $S(x,y) = c$ has the gradient of T as a tangent vector at each point where the analytic function $T(x,y) + iS(x,y)$ is conformal. The curves $S(x,y) = c$ are called *lines of flow*.

If the normal derivative dT/dn is zero along any part of the boundary of the sheet, the flux of heat across that part is zero. That is, the part is thermally insulated and is therefore a line of flow.

The function T may also denote the concentration of a substance that is diffusing through a solid. In that case, K is the diffusion constant. The above discussion and the derivation of equation (4) apply as well to steady-state diffusion.

82. Steady Temperatures in a Half Plane

Let us find a formula for the steady temperatures $T(x,y)$ in a thin semi-infinite plate $y \geqq 0$ whose faces are insulated and whose edge $y = 0$ is kept at temperature zero except for the portion $-1 < x < 1$, $y = 0$ where it is kept at temperature unity (Fig. 61). The function $T(x,y)$ is to be bounded; this condition is natural if we consider the given plate as the limiting case of the plate $0 \leqq y \leqq y_0$ whose upper edge is kept at a fixed temperature as y_0 is increased. In fact, it would be physically reasonable to stipulate that $T(x,y)$ approach zero as y tends to infinity.

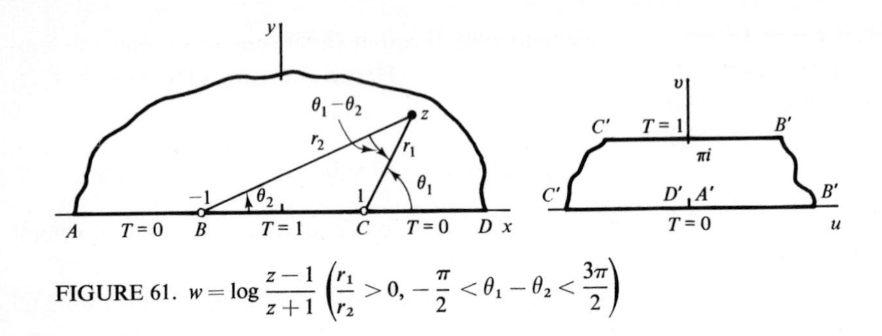

FIGURE 61. $w = \log \dfrac{z-1}{z+1} \left(\dfrac{r_1}{r_2} > 0, -\dfrac{\pi}{2} < \theta_1 - \theta_2 < \dfrac{3\pi}{2} \right)$

The boundary value problem to be solved can be written

$$T_{xx}(x,y) + T_{yy}(x,y) = 0 \qquad (-\infty < x < \infty, y > 0), \tag{1}$$

$$T(x,0) = \begin{cases} 1 & \text{when} \quad |x| < 1, \\ 0 & \text{when} \quad |x| > 1; \end{cases} \tag{2}$$

also, $|T(x,y)| < M$ where M is some positive constant.

This is a Dirichlet problem for the upper half of the xy plane. Our method of solution will be to obtain a new Dirichlet problem for a region in the uv plane. That region will be the image of the half plane under a transformation which is analytic in the domain $y > 0$ and which is conformal along the boundary $y = 0$ except at the points $(\pm 1,0)$ where it is undefined. It will be a simple matter to discover a bounded harmonic function satisfying the new problem. The two theorems of the previous chapter will then be applied to transform the solution of the problem in the uv plane into a solution of the original problem in the xy plane. Specifically, a harmonic function of u and v will be transformed into a harmonic function of x and y, and the boundary conditions in the uv plane will be preserved on corresponding portions of the boundary in the xy plane. There should be no confusion if we use the same symbol T to denote the different temperature functions in the two planes.

Let us write $z - 1 = r_1 \exp(i\theta_1)$ and $z + 1 = r_2 \exp(i\theta_2)$ where $-\pi/2 < \theta_k < 3\pi/2$ $(k = 1, 2)$. The transformation

$$w = \log \frac{z-1}{z+1} = \operatorname{Log} \frac{r_1}{r_2} + i(\theta_1 - \theta_2) \qquad \left(\frac{r_1}{r_2} > 0, -\frac{\pi}{2} < \theta_1 - \theta_2 < \frac{3\pi}{2} \right) \tag{3}$$

is defined on the upper half plane $y \geqq 0$, except for the points $z = \pm 1$, since $0 \leqq \theta_1 - \theta_2 \leqq \pi$ in that region (Fig. 61). Now the value of the logarithm is the principal value when $0 \leqq \theta_1 - \theta_2 \leqq \pi$, and we note from Fig. 19 of Appendix 2 that the upper half plane $y > 0$ is mapped onto the strip $0 < v < \pi$ in the w plane. Indeed, it was that figure which suggested transformation (3) here.

The segment of the x axis between $z = -1$ and $z = 1$, where $\theta_1 - \theta_2 = \pi$, is mapped onto the upper edge of the strip; and the rest of the x axis, where $\theta_1 - \theta_2 = 0$, is mapped onto the lower edge. The required analyticity and conformality conditions are evidently satisfied by transformation (3).

A bounded harmonic function of u and v that is zero on the edge $v = 0$ of the strip and unity on the edge $v = \pi$ is clearly

$$T = \frac{1}{\pi} v; \tag{4}$$

it is harmonic since it is the imaginary part of the entire function w/π. Changing to x and y coordinates by means of the equation

$$w = \log \left| \frac{z-1}{z+1} \right| + i \arg \frac{z-1}{z+1}, \tag{5}$$

we find that

$$v = \arg \left(\frac{x - 1 + iy}{x + 1 + iy} \right) = \arg \left[\frac{x^2 + y^2 - 1 + i2y}{(x+1)^2 + y^2} \right],$$

or

$$v = \arctan \left(\frac{2y}{x^2 + y^2 - 1} \right).$$

The range of the arctangent function here is from 0 to π since

$$\arg \frac{z-1}{z+1} = \theta_1 - \theta_2$$

and $0 \leqq \theta_1 - \theta_2 \leqq \pi$. Expression (4) now takes the form

$$T = \frac{1}{\pi} \arctan \left(\frac{2y}{x^2 + y^2 - 1} \right) \qquad (0 \leqq \arctan t \leqq \pi). \tag{6}$$

Since function (4) is harmonic in the strip $0 < v < \pi$ and since transformation (3) is analytic in the half plane $y > 0$, we may apply the theorem in Sec. 79 to conclude that function (6) is harmonic in that half plane. The boundary conditions for the two harmonic functions are the same on corresponding parts of the boundaries because they are of the type $T = c$ treated in the theorem of Sec. 80. The bounded function (6) is therefore the desired solution to the original problem. We can, of course, verify directly that function (6) satisfies Laplace's equation and has values tending to those indicated in Fig. 61 as the point (x,y) approaches the x axis from above.

The isotherms $T(x,y) = c$ $(0 < c < 1)$ are the circles

$$x^2 + y^2 - \frac{2}{\tan \pi c} y = 1$$

with centers on the y axis and passing through the points $(\pm 1,0)$.

Finally, we note that since the product of a harmonic function by a constant is also harmonic, the function

$$T = \frac{T_0}{\pi} \arctan \left(\frac{2y}{x^2 + y^2 - 1} \right) \qquad (0 \leqq \arctan t \leqq \pi)$$

represents the steady temperatures in the given half plane when the temperature unity along the edge $-1 < x < 1$, $y = 0$ is replaced by a fixed temperature T_0.

83. A Related Problem

Consider a semi-infinite slab in three-space formed by the planes $x = \pm\pi/2$ and $y = 0$ when the first two surfaces are kept at temperature zero and the last at temperature unity. We wish to find an expression for the temperature $T(x,y)$ at any interior point of the slab. The problem is also that of finding temperatures in a thin plate having the form of a semi-infinite strip $-\pi/2 \leqq x \leqq \pi/2$, $y \geqq 0$ when the faces of the plate are perfectly insulated (Fig. 62).

The boundary value problem to be solved here is

$$T_{xx}(x,y) + T_{yy}(x,y) = 0 \qquad \left(-\frac{\pi}{2} < x < \frac{\pi}{2}, y > 0\right), \tag{1}$$

$$T\left(-\frac{\pi}{2},y\right) = T\left(\frac{\pi}{2},y\right) = 0 \qquad (y > 0), \tag{2}$$

$$T(x,0) = 1 \qquad \left(-\frac{\pi}{2} < x < \frac{\pi}{2}\right), \tag{3}$$

where $T(x,y)$ is bounded.

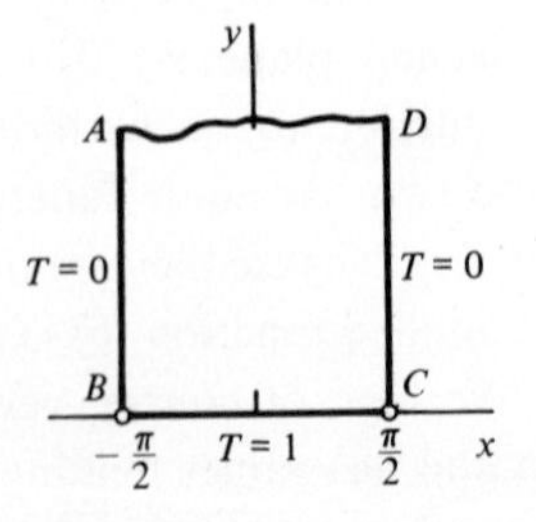

FIGURE 62

In view of Sec. 39, as well as Fig. 9 of Appendix 2, the mapping

$$w = \sin z \tag{4}$$

transforms this boundary value problem into the one posed in the previous section (Fig. 61). Hence, recalling solution (6) in that section, we write

$$T = \frac{1}{\pi}\arctan\left(\frac{2v}{u^2+v^2-1}\right) \qquad (0 \leqq \arctan t \leqq \pi). \tag{5}$$

The change of variables indicated in equation (4) can be written

$$u = \sin x \cosh y, \qquad v = \cos x \sinh y;$$

and the harmonic function (5) thus becomes

$$T = \frac{1}{\pi}\arctan\left(\frac{2\cos x \sinh y}{\sin^2 x \cosh^2 y + \cos^2 x \sinh^2 y - 1}\right).$$

The denominator here reduces to $\sinh^2 y - \cos^2 x$, and the quotient can be written

$$\frac{2\cos x \sinh y}{\sinh^2 y - \cos^2 x} = \frac{2\cos x/\sinh y}{1-(\cos x/\sinh y)^2} = \tan 2\alpha$$

where $\tan\alpha = \cos x/\sinh y$. Thus $\pi T = 2\alpha$, and our formula for T becomes

$$T = \frac{2}{\pi}\arctan\left(\frac{\cos x}{\sinh y}\right) \qquad \left(0 \leqq \arctan t \leqq \frac{\pi}{2}\right). \tag{6}$$

The arctangent function here has the range 0 to $\pi/2$ since its argument is nonnegative.

Now, since $\sin z$ is entire and function (5) is harmonic in the half plane $v > 0$, function (6) is harmonic in the strip $-\pi/2 < x < \pi/2$, $y > 0$. Also, function (5) satisfies the boundary conditions $T = 1$ when $|u| < 1$ and $v = 0$ and $T = 0$ when $|u| > 1$ and $v = 0$. Function (6) thus satisfies boundary conditions (2) and (3). Moreover, $|T(x,y)| \leqq 1$ throughout the strip. Formula (6) is therefore the temperature formula sought.

The isotherms $T(x,y) = c$ in the slab are the surfaces

$$\cos x = \tan\frac{\pi c}{2}\sinh y,$$

each of which passes through the points $(\pm\pi/2,0)$ in the xy plane. If K is the thermal conductivity, the flux of heat into the slab through the surface lying in the plane $y = 0$ is

$$-KT_y(x,0) = \frac{2K}{\pi\cos x} \qquad \left(-\frac{\pi}{2} < x < \frac{\pi}{2}\right);$$

the flux outward through the surface lying in the plane $x = \pi/2$ is

$$-KT_x\left(\frac{\pi}{2},y\right) = \frac{2K}{\pi \sinh y} \qquad (y > 0).$$

The boundary value problem posed in this section can also be solved by the method of separation of variables. That method is more direct, but it gives the solution in the form of an infinite series.[1]

84. Temperatures in a Quadrant with Part of One Boundary Insulated

Let us find the steady temperatures in a thin plate having the form of a quadrant if a segment at the end of one edge is insulated, if the rest of that edge is kept at a fixed temperature, and if the second edge is kept at another fixed temperature. The surfaces are insulated, and so the problem is two-dimensional.

The temperature scale and the unit of length can be chosen so that the boundary value problem for the temperature function T becomes

$$T_{xx}(x,y) + T_{yy}(x,y) = 0 \qquad (x > 0, y > 0), \tag{1}$$

$$\begin{cases} T_y(x,0) = 0 & \text{when} \quad 0 < x < 1, \\ T(x,0) = 1 & \text{when} \quad x > 1, \end{cases} \tag{2}$$

$$T(0,y) = 0 \qquad (y > 0), \tag{3}$$

where $T(x,y)$ is bounded in the quadrant. The plate and its boundary conditions are shown in Fig. 63.

Conditions (2) prescribe the value of the normal derivative of the function T over a part of a boundary line and the value of the function itself over the rest of that line. The separation of variables method mentioned at the end of the previous section is not adapted to such problems with different types of conditions along the same boundary line.

As indicated in Fig. 10 of Appendix 2, the transformation

$$z = \sin w \tag{4}$$

is a one to one mapping of the strip $0 \leqq u \leqq \pi/2$, $v \geqq 0$ onto the quadrant $x \geqq 0$, $y \geqq 0$. Observe now that the existence of an inverse is ensured by the fact that the given transformation is both one to one and onto. Since $\sin w$ is conformal throughout the strip except at the point $w = \pi/2$, the inverse transformation must be conformal throughout the quadrant except at the point $z = 1$.

[1] Essentially the same problem is treated in R. V. Churchill "Fourier Series and Boundary Value Problems," 2d ed., Exercises 3 and 4, pp. 150–151, 1963. Also, a short discussion of the uniqueness of solutions to boundary value problems will be found in Chap. 10 of that book.

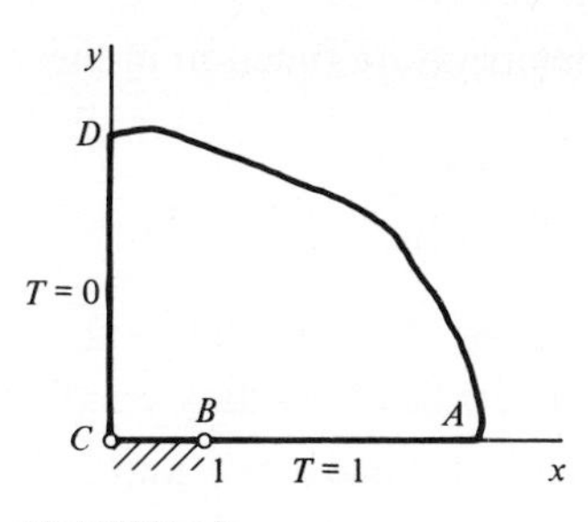

FIGURE 63

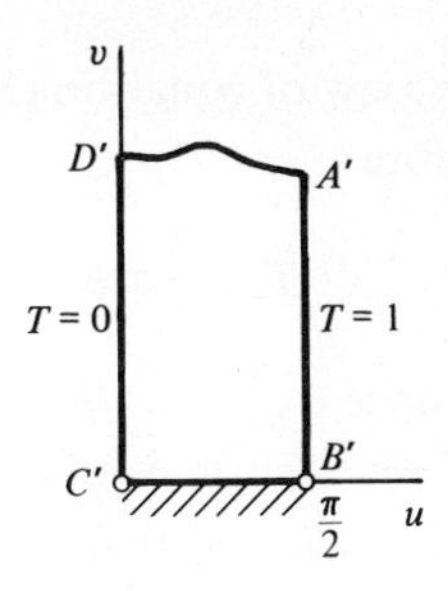

FIGURE 64

That inverse transformation maps the segment $0 < x < 1$, $y = 0$ of the boundary of the quadrant onto the base of the strip and the rest of the boundary onto the sides of the strip as shown in Fig. 64.

Since the inverse of transformation (4) is conformal in the quadrant, except when $z = 1$, the solution to the given problem can be obtained by finding a function that is harmonic in the strip and satisfies the boundary conditions given in Fig. 64. Observe that those boundary conditions are of the types $T = c$ and $dT/dn = 0$.

The required temperature function T for the new boundary value problem is clearly

$$T = \frac{2}{\pi} u, \tag{5}$$

the function $2u/\pi$ being the real part of the entire function $2w/\pi$. We must now express T in terms of x and y.

In order to obtain u in terms of x and y, we first note that, according to equation (4),

$$x = \sin u \cosh v, \qquad y = \cos u \sinh v; \tag{6}$$

hence

$$\frac{x^2}{\sin^2 u} - \frac{y^2}{\cos^2 u} = 1. \tag{7}$$

In solving for u it is convenient to observe that, for each fixed u, hyperbola (7) has foci at the points $(\pm 1,0)$ in the xy plane and transverse axis of length $2 \sin u$. Thus the difference of the distances between the foci and a point (x,y) lying on the part of the hyperbola in the first quadrant is

$$\sqrt{(x+1)^2 + y^2} - \sqrt{(x-1)^2 + y^2} = 2 \sin u.$$

In view of equation (5), the required temperature function in the xy plane is therefore

$$T = \frac{2}{\pi} \arcsin \frac{1}{2} [\sqrt{(x+1)^2 + y^2} - \sqrt{(x-1)^2 + y^2}] \tag{8}$$

where, since $0 \leqq u \leqq \pi/2$, the arcsine function has the range 0 to $\pi/2$.

If we wish to verify that this function satisfies boundary conditions (2), we must remember that $\sqrt{(x-1)^2}$ denotes $x - 1$ when $x > 1$ and $1 - x$ when $0 < x < 1$, the square roots being positive. Note too that the temperature at any point along the insulated part of the lower edge of the plate is

$$T(x,0) = \frac{2}{\pi} \arcsin x.$$

It can be seen from equation (5) that the isotherms $T(x,y) = c$ are the parts of the confocal hyperbolas (7), where $u = \pi c/2$, which lie in the first quadrant. Since the function $2v/\pi$ is a harmonic conjugate of function (5), the lines of flow are quarters of the confocal ellipses obtained by holding v constant in equations (6).

EXERCISES

1. In the problem of the semi-infinite plate shown on the left in Fig. 61, obtain a harmonic conjugate of the temperature function $T(x,y)$ from equation (5), Sec. 82, and find the lines of flow of heat. Show that they consist of the upper half of the y axis and the upper halves of certain circles on either side of that axis, the circles having their centers on the segment AB or CD of the x axis.

2. Show that if the function T of Sec. 82 is not required to be bounded, the harmonic function (4) of that section can be replaced by the harmonic function

$$T = \text{Im}\left(\frac{1}{\pi} w + A \cosh w\right) = \frac{1}{\pi} v + A \sinh u \sin v$$

where A is an arbitrary real constant. Conclude that the solution of the Dirichlet problem for the strip in the uv plane (Fig. 61) would not then be unique.

3. Suppose that the condition that T be bounded is omitted from the problem for temperatures in the semi-infinite slab of Sec. 83 (Fig. 62). Show that an infinite number of solutions are then possible by noting the effect of adding to the solution found there the imaginary part of the function $A \sin z$, where A is an arbitrary real constant.

4. Use the function Log z to find a formula for the bounded steady temperatures in a plate having the form of a quadrant $x \geqq 0$, $y \geqq 0$ if its faces are perfectly insulated and its edges have temperatures $T(x,0) = 0$ and $T(0,y) = 1$ (Fig. 65). Find the isotherms and lines of flow, and draw some of them.

Ans. $T = (2/\pi) \arctan (y/x)$.

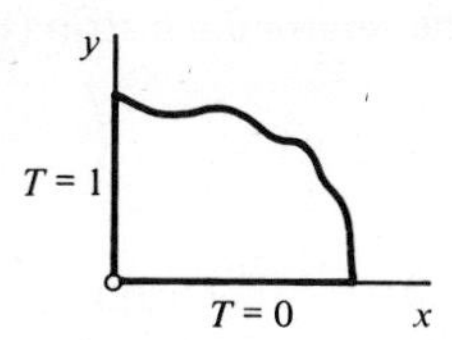

FIGURE 65

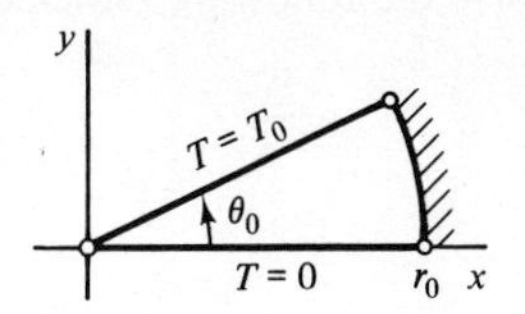

FIGURE 66

5. Find the steady temperatures in a solid whose shape is that of a long cylindrical wedge if its boundary planes $\theta = 0$ and $\theta = \theta_0$ are kept at constant temperatures zero and T_0, respectively, and its surface $r = r_0$ is perfectly insulated (Fig. 66).

Ans. $T = (T_0/\theta_0) \arctan (y/x)$.

6. Find the bounded steady temperatures $T(x,y)$ in the semi-infinite solid $y \geqq 0$ if $T = 0$ on the part $x < -1$, $y = 0$ of the boundary, if $T = 1$ on the part $x > 1$, $y = 0$, and if the strip $-1 < x < 1$, $y = 0$ of the boundary is insulated (Fig. 67).

$$Ans. \quad T = \frac{1}{2} + \frac{1}{\pi} \arcsin \frac{1}{2} [\sqrt{(x+1)^2 + y^2} - \sqrt{(x-1)^2 + y^2}]$$

$$\left(-\frac{\pi}{2} \leqq \arcsin t \leqq \frac{\pi}{2}\right).$$

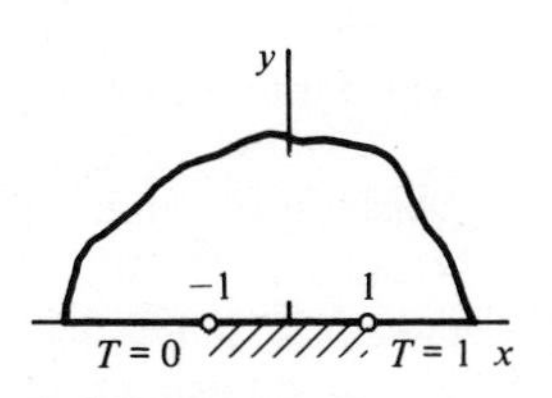

FIGURE 67

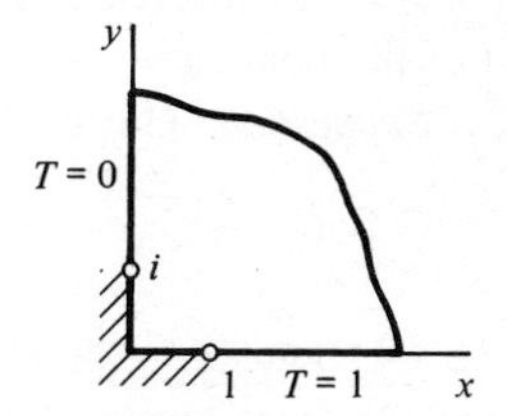

FIGURE 68

7. Find the bounded steady temperatures in the solid $x \geqq 0$, $y \geqq 0$ when the boundary surfaces are kept at fixed temperatures except for insulated strips of equal width at the corner, as shown in Fig. 68.

$$Ans. \quad T = \frac{1}{2} + \frac{1}{\pi} \arcsin \frac{1}{2} [\sqrt{(x^2 - y^2 + 1)^2 + 4x^2y^2} - \sqrt{(x^2 - y^2 - 1)^2 + 4x^2y^2}]$$

$$\left(-\frac{\pi}{2} \leqq \arcsin t \leqq \frac{\pi}{2}\right).$$

8. Solve the following Dirichlet problem for the semi-infinite strip (Fig. 69):

$$H_{xx}(x,y) + H_{yy}(x,y) = 0 \qquad \left(0 < x < \frac{\pi}{2}, y > 0\right),$$

$$H(x,0) = 0 \qquad \left(0 < x < \frac{\pi}{2}\right),$$

$$H(0,y) = 1, \qquad H\left(\frac{\pi}{2},y\right) = 0 \qquad (y > 0),$$

where $0 \leqq H(x,y) \leqq 1$.

Suggestion: This problem can be transformed into the one of Exercise 4.

Ans. $H = \dfrac{2}{\pi} \arctan \left(\dfrac{\tanh y}{\tan x}\right)$.

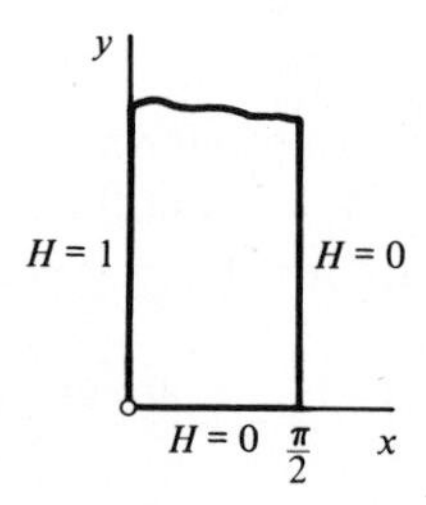

FIGURE 69

9. Derive a formula for the temperatures $T(r,\theta)$ in a semicircular plate $r \leqq 1$, $0 \leqq \theta \leqq \pi$ with insulated faces if $T = 1$ along the radial edge $\theta = 0$ and $T = 0$ on the rest of the boundary.

Suggestion: This problem can be transformed into the one of Exercise 8.

Ans. $T = \dfrac{2}{\pi} \arctan \left(\dfrac{1-r}{1+r} \cot \dfrac{\theta}{2}\right)$.

10. Solve the boundary value problem for the plate $X \geqq 0$, $Y \geqq 0$ in the Z plane where the faces are insulated and the boundary conditions are those indicated in Fig. 70.

Suggestion: Using the mapping $z = i/Z$, transform this problem into the one posed in Sec. 84 (Fig. 63).

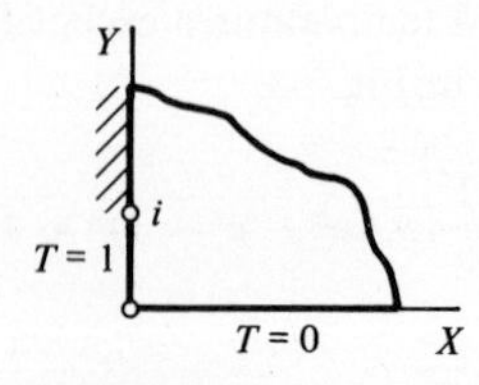

FIGURE 70

11. The portions $x < 0$, $y = 0$ and $x < 0$, $y = \pi$ of the edges of an infinite plate $0 \leqq y \leqq \pi$ are thermally insulated, as are the faces of the plate. The conditions $T(x,0) = 1$ and $T(x,\pi) = 0$ are maintained when $x > 0$ (Fig. 71). Find the steady temperatures in the plate.

Suggestion: This problem can be transformed into the one of Exercise 6.

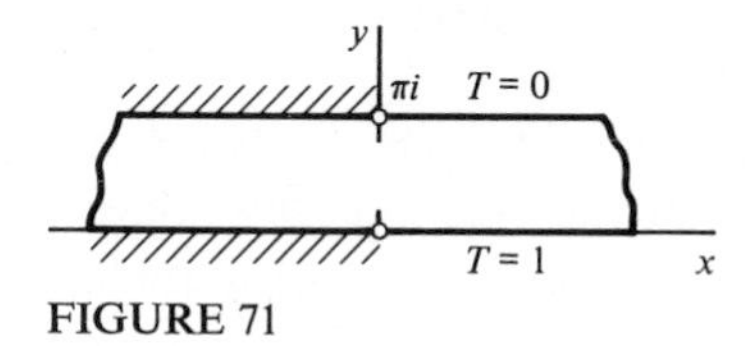

FIGURE 71

12. A thin plate with insulated faces has the semi-elliptical shape shown in the uv plane of Fig. 11, Appendix 2. The temperature on the elliptical part of its boundary is $T = 1$. The temperature along the line segment $-1 < u < 1$, $v = 0$ is $T = 0$, and the rest of the boundary along the u axis is insulated. Find the lines of flow of heat.

13. According to Exercises 11 and 12 of Sec. 55, if a function $f(z) = u(x,y) + iv(x,y)$ is continuous in a closed bounded region R and analytic and not constant in the interior of R, then the function $u(x,y)$ reaches its maximum and minimum values on the boundary of R, never in the interior. By interpreting $u(x,y)$ as a steady temperature, state a physical reason why that property of maximum and minimum values should hold true.

85. Electrostatic Potential

In an electrostatic force field the *field intensity* at a point is a vector representing the force exerted on a unit positive charge placed at that point. The electrostatic *potential* is a scalar function of the space coordinates such that at each point its directional derivative in any direction is the negative of the component of the field intensity in that direction.

For two stationary charged particles the magnitude of the force of attraction or repulsion exerted by one particle on the other is directly proportional to the product of the charges and inversely proportional to the square of the distance between those particles. From this inverse-square law, it can be shown that the potential at a point due to a single particle in space is inversely proportional to the distance between the point and the particle. In any region free of charges the potential due to a distribution of charges outside that region can then be shown to satisfy Laplace's equation for three-dimensional space.

If conditions are such that the potential V is the same in all planes parallel to the xy plane, then in regions free from charges V is a harmonic function of just the two variables x and y:

$$V_{xx}(x,y) + V_{yy}(x,y) = 0.$$

The field intensity vector at each point is parallel to the xy plane with x and y components $-V_x(x,y)$ and $-V_y(x,y)$, respectively. That vector is therefore the negative of the gradient of $V(x,y)$.

A surface along which $V(x,y)$ is constant is an equipotential surface. The tangential component of the field intensity vector at a point on a conducting surface is zero in the static case since charges are free to move on such a surface. Hence $V(x,y)$ is constant along the surface of a conductor and that surface is an *equipotential*.

If U is a harmonic conjugate of V, the curves $U(x,y) = c$ in the xy plane are called *flux lines*. When such a curve intersects an equipotential curve at a point where the derivative of the analytic function $V(x,y) + iU(x,y)$ is not zero, the two curves are orthogonal at that point and the field intensity is tangent to the flux line there.

Boundary value problems for the potential V are the same mathematical problems as those for steady temperatures T; and, as in the case of steady temperatures, the methods of complex variables are limited to two-dimensional problems. The problem posed in Sec. 83 (Fig. 62), for instance, can be interpreted as that of finding the two-dimensional electrostatic potential in the empty space $-\pi/2 < x < \pi/2$, $y > 0$ formed by the conducting planes $x = \pm\pi/2$ and $y = 0$, insulated at their intersections, when the first two surfaces are kept at potential zero and the third at potential unity. Problems of this type arise in electronics. If the space charge inside a vacuum tube is small, the space is sometimes considered free of charge and the potential there can be assumed to satisfy Laplace's equation.

The potential in the steady flow of electricity in a plane conducting sheet is also a harmonic function at points free from sources and sinks. Gravitational potential is a further example of a harmonic function in physics.

86. Potential in a Cylindrical Space

A long hollow circular cylinder is made out of a thin sheet of conducting material, and the cylinder is split along two of its elements to form two equal parts. These parts are separated by slender strips of insulating material and are used as electrodes, one of which is grounded at potential zero and the other kept at a different fixed potential. We take the coordinate axes and units of length and potential difference as indicated in Fig. 72. We then interpret the electrostatic potential $V(x,y)$ over any cross section of the enclosed space that is distant from the ends of the cylinder as a harmonic function inside the circle $x^2 + y^2 = 1$ in the xy plane; also, $V = 0$ on the upper half of the circle and $V = 1$ on the lower half.

A linear fractional transformation that maps the upper half plane onto the interior of the unit circle centered at the origin, the positive real axis onto the upper half of the circle, and the negative real axis onto the lower half of the circle was verified in Exercise 11, Sec. 34. The result is given in Fig. 13 of Appendix 2; interchanging z and w there, we find that the inverse of the transformation

$$z = \frac{i - w}{i + w} \tag{1}$$

gives us a new problem for V in a half plane, indicated in Fig. 73.

Observe now that the imaginary part of the function

$$\frac{1}{\pi} \operatorname{Log} w = \frac{1}{\pi} \operatorname{Log} \rho + \frac{i}{\pi} \phi \qquad (\rho > 0, 0 \leqq \phi \leqq \pi) \tag{2}$$

is a bounded function of u and v that assumes the required constant values on the two parts $\phi = 0$ and $\phi = \pi$ of the u axis. The desired harmonic function for the half plane is therefore

$$V = \frac{1}{\pi} \arctan \frac{v}{u}, \tag{3}$$

where the values of the arctangent function range between 0 and π.

The inverse of transformation (1) is

$$w = i\frac{1 - z}{1 + z}, \tag{4}$$

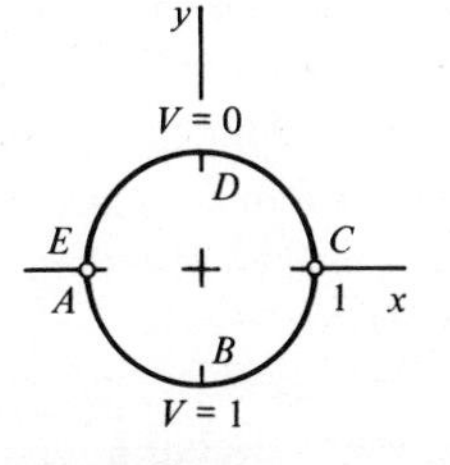

FIGURE 72

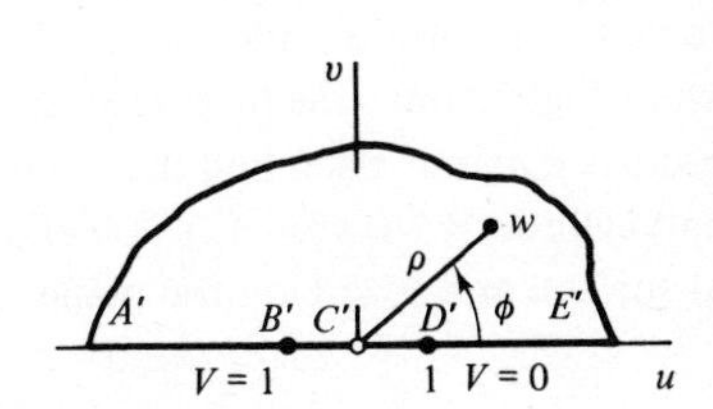

FIGURE 73

from which u and v can be expressed in terms of x and y. Equation (3) then becomes

$$V = \frac{1}{\pi} \arctan \left(\frac{1 - x^2 - y^2}{2y} \right) \qquad (0 \leqq \arctan t \leqq \pi). \tag{5}$$

Function (5) is the potential function for the space enclosed by the cylindrical electrodes since it is harmonic inside the circle and it assumes the required values on the semicircles. If we wish to verify this solution we must note that

$$\lim_{t \to 0} \arctan t = 0 \qquad (t > 0)$$

and

$$\lim_{t \to 0} \arctan t = \pi \qquad (t < 0).$$

The equipotential curves $V(x,y) = c$ in the circular region are arcs of the circles

$$x^2 + y^2 + 2y \tan \pi c = 1,$$

each circle passing through the points $(\pm 1,0)$. Also, the segment of the x axis between those points is the equipotential $V(x,y) = 1/2$. A harmonic conjugate U of V is $(-1/\pi) \operatorname{Log} \rho$, the imaginary part of the function $(-i/\pi) \operatorname{Log} w$. In view of equation (4), U may be written

$$U = -\frac{1}{\pi} \operatorname{Log} \left| \frac{1 - z}{1 + z} \right|.$$

From this equation it can be seen that the flux lines $U(x,y) = c$ are arcs of circles with centers on the x axis. The segment of the y axis between the electrodes is also a flux line.

EXERCISES

1. The harmonic function (3) of Sec. 86 is bounded in the half plane $v \geqq 0$ and satisfies the boundary conditions indicated in Fig. 73. Show that if the imaginary part of Ae^w, where A is any real constant, is added to that function, the resulting function satisfies all the requirements except the boundedness condition.
2. Prove that transformation (4) of Sec. 86 maps the upper half of the circular region shown in Fig. 72 onto the first quadrant of the w plane and the diameter CE onto the positive v axis. Then find the electrostatic potential V in the space formed by the half cylinder $x^2 + y^2 = 1$, $y \geqq 0$ and the plane $y = 0$ when $V = 0$ on the cylindrical surface and $V = 1$ on the plane surface (Fig. 74).

Ans. $V = \dfrac{2}{\pi} \arctan \left(\dfrac{1 - x^2 - y^2}{2y} \right)$.

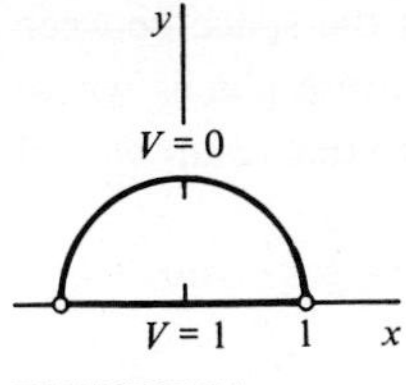

FIGURE 74

3. Find the electrostatic potential $V(r,\theta)$ in the space $0 < r < 1$, $0 < \theta < \pi/4$ formed by the half planes $\theta = 0$ and $\theta = \pi/4$ and the portion $0 \leqq \theta \leqq \pi/4$ of the cylindrical surface $r = 1$ when $V = 1$ on the plane boundaries and $V = 0$ on the cylindrical boundary. (See Exercise 2.) Verify that your function satisfies these boundary conditions.
4. Note that all branches of $\log z$ have the same real component which is harmonic everywhere except at the origin. Then write a formula for the electrostatic potential $V(x,y)$ in the space between two coaxial conducting cylindrical surfaces $x^2 + y^2 = 1$ and $x^2 + y^2 = r_0^2$ ($r_0 \neq 1$) if $V = 0$ on the first surface and $V = 1$ on the second.

$$\textit{Ans.}\quad V = \frac{\text{Log}\,(x^2 + y^2)}{2\,\text{Log}\,r_0}.$$

5. Find the bounded electrostatic potential $V(x,y)$ in the space $y > 0$ bounded by an infinite conducting plane $y = 0$ one strip $(-a < x < a,\ y = 0)$ of which is insulated from the rest of the plane and kept at potential $V = 1$, while $V = 0$ on the rest (Fig. 75). Verify that your function satisfies the boundary conditions.

$$\textit{Ans.}\quad V = \frac{1}{\pi}\arctan\left(\frac{2ay}{x^2 + y^2 - a^2}\right)\quad (0 \leqq \arctan t \leqq \pi).$$

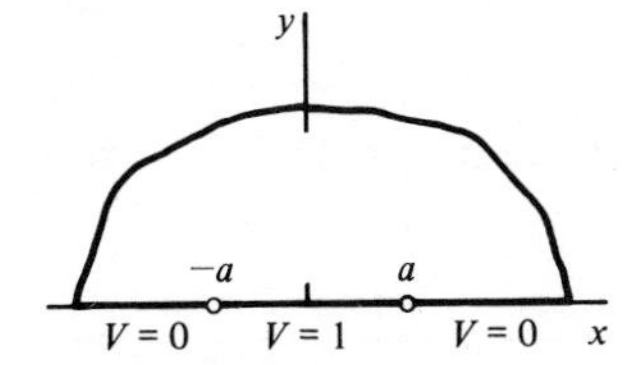

FIGURE 75

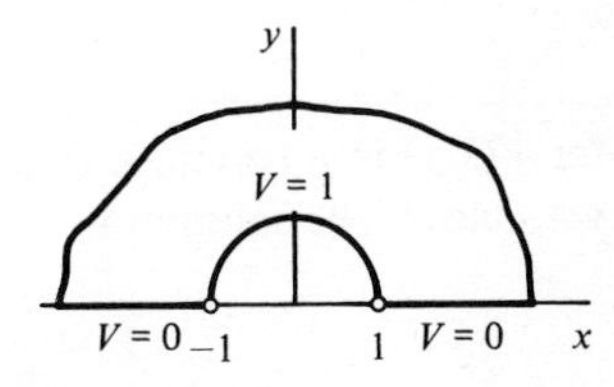

FIGURE 76

6. Derive a formula for the electrostatic potential in the space indicated in Fig. 76, bounded by two half planes and a half cylinder, when $V = 1$ on the cylindrical surface and $V = 0$ on the plane surfaces. Draw some of the equipotential curves in the xy plane.

$$\textit{Ans.}\quad V = \frac{2}{\pi}\arctan\left(\frac{2y}{x^2 + y^2 - 1}\right).$$

7. Find the potential V in the space between the planes $y = 0$ and $y = \pi$ if $V = 0$ on the part of each of those planes where $x > 0$ and $V = 1$ on the parts where $x < 0$ (Fig. 77). Check your result with the boundary conditions.

$$\textit{Ans.}\quad V = \frac{1}{\pi}\arctan\left(\frac{\sin y}{\sinh x}\right) \quad (0 \leqq \arctan t \leqq \pi).$$

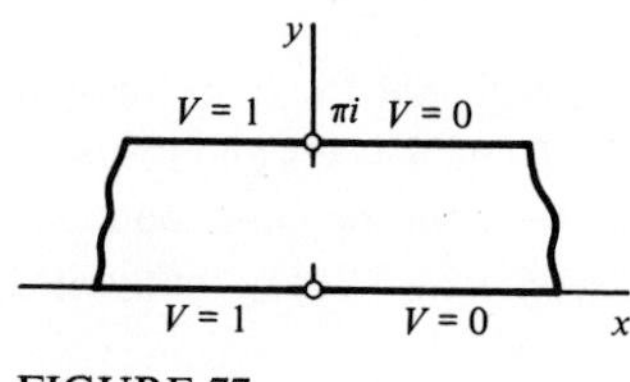

FIGURE 77

8. Derive a formula for the electrostatic potential V in the space interior to a long cylinder $r = 1$ if $V = 0$ on the first quadrant ($r = 1, 0 < \theta < \pi/2$) of the cylindrical surface and $V = 1$ on the rest ($r = 1, \pi/2 < \theta < 2\pi$) of that surface. (See Fig. 24 and Exercise 14, Sec. 34.) Show that $V = 3/4$ on the axis of the cylinder. Check your formula with the boundary conditions.

9. Using Fig. 20 of Appendix 2, find a temperature function $T(x,y)$ which is harmonic in the shaded domain of the xy plane shown there and which assumes the values $T = 0$ along the semicircle ABC and $T = 1$ along the line segment DEF. Verify that your function satisfies the required boundary conditions. (See Exercise 2.)

10. The Dirichlet problem

$$V_{xx}(x,y) + V_{xx}(x,y) = 0 \qquad (0 < x < a, 0 < y < b),$$
$$V(x,0) = 0, \qquad V(x,b) = 1 \qquad (0 < x < a),$$
$$V(0,y) = V(a,y) = 0 \qquad (0 < y < b)$$

for $V(x,y)$ in a rectangle (Fig. 78) can be solved by the method of separation of variables.[1] The solution is

$$V = \frac{4}{\pi}\sum_{n=1}^{\infty}\frac{\sinh(m\pi y/a)}{m\sinh(m\pi b/a)}\sin\frac{m\pi x}{a} \qquad (m = 2n - 1).$$

Accepting this formula, find the potential $V(r,\theta)$ in the space $1 < r < r_0, 0 < \theta < \pi$ if $V = 1$ on the part of the boundary where $\theta = \pi$ and $V = 0$ on the rest of the boundary (Fig. 79).

$$\textit{Ans.}\quad V = \frac{4}{\pi}\sum_{n=1}^{\infty}\frac{\sinh\alpha_n\theta}{\sinh\alpha_n\pi}\frac{\sin(\alpha_n \operatorname{Log} r)}{2n-1} \qquad \left[\alpha_n = \frac{(2n-1)\pi}{\operatorname{Log} r_0}\right].$$

[1] See R. V. Churchill, "Fourier Series and Boundary Value Problems," 2d ed., pp. 147–148, 1963.

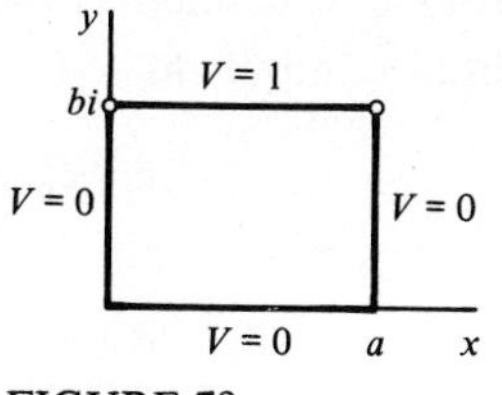

FIGURE 78

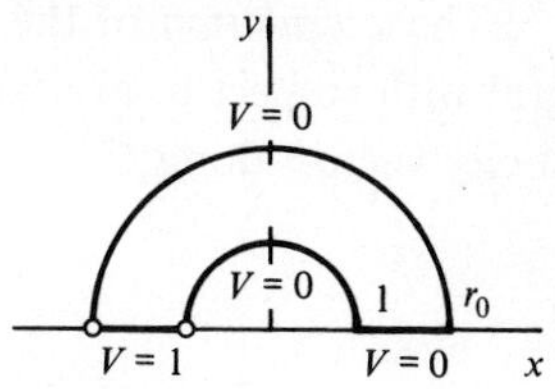

FIGURE 79

11. With the aid of the formula for $V(x,y)$ in the rectangle of Exercise 10, find the potential function $V(r,\theta)$ for the space $1 < r < r_0$, $0 < \theta < \pi$ if $V = 1$ on the part of the boundary $r = r_0$, $0 < \theta < \pi$ and $V = 0$ on the rest of the boundary (Fig. 80).

$$\textit{Ans.}\quad V = \frac{4}{\pi}\sum_{n=1}^{\infty}\frac{r^m - r^{-m}}{r_0^{\,m} - r_0^{\,-m}}\,\frac{\sin m\theta}{m} \quad (m = 2n - 1).$$

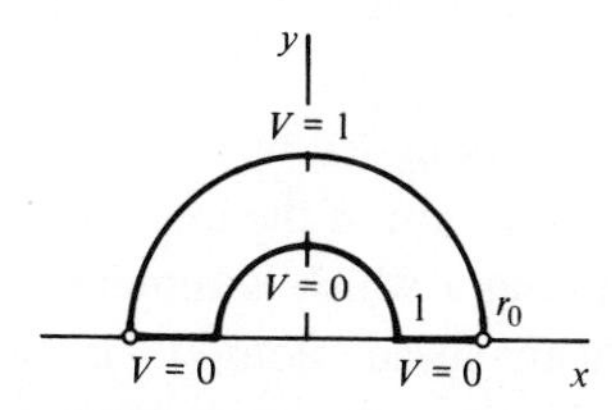

FIGURE 80

87. Two-dimensional Fluid Flow

Harmonic functions play an important role in hydrodynamics and aerodynamics. Again, we consider only the two-dimensional steady-state type of problem. That is, the motion of the fluid is assumed to be the same in all planes parallel to the xy plane, the velocity being parallel to that plane and independent of time. It is then sufficient to consider the motion of a sheet of fluid in the xy plane.

We let the vector representing the complex number

$$V = p + iq$$

denote the velocity of a particle of the fluid at any point (x,y); hence the x and y components of the velocity vector are $p(x,y)$ and $q(x,y)$, respectively. At points interior to a region of flow in which no sources or sinks of the fluid occur, the functions $p(x,y)$ and $q(x,y)$ and their first partial derivatives are assumed to be continuous.

The *circulation* of the fluid along any contour C is defined as the line integral with respect to arc length σ of the tangential component $V_T(x,y)$ of the velocity vector along C:

$$\int_C V_T(x,y)\,d\sigma. \tag{1}$$

The ratio of the circulation along C to the length of C is therefore a mean speed of the fluid along that contour. It is shown in the calculus of real variables that such an expression as integral (1) can be written[1]

$$\int_C p(x,y)\,dx + q(x,y)\,dy. \tag{2}$$

When C is a simple closed contour lying in a simply connected domain of flow containing no sources or sinks, Green's theorem allows us to write

$$\int_C p(x,y)\,dx + q(x,y)\,dy = \iint_R [q_x(x,y) - p_y(x,y)]\,dx\,dy, \tag{3}$$

where R is the closed region bounded by C.

For a physical interpretation of the integrand on the right in equation (3), let C denote a circle of radius r which is centered at a point (x_0,y_0) and taken counterclockwise. A mean speed along C is then found by dividing the circulation by $2\pi r$, and the corresponding mean angular velocity of the fluid about the axis of the circle is found by dividing that mean speed by r:

$$\frac{1}{\pi r^2}\iint_R \frac{1}{2}[q_x(x,y) - p_y(x,y)]\,dx\,dy.$$

This expression represents a mean value of the function

$$\omega(x,y) = \tfrac{1}{2}[q_x(x,y) - p_y(x,y)] \tag{4}$$

over the circular region R bounded by C. Its limit as r tends to zero is the value of ω at the point (x_0,y_0). Hence the function $\omega(x,y)$, called the *rotation* of the fluid, represents the limiting angular velocity of a circular element of the fluid as the circle shrinks to its center, the point (x,y).

If $\omega(x,y) = 0$ at each point in some domain, the flow is *irrotational* in that domain. We consider only irrotational flows here, and we also assume that the fluid is *incompressible* and *free from viscosity*.

[1] For those properties of line integrals in the calculus of real variables that are used in this and the following section, see, for instance, W. Kaplan, "Advanced Calculus," 2d ed., pp. 293ff, 1973.

Let D be a simply connected domain in which the flow is irrotational. If C is any simple closed contour in D, it follows from equation (3) that the circulation around C is zero; that is,

$$\int_C p(x,y)\,dx + q(x,y)\,dy = 0.$$

Consequently, if (x_0,y_0) is any fixed point in D, we can define the function

$$\phi(x,y) = \int_{(x_0,\,y_0)}^{(x,y)} p(r,t)\,dr + q(r,t)\,dt \tag{5}$$

on the domain D. The variables of integration here are denoted by r and t to distinguish them from the upper limit of integration. The integral in equation (5) is independent of the path taken between the limits of integration so long as that path is a contour within D. For the difference between the integrals along two paths is the integral along a closed path, and the latter integral must be zero.

Since line integral (5) is independent of path, its integrand is the differential of the function $\phi(x,y)$; that is,

$$p(x,y) = \phi_x(x,y), \qquad q(x,y) = \phi_y(x,y). \tag{6}$$

The velocity vector $V = p + iq$ is then the gradient of ϕ; and the directional derivative of ϕ in any direction represents the component of the velocity of flow in that direction.

The function $\phi(x,y)$ is called the *velocity potential.* From equation (5) it is evident that $\phi(x,y)$ changes by an additive constant when the reference point (x_0,y_0) is changed. The level curves $\phi(x,y) = c$ are called *equipotentials.* Because it is the gradient of $\phi(x,y)$, the velocity vector V is normal to an equipotential at any point where V is not the zero vector.

Just as in the case of the flow of heat, the condition that the incompressible fluid enter or leave an element of volume only by flowing through the boundaries of that element requires that $\phi(x,y)$ must satisfy Laplace's equation

$$\phi_{xx}(x,y) + \phi_{yy}(x,y) = 0$$

in a domain where the fluid is free from sources or sinks. In view of equations (6) and the continuity of the functions p and q and their first partial derivatives, it follows that the first and second partial derivatives of ϕ are continuous in such a domain. Hence the velocity potential ϕ *is a harmonic function* in that domain.

88. The Stream Function

According to the previous section, the velocity vector

$$V = p(x,y) + iq(x,y) \tag{1}$$

for a simply connected domain in which the flow is irrotational can be written

$$V = \phi_x(x,y) + i\phi_y(x,y) \tag{2}$$

where ϕ is the velocity potential.

When the velocity vector is not the zero vector, it is normal to an equipotential passing through the point (x,y). If, moreover, $\psi(x,y)$ is a harmonic conjugate of $\phi(x,y)$, the velocity vector is tangent to a curve $\psi(x,y) = c$. The curves $\psi(x,y) = c$ are called the *streamlines* of the flow, and the function ψ is the *stream function*. In particular, a boundary across which fluid cannot flow is a streamline.

The analytic function

$$F(z) = \phi(x,y) + i\psi(x,y)$$

is called the *complex potential* of the flow. Note that

$$F'(z) = \phi_x(x,y) + i\psi_x(x,y),$$

or, in view of the Cauchy-Riemann equations,

$$F'(z) = \phi_x(x,y) - i\phi_y(x,y).$$

Expression (2) for the velocity thus becomes

$$V = \overline{F'(z)}. \tag{3}$$

The speed, or magnitude of the velocity, is given by the formula

$$|V| = |F'(z)|.$$

According to equation (3), Sec. 78, if ϕ is harmonic in a simply connected domain D, a harmonic conjugate of ϕ there can be written

$$\psi(x,y) = \int_{(x_0,y_0)}^{(x,y)} -\phi_t(r,t)\,dr + \phi_r(r,t)\,dt$$

where the integration is independent of path. With the aid of equations (6), Sec. 87, we can therefore write

$$\psi(x,y) = \int_C -q(r,t)\,dr + p(r,t)\,dt \tag{4}$$

where C is any contour in D from (x_0,y_0) to (x,y).

Now it is shown in the calculus of real variables that the right-hand side of equation (4) represents the integral with respect to arc length σ along C of the normal component $V_N(x,y)$ of the vector whose x and y components are $p(x,y)$ and $q(x,y)$, respectively. Thus formula (4) can be written

$$\psi(x,y) = \int_C V_N(x,y)\,d\sigma. \tag{5}$$

Physically, then, $\psi(x,y)$ represents the time rate of flow of the fluid across C. More precisely, $\psi(x,y)$ denotes the rate of flow, by volume, across a surface of unit height standing on the curve C perpendicular to the xy plane.

Since ϕ and ψ are harmonic functions in the xy plane, the results of Secs. 79 and 80 can be applied. That is, a transformation

$$z = f(w) = x(u,v) + iy(u,v),$$

where f is analytic, transforms $\phi(x,y)$ and $\psi(x,y)$ into harmonic functions of u and v. These new functions may be interpreted as velocity potential and stream function, respectively, for a flow in the new region in the uv plane. A streamline or natural boundary $\psi(x,y) = c$ in the xy plane is transformed into a streamline or natural boundary $\psi[x(u,v),y(u,v)] = c$ in the uv plane.

Under our assumptions of steady irrotational flow of fluids with uniform density ρ, it can be shown that the fluid pressure $P(x,y)$ satisfies the following special case of *Bernoulli's equation:*

$$\frac{P}{\rho} + \frac{1}{2}\,|V|^2 = \text{constant}. \tag{5}$$

Note that the pressure is greatest where the speed $|V|$ is least.

89. Flow around a Corner

When the complex potential is the function

$$F(z) = Az \tag{1}$$

where A is a positive real constant,

$$\phi(x,y) = Ax, \qquad \psi(x,y) = Ay. \tag{2}$$

The streamlines $\psi(x,y) = c$ are the horizontal lines $y = c/A$, and the velocity at any point is

$$V = \overline{F'(z)} = A.$$

Here a point (x_0,y_0) at which $\psi(x,y) = 0$ is any point on the x axis. If the point (x_0,y_0) is taken as the origin, then $\psi(x,y)$ is the rate of flow across any contour drawn from the origin to the point (x,y) (Fig. 81). The flow is uniform and to the right. It can be interpreted as the uniform flow in the upper half plane bounded by the x axis or as the uniform flow between two parallel lines $y = y_1$ and $y = y_2$.

To determine a flow in the quadrant $u \geqq 0$, $v \geqq 0$, we note that the transformation

$$z = w^2 \tag{3}$$

maps the quadrant onto the upper half of the xy plane, the boundary of the quadrant being mapped onto the entire x axis. Since $y = 2uv$, the stream function $\psi(x,y) = Ay$ for the flow in the half plane corresponds to the stream function

$$\psi(u,v) = 2Auv \tag{4}$$

for the flow in the quadrant. This function must, of course, be harmonic in the quadrant and have zero values on the boundary.

The streamlines in the quadrant are branches of the rectangular hyperbolas (Fig. 82)

$$2Auv = c.$$

The complex potential is the function $F(w) = Aw^2$, and the velocity of the fluid is

$$V = \overline{F'(w)} = 2A(u - iv).$$

The speed

$$|V| = 2A\sqrt{u^2 + v^2}$$

is directly proportional to the distance of the particle from the origin. The value of stream function (4) can be interpreted here as the rate of flow across a line segment extending from the origin to the point (u,v).

In such problems it is simplest to write first the complex potential as a function of the complex variable in the new region. The stream function and the velocity can then be obtained from that potential function.

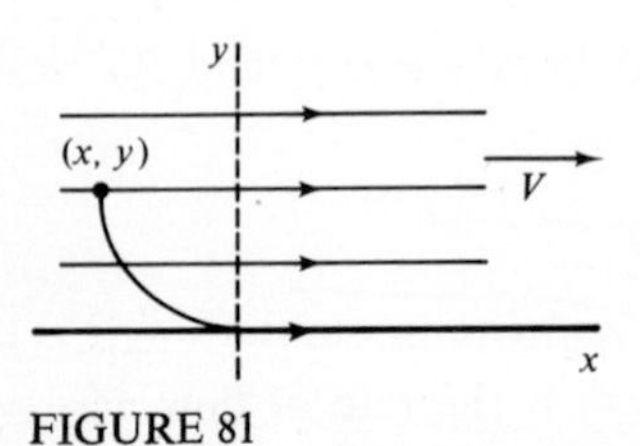

FIGURE 81

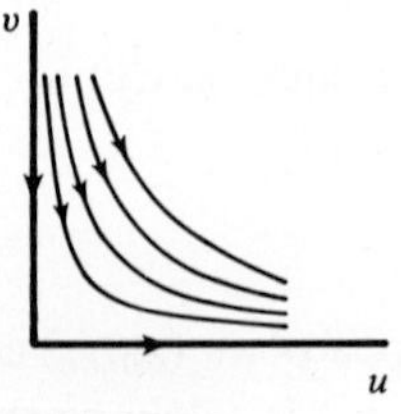

FIGURE 82

The function ψ characterizes a definite flow in a region. The question of whether just one such function exists corresponding to a given region, except possibly for a constant factor or an additive constant, will not be examined here. In some of the examples to follow, where the velocity is uniform far from the obstruction, or in Chap. 10, where sources and sinks are involved, the physical situation indicates that the flow is uniquely determined by the conditions given in the problem.

A harmonic function is not always uniquely determined, even up to a constant factor, by simply prescribing its values on the boundary of a region. We saw above, for example, that the function $\psi(x,y) = Ay$ is harmonic in the half plane $y > 0$ and has zero values on the boundary. The function $\psi_1(x,y) = Be^x \sin y$ also satisfies those conditions. However, the streamline $\psi_1(x,y) = 0$ consists not only of the line $y = 0$ but also of the lines $y = n\pi$ $(n = 1, 2, \ldots)$. Here the function $F_1(z) = Be^z$ is the complex potential for the flow in the strip between the lines $y = 0$ and $y = \pi$, both boundaries making up the streamline $\psi_1(x,y) = 0$; if $B > 0$, the fluid flows to the right along the lower boundary and to the left along the upper one.

90. Flow around a Cylinder

Let a long circular cylinder of unit radius be placed in a large body of fluid flowing with a uniform velocity, the axis of the cylinder being perpendicular to the direction of flow. To determine the steady flow around the cylinder, we represent the cylinder by the circle $x^2 + y^2 = 1$ and let the flow distant from it be parallel to the x axis (Fig. 83). Symmetry shows that the part of the x axis exterior to the circle may be treated as a boundary; so we need to consider only the upper part of the figure as the region of flow.

The boundary of this region of flow, consisting of the upper semicircle and the parts of the x axis exterior to the circle, is mapped onto the entire u axis by the transformation

$$w = z + \frac{1}{z}. \tag{1}$$

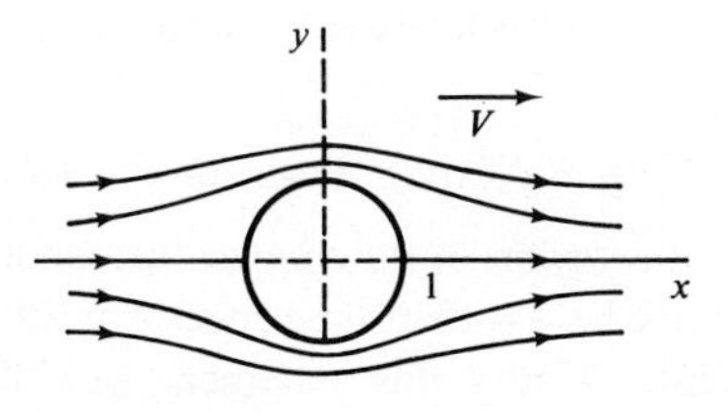

FIGURE 83

The region is mapped onto the half plane $v \geqq 0$, as indicated in Fig. 17, Appendix 2. The complex potential for a uniform flow in that half plane is

$$F(w) = Aw,$$

where A is a real constant. Hence the complex potential for the region about the circle is

$$F(z) = A\left(z + \frac{1}{z}\right). \tag{2}$$

The velocity

$$V = A\left(1 - \frac{1}{\bar{z}^2}\right) \tag{3}$$

approaches A as $|z|$ increases; that is, the flow is nearly uniform and parallel to the x axis at points distant from the circle.

From formula (3) we see that $V(\bar{z}) = \overline{V(z)}$; hence that formula also represents velocities of flow in the lower region, the lower semicircle being a streamline.

According to formula (2), the stream function for the given problem is, in polar coordinates,

$$\psi(r,\theta) = A\left(r - \frac{1}{r}\right)\sin\theta. \tag{4}$$

The streamlines

$$A\left(r - \frac{1}{r}\right)\sin\theta = c$$

are symmetric to the y axis and have asymptotes parallel to the x axis. Note that when $c = 0$, the streamline consists of the circle $r = 1$ and the parts of the x axis where $|x| \geqq 1$.

EXERCISES

1. State why the components of velocity can be obtained from the stream function by the formulas

$$p(x,y) = \psi_y(x,y), \qquad q(x,y) = -\psi_x(x,y).$$

2. At an interior point of a region of flow under the conditions we have assumed, the fluid pressure cannot be less than the pressure at all other points in a neighborhood of that point. Justify this statement with the aid of statements in Secs. 54 and 88.

3. For the flow around a corner described in Sec. 89, at what point of the region $x \geqq 0$, $y \geqq 0$ is the fluid pressure greatest?
4. Show that the speed of the fluid at points on the cylindrical surface in Sec. 90 is $2|A \sin \theta|$ and that the fluid pressure on the cylinder is greatest at the points $z = \pm 1$ and least at the points $z = \pm i$.
5. Write the complex potential for the flow around a cylinder $r = r_0$ when the velocity V approaches a real constant A as the point recedes from the cylinder.
6. Obtain the stream function $\psi(r,\theta) = Ar^4 \sin 4\theta$ for a flow in the angular region $0 \leqq \theta \leqq \pi/4$ (Fig. 84), and trace one or two of the streamlines in the interior of the region.

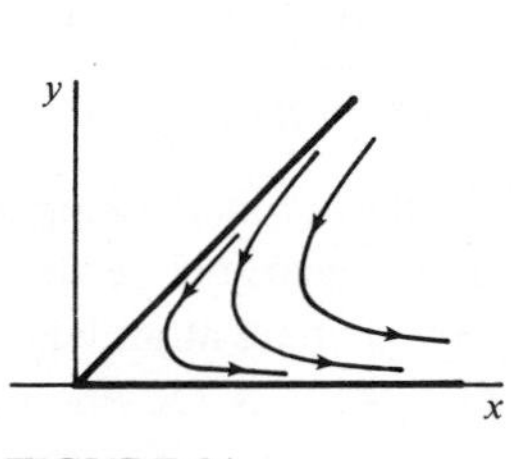

FIGURE 84

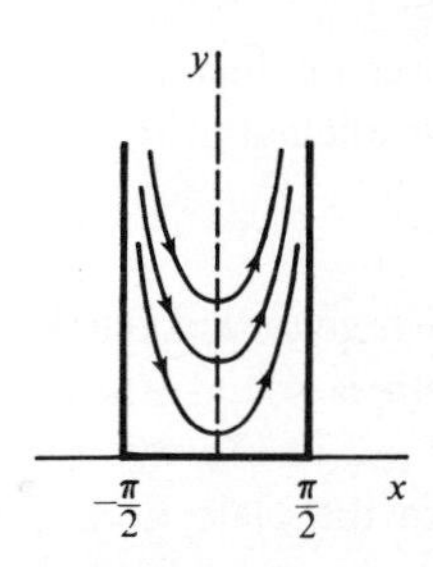

FIGURE 85

7. Obtain the complex potential $F(z) = A \sin z$ for a flow inside the semi-infinite region $-\pi/2 \leqq x \leqq \pi/2$, $y \geqq 0$ (Fig. 85). Write the equations of the streamlines.
8. Show that if the velocity potential is $\phi(r,\theta) = A \text{ Log } r$ $(A > 0)$ for flow in the region $r \geqq r_0$, the streamlines are the half lines $r \geqq r_0$, $\theta = c$ and the rate of flow outward through each complete circle about the origin is $2\pi A$, corresponding to a source of that strength at the origin.
9. Obtain the complex potential $F(z) = A(z^2 + z^{-2})$ for a flow in the region $r \geqq 1$, $0 \leqq \theta \leqq \pi/2$. Write formulas for V and ψ. Note how the speed $|V|$ varies along the boundary of the region, and verify that $\psi(x,y) = 0$ on the boundary.
10. Suppose that the flow at an infinite distance from the cylinder of unit radius in Sec. 90 is uniform in a direction making an angle α with the x axis; that is,

$$\lim_{|z| \to \infty} V = A \exp (i\alpha) \qquad (A > 0).$$

Find the complex potential.

Ans. $F(z) = A[z \exp (-i\alpha) + z^{-1} \exp (i\alpha)]$.

11. The transformation $z = w + 1/w$ maps the circle $|w| = 1$ onto the line segment joining the points $z = 2$ and $z = -2$, and it maps the domain outside that circle onto the rest of the z plane. (See Exercises 18 and 19, Sec. 41.) Write

$$z - 2 = r_1 \exp (i\theta_1), \qquad z + 2 = r_2 \exp (i\theta_2),$$

and

$$(z^2-4)^{1/2} = \sqrt{r_1 r_2}\exp\frac{i(\theta_1+\theta_2)}{2} \qquad (0 \leqq \theta_1 < 2\pi, 0 \leqq \theta_2 < 2\pi);$$

the function $(z^2-4)^{1/2}$ is then single-valued and analytic everywhere except on the branch cut consisting of the segment of the x axis between the points $z = \pm 2$. Show that the inverse of the transformation $z = w + 1/w$, such that $|w| > 1$ for every point z not on the branch cut, can be written

$$w = \frac{1}{2}[z + (z^2-4)^{1/2}] = \frac{1}{4}\left(\sqrt{r_1}\exp\frac{i\theta_1}{2} + \sqrt{r_2}\exp\frac{i\theta_2}{2}\right)^2.$$

Thus the transformation and that inverse establish a one to one correspondence between points in the two domains.

12. With the aid of the results found in Exercises 10 and 11, derive the formula

$$F(z) = A[z\cos\alpha - i(z^2-4)^{1/2}\sin\alpha]$$

which gives the complex potential for the steady flow around a long plate whose width is 4 and whose cross section is the line segment joining the two points $z = \pm 2$ in Fig. 86, assuming that the velocity of the fluid at an infinite distance from the plate is $A\exp(i\alpha)$. The branch $(z^2-4)^{1/2}$ is the one described in Exercise 11, and $A > 0$.

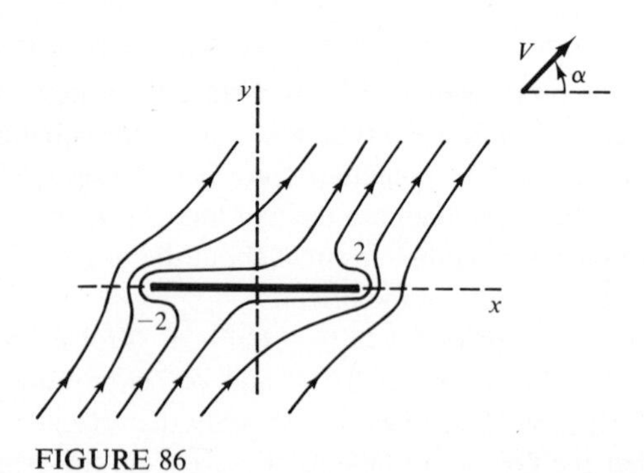

FIGURE 86

13. Show that if $\sin\alpha \neq 0$ in Exercise 12, the speed of the fluid along the line segment joining the points $z = \pm 2$ is infinite at the ends and is equal to $A|\cos\alpha|$ at the midpoint.

14. For the sake of simplicity, suppose that $0 < \alpha \leqq \pi/2$ in Exercise 12. Then show that the velocity of the fluid along the upper side of the line segment representing the plate in Fig. 86 is zero at the point $x = 2\cos\alpha$ and that the velocity along the lower side of the segment is zero at the point $x = -2\cos\alpha$.

15. A circle with its center at a point x_0 $(0 < x_0 < 1)$ on the x axis and passing through the point $z = -1$ is subjected to the transformation $w = z + 1/z$. Individual nonzero points $z = \exp(i\theta)$ can be mapped geometrically by adding the vector $r^{-1} \exp(-i\theta)$ to the vector z. Indicate by mapping some points that the image of the circle is a profile of the type shown in Fig. 87 and that points exterior to the circle map onto points exterior to the profile. This is a special case of the profile of a *Joukowski airfoil.* (See also Exercises 16 and 17 below.)

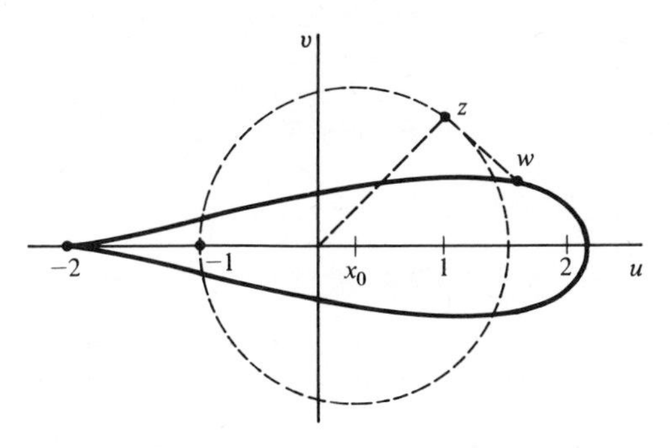

FIGURE 87

16. (*a*) Show that the mapping of the circle in Exercise 15 is conformal except at the point $z = -1$. (*b*) Let the complex numbers

$$t = \lim_{\Delta z \to 0} \frac{\Delta z}{|\Delta z|}, \qquad \tau = \lim_{\Delta z \to 0} \frac{\Delta w}{|\Delta w|}$$

represent unit vectors tangent to a directed arc at $z = -1$ and that arc's image, respectively, under the transformation $w = z + 1/z$. Show that $\tau = -t^2$ and hence that the Joukowski profile of Fig. 87 has a cusp at the point $w = -2$, the angle between the tangents at the cusp being zero.

17. The inverse of the transformation $w = z + 1/z$ used in Exercise 15 is given, with z and w interchanged, in Exercise 11. Find the complex potential for the flow around the airfoil introduced in Exercise 15 when the velocity V of the fluid at an infinite distance from the origin is a real constant A.

18. Note that under the transformation

$$w = e^z + z$$

both the positive and negative parts of the line $y = \pi$ are mapped onto the half line $u \leqq -1$, $v = \pi$. Similarly, $y = -\pi$ is mapped onto $u \leqq -1$, $v = -\pi$; and the strip $-\pi \leqq y \leqq \pi$ is mapped onto the w plane. Also note that the change of directions, arg (dw/dz), under this transformation approaches zero as x tends to $-\infty$. Show that the streamlines of a fluid flowing through the open channel formed by the half lines in the w plane (Fig. 88) are the images of the lines $y = c$ in the strip. These streamlines also represent the equipotential curves of the electrostatic field near the edge of a parallel-plate capacitor.

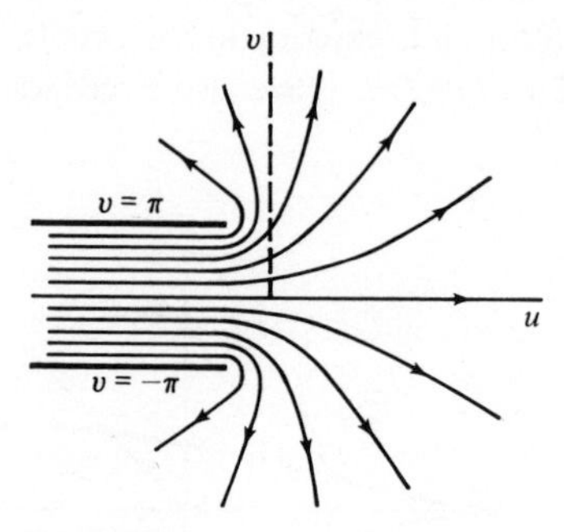

FIGURE 88

10

THE SCHWARZ-CHRISTOFFEL TRANSFORMATION

In this chapter we construct a transformation, known as the Schwarz-Christoffel transformation, which maps the x axis and the upper half of the z plane onto a given simple closed polygon and its interior in the w plane. Applications are made to the solution of problems in fluid flow and electrostatic potential theory.

91. Mapping the Real Axis onto a Polygon

We represent the unit vector tangent to a smooth directed arc C at a point z_0 by the complex number t. Let the number τ denote the unit vector tangent to the image Γ of C at the corresponding point w_0 under a transformation $w = f(z)$. We assume that f is analytic at z_0 and that $f'(z_0) \neq 0$. According to Sec. 76,

$$\arg \tau = \arg t + \arg f'(z_0). \tag{1}$$

In particular, if C is a segment of the x axis with positive sense to the right, $t = 1$ and $\arg t = 0$ at each point $z_0 = x$ on C. Equation (1) then becomes

$$\arg \tau = \arg f'(x). \tag{2}$$

If $f'(z)$ has a constant argument along that segment, it follows that $\arg \tau$ is constant; that is, the image Γ of C is also a segment of a straight line.

Let us now construct a transformation $w = f(z)$ that maps the whole x axis onto a polygon of n sides where $x_1, x_2, \ldots, x_{n-1}$ and $z = \infty$ are the points on that axis whose images are to be the vertices of the polygon and where

$$x_1 < x_2 < \cdots < x_{n-1}.$$

The vertices are the points $w_j = f(x_j)$ $(j = 1, 2, \ldots, n-1)$ and $w_n = f(\infty)$. The function f should be such that $\arg f'(z)$ jumps from one constant value to another at the points $z = x_j$ as the point z traces the x axis.

If the function f is chosen such that

$$f'(z) = A(z - x_1)^{-k_1}(z - x_2)^{-k_2} \cdots (z - x_{n-1})^{-k_{n-1}}, \tag{3}$$

where A is a complex constant and each k_j is a real constant, the argument of $f'(z)$ changes in the prescribed manner as z describes the real axis; for the argument of function (3) can be written

$$\begin{aligned} \arg f'(z) = \arg A - k_1 \arg (z - x_1) \\ - k_2 \arg (z - x_2) - \cdots - k_{n-1} \arg (z - x_{n-1}). \end{aligned} \tag{4}$$

When $z = x$ and $x < x_1$,

$$\arg (z - x_1) = \arg (z - x_2) = \cdots = \arg (z - x_{n-1}) = \pi.$$

When $x_1 < x < x_2$, $\arg (z - x_1) = 0$ and each of the other arguments is π. According to equation (4), then, $\arg f'(z)$ increases abruptly by the angle $k_1\pi$ as z moves to the right through the point $z = x_1$. It again jumps in value, by the amount $k_2\pi$, as z passes through the point x_2, etc.

In view of equation (2), the unit vector τ is constant in direction as z moves from x_{j-1} to x_j; w then moves in that fixed direction along a straight line. The direction of τ changes abruptly, by the angle $k_j\pi$, at the image point w_j of x_j (Fig. 89). Those angles $k_j\pi$ are the exterior angles of the polygon described by the point w.

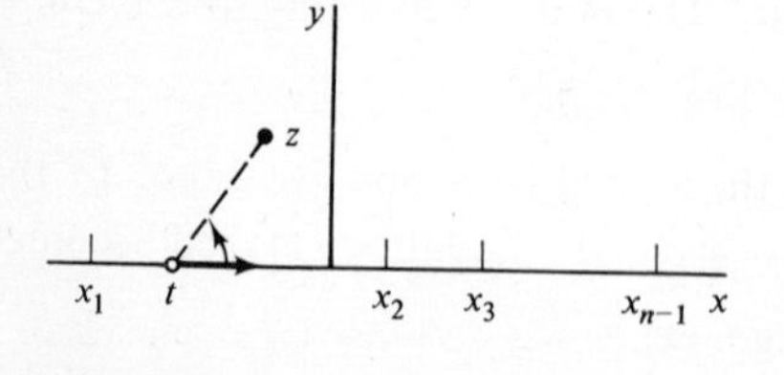

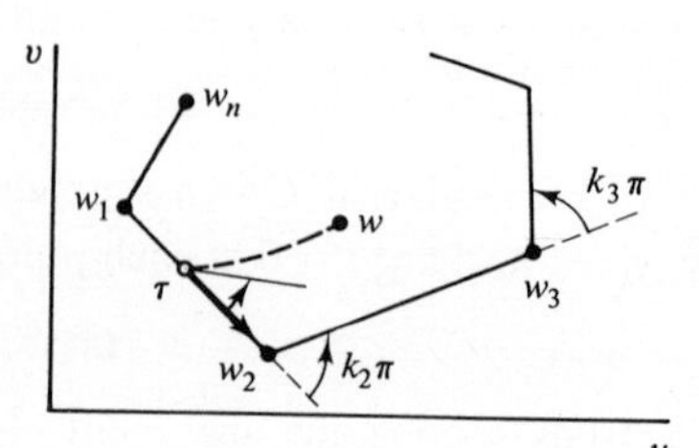

FIGURE 89

The exterior angles can be limited to angles between $-\pi$ and π; that is, $-1 < k_j < 1$. We assume that the sides of the polygon never cross one another and that the polygon is given a positive, or counterclockwise, orientation. The sum of the exterior angles of a *closed* polygon is then 2π; and the exterior angle at the vertex w_n, the image of the point $z = \infty$, can be written

$$k_n \pi = 2\pi - (k_1 + k_2 + \cdots + k_{n-1})\pi.$$

Thus the numbers k_j must necessarily satisfy the conditions

$$k_1 + k_2 + \cdots + k_{n-1} + k_n = 2, \qquad -1 < k_j < 1 \qquad (j = 1, 2, \ldots, n). \tag{5}$$

Note that $k_n = 0$ if

$$k_1 + k_2 + \cdots + k_{n-1} = 2. \tag{6}$$

In that case the direction of τ does not change at the point w_n; so w_n is not a vertex, and the polygon has $n - 1$ sides.

The existence of a mapping function f whose derivative is given by formula (3) will now be established.

92. The Schwarz-Christoffel Transformation

In our formula

$$f'(z) = A(z - x_1)^{-k_1}(z - x_2)^{-k_2} \cdots (z - x_{n-1})^{-k_{n-1}} \tag{1}$$

for the derivative of a function that is to map the x axis onto a polygon, let the factors $(z - x_j)^{-k_j}$ represent branches of power functions with branch cuts extending below that axis. To be specific, write

$$(z - x_j)^{-k_j} = |z - x_j|^{-k_j} \exp(-ik_j\theta_j) \qquad \left(-\frac{\pi}{2} < \theta_j < \frac{3\pi}{2}\right) \tag{2}$$

where $\theta_j = \arg(z - x_j)$ and $j = 1, 2, \ldots, n - 1$. Then $f'(z)$ is analtyic everywhere in the half plane $y \geqq 0$ except at the $n - 1$ branch points x_j.

If z_0 is a point in that region of analyticity, denoted here by R, then the function

$$F(z) = \int_{z_0}^{z} f'(s)\, ds \tag{3}$$

is single-valued and analytic throughout that same region, where the path of integration from z_0 to z is any contour lying within R. Moreover, $F'(z) = f'(z)$.

In order to define the function F at the point $z = x_1$ so that it is continuous there, we note that $(z - x_1)^{-k_1}$ is the only factor in expression (1) that is not analytic at x_1. Hence if $\phi(z)$ denotes the product of the rest of the factors in that expression, $\phi(z)$ is analytic at x_1 and is represented throughout an open disk $|z - x_1| < r_1$ by its Taylor series about x_1. So we can write

$$\begin{aligned} f'(z) &= (z - x_1)^{-k_1}\phi(z) \\ &= (z - x_1)^{-k_1}\left[\phi(x_1) + \phi'(x_1)(z - x_1) + \frac{\phi''(x_1)}{2!}(z - x_1)^2 + \cdots\right], \end{aligned}$$

or

$$f'(z) = \phi(x_1)(z - x_1)^{-k_1} + (z - x_1)^{1-k_1}\psi(z) \tag{4}$$

where ψ is analytic, and therefore continuous, throughout the entire open disk. Since $1 - k_1 > 0$, the last term on the right in equation (4) thus represents a continuous function of z throughout the upper half of the disk, where $\operatorname{Im} z \geqq 0$, if we assign that term the value zero at $z = x_1$. It follows that the integral

$$\int_{Z_1}^{z} (s - x_1)^{1-k_1}\psi(s)\,ds$$

of the last term along a contour from Z_1 to z, where Z_1 and the contour lie in the half disk, is a continuous function of z at $z = x_1$. The integral

$$\int_{Z_1}^{z} (s - x_1)^{-k_1}\,ds = \frac{1}{1 - k_1}\left[(z - x_1)^{1-k_1} - (Z_1 - x_1)^{1-k_1}\right]$$

along the same path also represents a continuous function of z at x_1 if we define the value of the integral there as its limit as z approaches x_1 in the half disk. The integral of function (4) along the stated path from Z_1 to z is therefore continuous at $z = x_1$, and so then is integral (3) since it can be written as an integral along a contour in R from z_0 to Z_1 plus the integral from Z_1 to z.

The above argument applies at each of the $n - 1$ points x_j to make F continuous throughout the region $y \geqq 0$.

From equation (1) we can show that for a sufficiently large positive number R a positive constant M exists such that if $\operatorname{Im} z \geqq 0$,

$$|f'(z)| < \frac{M}{|z|^{2-k_n}} \qquad \text{whenever} \qquad |z| > R. \tag{5}$$

Since $2 - k_n > 1$, this order property of the integrand in equation (3) ensures the existence of the limit of the integral there as z tends to infinity; that is, a number W_n exists such that

$$\lim_{z \to \infty} F(z) = W_n \qquad (\operatorname{Im} z \geqq 0). \tag{6}$$

Details of the argument are left to Exercises 10 and 11, Sec. 95.

Our mapping function whose derivative is given by formula (1) can be written $f(z) = F(z) + B$, where B is a complex constant. The resulting transformation,

$$(7) \qquad w = A \int_{z_0}^{z} (s - x_1)^{-k_1}(s - x_2)^{-k_2} \cdots (s - x_{n-1})^{-k_{n-1}}\, ds + B,$$

is the *Schwarz-Christoffel transformation*, named in honor of the two German mathematicians H. A. Schwarz (1843–1921) and E. B. Christoffel (1829–1900) who discovered it independently.

Transformation (7) is continuous throughout the half plane $y \geqq 0$; it is conformal there except at the points x_j. We have assumed that the numbers k_j satisfy conditions (5), Sec. 91. In addition, we suppose that the constants x_j and k_j are such that the sides of the polygon do not cross, so that the polygon is a simple closed contour. Then, according to Sec. 91, as the point z describes the x axis in the positive direction, its image w describes the polygon P in the positive sense, and there is a one to one correspondence between points on that axis and points on P. According to condition (6), the image w_n of the point $z = \infty$ exists, it being $w_n = W_n + B$.

If z is an interior point of the upper half plane $y \geqq 0$ and x_0 is any point other than one of the x_j on the x axis, then the angle from the vector t at x_0 up to the line segment joining x_0 and z is positive and less than π (Fig. 89). At the image w_0 of x_0, the corresponding angle from the vector τ to the image of the line segment joining x_0 and z has that same value. Thus the images of interior points in the half plane lie to the left of the sides of the polygon taken counterclockwise. A proof that the transformation establishes a one to one correspondence between the interior points of the half plane and the points within the polygon is left to the exercises.

Given a definite polygon P, let us examine the number of constants in the Schwarz-Christoffel transformation that must be determined in order to map the x axis onto P. For this purpose we may write $z_0 = 0$, $A = 1$, and $B = 0$ and simply require that the x axis be mapped onto some polygon P' similar to P. The size and position of P' can then be adjusted to match those of P by introducing the appropriate constants A and B.

The numbers k_j are all determined from the exterior angles at the vertices of P. The $n - 1$ constants x_j remain to be chosen. The image of the x axis is some polygon P' which has the same angles as P. But if P' is to be similar to P, then $n - 2$ connected sides must have a common ratio to the corresponding sides of P; this condition is expressed by means of $n - 3$ equations in the $n - 1$ real unknowns x_j. Thus *two of the numbers* x_j, *or two relations between them, can be chosen arbitrarily*, provided those $n-3$ equations in the remaining $n-3$ unknowns have real-valued solutions.

When a finite point $z = x_n$ on the x axis, instead of the point at infinity, represents the point whose image is the vertex w_n, it follows from the preceding section that the Schwarz-Christoffel transformation takes the form

$$w = A \int_{z_0}^{z} (s - x_1)^{-k_1}(s - x_2)^{-k_2} \cdots (s - x_n)^{-k_n}\, ds + B \tag{8}$$

where $k_1 + k_2 + \cdots + k_n = 2$. The exponents k_j are determined from the exterior angles of the polygon. But in this case there are n real constants x_j which must satisfy the $n-3$ equations noted above. Thus *three of the numbers x_j, or three conditions on those n numbers, can be chosen arbitrarily* in transformation (8) of the x axis onto a given polygon.

93. Triangles and Rectangles

The Schwarz-Christoffel transformation is written in terms of the points x_j and not in terms of their images, the vertices of the polygon. Not more than three of those points can be chosen arbitrarily; so when the given polygon has more than three sides, some of the points x_j must be determined in order to make the given polygon, or any polygon similar to it, be the image of the x axis. The selection of conditions for the determination of those constants, conditions that are convenient to use, often requires ingenuity.

Another limitation in using the transformation is due to the integration that is involved. Often the integral cannot be evaluated in terms of a finite number of elementary functions. In such cases the solution of problems by means of it can become quite involved.

If the polygon is a triangle with vertices at the points w_1, w_2, and w_3 (Fig. 90), the transformation can be written

$$w = A \int_{z_0}^{z} (s - x_1)^{-k_1}(s - x_2)^{-k_2}(s - x_3)^{-k_3}\, ds + B \tag{1}$$

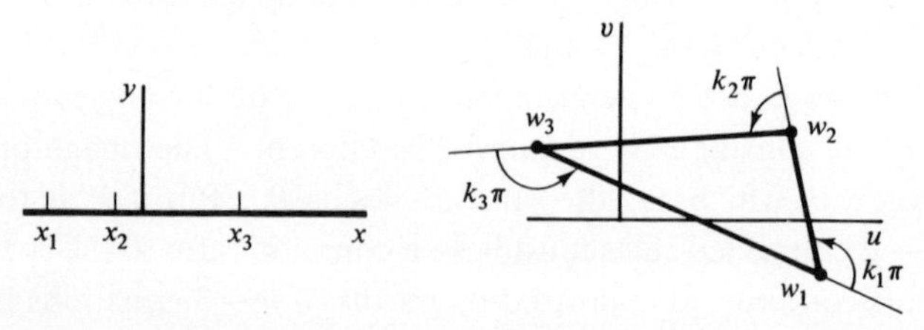

FIGURE 90

where $k_1 + k_2 + k_3 = 2$. In terms of the interior angles θ_j,

$$k_j = 1 - \frac{1}{\pi}\theta_j \qquad (j = 1, 2, 3).$$

Here we have taken all three points x_j as finite points on the x axis. Arbitrary values can be assigned to each of them. The complex constants A and B, associated with the size and position of the triangle, can be determined so that the upper half plane is mapped onto the given triangular region.

If we take the vertex w_3 as the image of the point at infinity, the transformation becomes

$$(2) \qquad w = A\int_{z_0}^{z}(s - x_1)^{-k_1}(s - x_2)^{-k_2}\,ds + B,$$

where arbitrary real values can be assigned to x_1 and x_2.

The integrals in equations (1) and (2) do not represent elementary functions unless the triangle is degenerate with one or two of its vertices at infinity. The integral in equation (2) becomes an elliptic integral when the triangle is equilateral or when it is a right triangle with one of its angles equal to either $\pi/3$ or $\pi/4$.

For an equilateral triangle, $k_1 = k_2 = k_3 = 2/3$. It is convenient to write $x_1 = -1$, $x_2 = 1$, and $x_3 = \infty$ and to use equation (2) where $z_0 = 1$, $A = 1$, and $B = 0$. The transformation then becomes

$$(3) \qquad w = \int_1^z (s + 1)^{-2/3}(s - 1)^{-2/3}\,ds.$$

The image of the point $z = 1$ is clearly $w = 0$; that is, $w_2 = 0$. When $z = -1$ in the integral, we can write $s = x$; then $-1 < x < 1$ and $x + 1 > 0$ and $\arg(x + 1) = 0$, while $|x - 1| = 1 - x$ and $\arg(x - 1) = \pi$. Hence

$$(4) \qquad \begin{aligned} w_1 &= \int_1^{-1}(x + 1)^{-2/3}(1 - x)^{-2/3}\exp\left(-\frac{2\pi i}{3}\right)dx \\ &= \exp\left(\frac{\pi i}{3}\right)\int_{-1}^{1}\frac{dx}{(1 - x^2)^{2/3}}. \end{aligned}$$

The last integral here reduces to the one used in defining the beta function (Exercise 9, Sec. 75). Let b denote its value, which is positive:

$$(5) \qquad b = 2\int_0^1 \frac{dx}{(1 - x^2)^{2/3}} = B(\tfrac{1}{2}, \tfrac{1}{3}).$$

The vertex w_1 is therefore the point (Fig. 91)

$$(6) \qquad w_1 = b\exp\frac{\pi i}{3}.$$

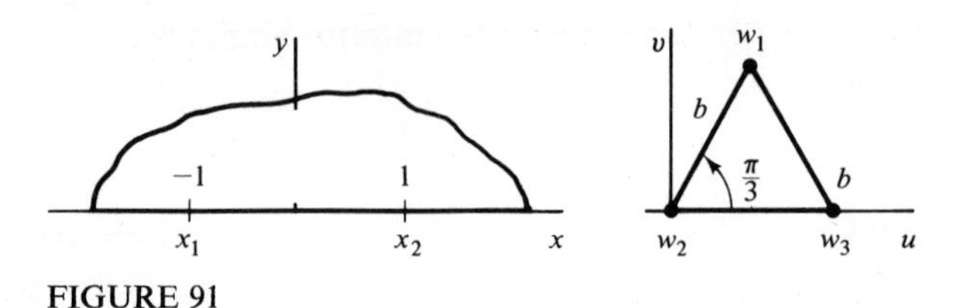

FIGURE 91

The vertex w_3 is on the positive u axis because

$$w_3 = \int_1^\infty (x+1)^{-2/3}(x-1)^{-2/3}\,dx = \int_1^\infty \frac{dx}{(x^2-1)^{2/3}}.$$

But the value of w_3 is also represented by integral (3) when z tends to infinity along the negative x axis; that is,

$$w_3 = \int_1^{-1} (|x+1|\,|x-1|)^{-2/3} \exp\left(-\frac{2\pi i}{3}\right) dx$$
$$+ \int_{-1}^{-\infty} (|x+1|\,|x-1|)^{-2/3} \exp\left(-\frac{4\pi i}{3}\right) dx.$$

In view of equation (4), then,

$$w_3 = w_1 + \exp\left(-\frac{4\pi i}{3}\right) \int_{-1}^{-\infty} (|x+1|\,|x-1|)^{-2/3}\,dx$$
$$= b \exp\frac{\pi i}{3} + \exp\left(-\frac{\pi i}{3}\right) \int_1^\infty \frac{dx}{(x^2-1)^{2/3}},$$

or

$$w_3 = b \exp\frac{\pi i}{3} + w_3 \exp\left(-\frac{\pi i}{3}\right).$$

Solving for w_3, we find that

(7) $$w_3 = b.$$

We have thus verified that the image of the x axis is the equilateral triangle of side b shown in Fig. 91. We can see also that $w = (b/2)\exp(\pi i/3)$ when $z = 0$.

When the polygon is a rectangle, each $k_j = 1/2$. If we choose ± 1 and $\pm a$ as the points x_j whose images are the vertices and write

(8) $$g(z) = (z+a)^{-1/2}(z+1)^{-1/2}(z-1)^{-1/2}(z-a)^{-1/2}$$

where $0 \leqq \arg(z - x_j) \leqq \pi$, the Schwarz-Christoffel transformation becomes

(9) $$w = -\int_0^z g(s)\,ds$$

except for a transformation $W = Aw + B$ to adjust the size and position of the rectangle. Integral (9) is a constant times the elliptic integral

$$\int_0^z (1 - s^2)^{-1/2}(1 - k^2s^2)^{-1/2}\,ds \qquad \left(k = \frac{1}{a}\right);$$

but form (8) of the integrand indicates more clearly the appropriate branches of the irrational functions involved.

Let us locate the vertices of the rectangle when $a > 1$. As shown in Fig. 92, $x_1 = -a$, $x_2 = -1$, $x_3 = 1$, and $x_4 = a$. All four vertices can be described in terms of two positive numbers b and c which depend on the value of a in the following manner:

$$b = \int_0^1 |g(x)|\,dx = \int_0^1 \frac{dx}{\sqrt{(1 - x^2)(a^2 - x^2)}}, \tag{10}$$

$$c = \int_1^a |g(x)|\,dx = \int_1^a \frac{dx}{\sqrt{(x^2 - 1)(a^2 - x^2)}}. \tag{11}$$

When $-1 < x < 0$, then $\arg(x + a) = \arg(x + 1) = 0$ and $\arg(x - 1) = \arg(x - a) = \pi$; hence

$$g(x) = [\exp(-\pi i/2)]^2\,|g(x)| = -|g(x)|.$$

When $-a < x < -1$, then $g(x) = [\exp(-\pi i/2)]^3\,|g(x)| = i|g(x)|$. Thus

$$w_1 = -\int_0^{-a} g(x)\,dx = -\int_0^{-1} g(x)\,dx - \int_{-1}^{-a} g(x)\,dx$$
$$= \int_0^{-1} |g(x)|\,dx - i\int_{-1}^{-a} |g(x)|\,dx = -b + ic.$$

It is left to the exercises to show that

$$w_2 = -b, \qquad w_3 = b, \qquad w_4 = b + ic. \tag{12}$$

The position and dimensions of the rectangle are shown in Fig. 92.

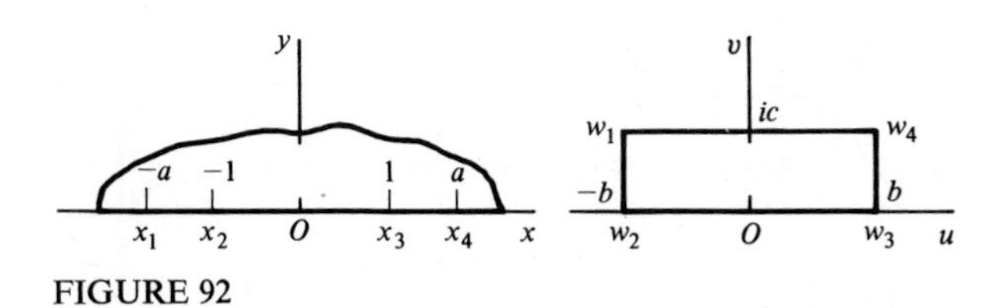

FIGURE 92

94. Degenerate Polygons

We now apply the Schwarz-Christoffel transformation to some degenerate polygons for which the integrals represent elementary functions. For the purpose of illustration, we begin with some known transformations.

First, let us map the half plane $y \geqq 0$ onto the semi-infinite strip

$$-\frac{\pi}{2} \leqq u \leqq \frac{\pi}{2}, \qquad v \geqq 0.$$

We consider the strip as the limiting form of a triangle with vertices w_1, w_2, and w_3 (Fig. 93) as the imaginary part of w_3 tends to infinity.

The limiting values of the exterior angles are

$$k_1\pi = k_2\pi = \frac{\pi}{2}, \qquad k_3\pi = \pi.$$

We choose the points $x_1 = -1$, $x_2 = 1$, and $x_3 = \infty$ as the points whose images are the vertices. Then the derivative of the mapping function can be written

$$\frac{dw}{dz} = A(z+1)^{-1/2}(z-1)^{-1/2} = A'(1-z^2)^{-1/2}.$$

Hence $w = A' \sin^{-1} z + B$. If we write $A' = 1/a$ and $B = b/a$, then

$$z = \sin(aw - b).$$

This transformation from the w to the z plane satisfies the conditions $z = -1$ when $w = -\pi/2$ and $z = 1$ when $w = \pi/2$ if $a = 1$ and $b = 0$. The resulting transformation is

$$z = \sin w,$$

which we verified in Sec. 39 as one that maps the strip onto the half plane.

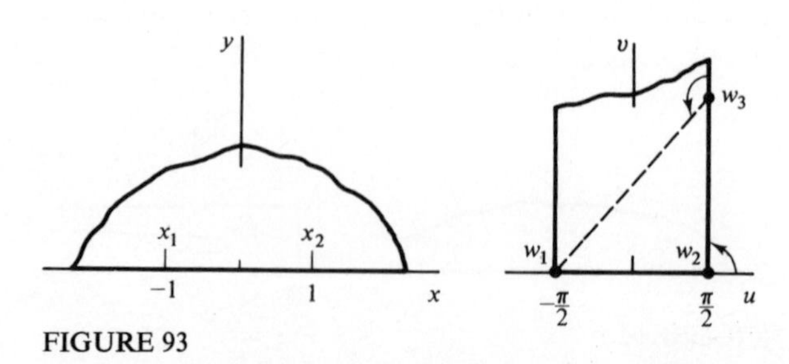

FIGURE 93

95. The Infinite Strip

Consider the strip $0 < v < \pi$ as the limiting form of a rhombus with vertices at the points $w_1 = \pi i$, w_2, $w_3 = 0$, and w_4, as the points w_2 and w_4 are moved infinitely far to the left and right, respectively (Fig. 94). In the limit the exterior angles become

$$k_1\pi = 0, \qquad k_2\pi = \pi, \qquad k_3\pi = 0, \qquad k_4\pi = \pi.$$

We leave x_1 to be determined and choose the values $x_2 = 0$, $x_3 = 1$, and $x_4 = \infty$. The derivative of the Schwarz-Christoffel mapping function then becomes

$$\frac{dw}{dz} = A(z - x_1)^0 z^{-1}(z-1)^0 = \frac{A}{z};$$

thus

$$w = A \operatorname{Log} z + B.$$

Now $B = 0$ because $w = 0$ when $z = 1$. The constant A must be real because the point w lies on the real axis when $z = x$ and $x > 0$. The point $w = \pi i$ is the image of the point $z = x_1$, where x_1 is a negative number; therefore

$$\pi i = A \operatorname{Log} x_1 = A \operatorname{Log} |x_1| + A\pi i.$$

By identifying real and imaginary parts here, we see that $|x_1| = 1$ and $A = 1$. Hence the transformation becomes

$$w = \operatorname{Log} z;$$

also $x_1 = -1$. We already know from Sec. 38 that this transformation maps the half plane onto the strip.

The procedure used here and in the preceding section is not rigorous because limiting values of angles and coordinates were not introduced in an orderly way. Limiting values were used whenever it seemed expedient to do so. But if we verify the mapping obtained, it is not essential that we justify the steps in our derivation of the mapping function. The formal method used here is shorter and less tedious than rigorous methods.

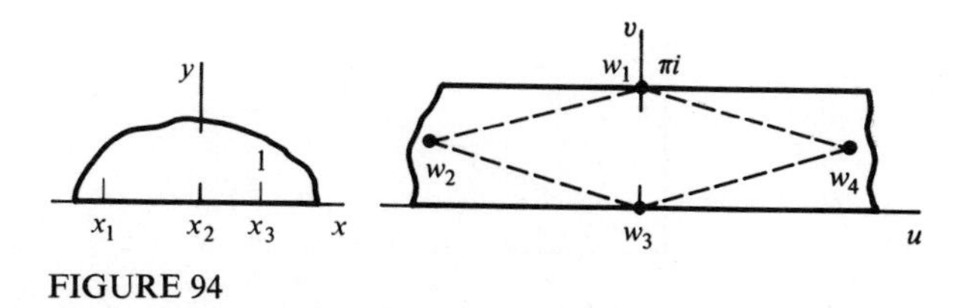

FIGURE 94

EXERCISES

1. In transformation (1), Sec. 93, write $B = z_0 = 0$ and

$$A = \exp\frac{3\pi i}{4}, \qquad x_1 = -1, \qquad x_2 = 0, \qquad x_3 = 1,$$

$$k_1 = \tfrac{3}{4}, \qquad k_2 = \tfrac{1}{2}, \qquad k_3 = \tfrac{3}{4}$$

to map the x axis onto an *isosceles right triangle.* Show that the vertices of that triangle are the points

$$w_1 = bi, \qquad w_2 = 0, \qquad w_3 = b,$$

where b is the positive constant

$$b = \int_0^1 (1 - x^2)^{-3/4} x^{-1/2}\, dx.$$

Also show that $2b = B(\frac{1}{4}, \frac{1}{4})$ where B is the beta function.

2. Obtain formulas (12) in Sec. 93 for the rest of the vertices of the rectangle shown in Fig. 92.

3. Show that when $0 < a < 1$ in formulas (8) and (9), Sec. 93, the vertices of the rectangle are those shown in Fig. 92 where b and c now have the values

$$b = \int_0^a |g(x)|\, dx, \qquad c = \int_a^1 |g(x)|\, dx.$$

4. Show that the special case

$$w = i\int_0^z (s+1)^{-1/2}(s-1)^{-1/2}s^{-1/2}\, ds$$

of the Schwarz-Christoffel transformation (7), Sec. 92, maps the x axis onto the *square* with vertices

$$w_1 = bi, \qquad w_2 = 0, \qquad w_3 = b, \qquad w_4 = b + ib$$

where the positive number b is given in terms of the beta function:

$$b = \tfrac{1}{2}B(\tfrac{1}{4}, \tfrac{1}{2}).$$

5. Use the Schwarz-Christoffel transformation to arrive at the transformation $w = z^m$ $(0 < m < 1)$ which maps the half plane $y \geqq 0$ onto the angular region $0 \leqq \arg w \leqq m\pi$ and the point $z = 1$ onto the point $w = 1$. Consider the angular region as the limiting case of the triangle shown in Fig. 95 as the angle α tends to 0.

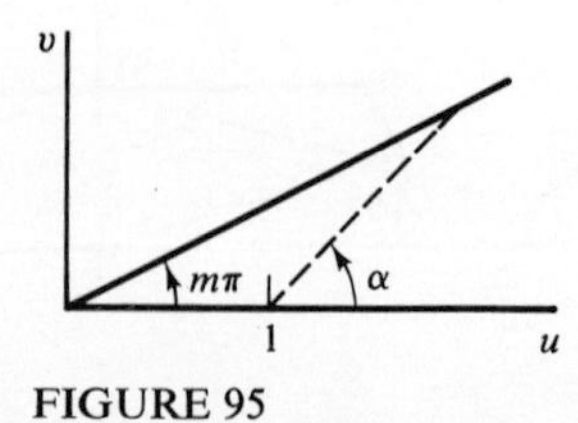

FIGURE 95

6. Refer to Fig. 26, Appendix 2. As the point z moves to the right along the negative real axis, its image point w is to move to the right along the entire u axis. As z describes the segment $0 \leqq x \leqq 1, y = 0$ of the positive real axis, its image point w is to move to the left along the half line $u \geqq 1$, $v = \pi i$; and as z moves to the right along that part of the positive real axis where $x \geqq 1$, its image point w is to move to the right along the same half line $u \geqq 1$, $v = \pi i$. Note the changes in direction of the motion of w at the images of the points $z = 0$ and $z = 1$. These changes suggest that the derivative of a mapping function be

$$f'(z) = k(z - 0)^{-1}(z - 1)$$

where k is some constant; thus obtain formally the mapping function

$$w = \pi i + z - \operatorname{Log} z$$

which can be verified as one that maps the half plane Re $z > 0$ as indicated in the figure.

7. As the point z moves to the right along that part of the negative real axis where $x \leqq -1$, its image point is to move to the right along the negative real axis in the w plane. As z moves on the real axis to the right along the segment $-1 \leqq x \leqq 0$, $y = 0$ and then along the segment $0 \leqq x \leqq 1$, $y = 0$, its image point w is to move in the direction of increasing v along the segment $u = 0$, $0 \leqq v \leqq 1$ and then in the direction of decreasing v along the same segment. Finally, as z moves to the right along that part of the positive real axis where $x \geqq 1$, its image point is to move to the right along the positive real axis in the w plane. Note the changes in the direction of the motion of w at the images of the points $z = -1$, $z = 0$, and $z = 1$. A mapping function whose derivative is

$$f'(z) = k(z + 1)^{-1/2}(z - 0)^{1}(z - 1)^{-1/2},$$

where k is some constant, is thus indicated. Obtain formally the mapping function

$$w = \sqrt{z^2 - 1}$$

where $0 < \arg \sqrt{z^2 - 1} < \pi$. By considering the successive mappings $Z = z^2$, $W = Z - 1$, and $w = \sqrt{W}$, verify that the resulting transformation maps the half plane Re $z > 0$ onto the half plane Im $w > 0$ with a cut along the segment $u = 0$, $0 < v \leqq 1$.

8. The inverse of the linear fractional transformation

$$Z = \frac{i - z}{i + z}$$

maps the unit disk $|Z| \leqq 1$ conformally, except at the point $Z = -1$, onto the half plane Im $z \geqq 0$. (See Fig. 13, Appendix 2.) Let Z_j be points on the circle $|Z| = 1$ whose images are the points $z = x_j$ ($j = 1, 2, \ldots, n$) which are used in the Schwarz-Christoffel transformation (8), Sec. 92. Show formally, without determining the branches of the irrational functions, that

$$\frac{dw}{dZ} = A'(Z - Z_1)^{-k_1}(Z - Z_2)^{-k_2} \cdots (Z - Z_n)^{-k_n}$$

where A' is a constant. Thus show that *the transformation*

$$w = A' \int_0^Z (S - Z_1)^{-k_1}(S - Z_2)^{-k_2} \cdots (S - Z_n)^{-k_n}\, dS + B$$

maps the interior of the circle $|Z| = 1$ *onto the interior of a polygon*, the vertices of the polygon being the images of the points Z_j on the circle.

9. In the integral of Exercise 8, let the numbers Z_j $(j = 1, 2, \ldots, n)$ be the nth roots of unity. Write $\omega = \exp(2\pi i/n)$ and $Z_1 = 1, Z_2 = \omega, \ldots, Z_n = \omega^{n-1}$. Let each of the numbers k_j $(j = 1, 2, \ldots, n)$ have the value $2/n$. The integral in Exercise 8 then becomes

$$w = A' \int_0^Z \frac{dS}{(S^n - 1)^{2/n}} + B.$$

Show that when $A' = 1$ and $B = 0$, this transformation maps the interior of the unit circle $|Z| = 1$ onto the interior of a regular polygon of n sides and that the center of the polygon is the point $w = 0$.

Suggestion: The image of each of the points Z_j $(j = 1, 2, \ldots, n)$ is a vertex of some polygon with an exterior angle of $2\pi/n$ at that vertex. Write

$$w_1 = \int_0^1 \frac{dS}{(S^n - 1)^{2/n}}$$

where the path of the integration is along the positive real axis from $Z = 0$ to $Z = 1$ and the principal value of the nth root of $(S^n - 1)^2$ is to be taken. Then show that the images of the points $Z_2 = \omega, \ldots, Z_n = \omega^{n-1}$ are the points $\omega w_1, \ldots, \omega^{n-1} w_1$, respectively. Thus verify that the polygon is regular and is centered at $w = 0$.

10. Obtain inequality (5), Sec. 92.

Suggestion: Let R be larger than any of the numbers $|x_j|$ $(j = 1, 2, \ldots, n - 1)$. Note that for R sufficiently large the inequalities $|z|/2 < |z - x_j| < 2|z|$ hold for each x_j when $|z| > R$. Then use equation (1), Sec. 92, along with conditions (5), Sec. 91.

11. In Sec. 92, use condition (5) and sufficient conditions for the existence of improper integrals of real-valued functions to show that $F(x)$ has some limit W_n as x tends to infinity, where $F(z)$ is defined by equation (3) of that section. Also show that the integral of $f'(z)$ over each arc of a semicircle $|z| = R$, Im $z \geqq 0$ approaches 0 as R tends to ∞. Then deduce that

$$\lim_{z \to \infty} F(z) = W_n \qquad (\text{Im } z \geqq 0),$$

as stated in equation (6) there.

12. According to Exercise 14, Sec. 71, the formula

$$N = \frac{1}{2\pi i} \int_C \frac{g'(z)}{g(z)}\, dz$$

can be used to determine the number of zeros of a function g interior to a positively oriented simple closed contour C when C lies in a simply connected domain D

throughout which $g(z)$ is analytic and $g'(z)$ is never zero. In that formula write $g(z) = f(z) - w_0$, where $f(z)$ is the Schwarz-Christoffel mapping function (7), Sec. 92, and the point w_0 is either interior to or exterior to the polygon P that is the image of the x axis; thus $f(x) \neq w_0$. Let the contour C consist of the upper half of a circle $|z| = R$ and a segment $-R < x < R$ of the x axis that contains all $n-1$ points x_j, except that a small segment about each point x_j is replaced by the upper half of a circle $|z - x_j| = r_j$ with that segment as its diameter. Then the number of points z interior to C such that $f(z) = w_0$ is

$$N_C = \frac{1}{2\pi i} \int_C \frac{f'(z)}{f(z) - w_0}\, dz.$$

Note that $f(z) - w_0$ approaches the nonzero point $W_n - w_0$ when $|z| = R$ and R tends to ∞, and recall the order property (5), Sec. 92, for $|f'(z)|$. Let the r_j tend to zero and prove that the number of points in the upper half of the z plane at which $f(z) = w_0$ is

$$N = \frac{1}{2\pi i} \lim_{R\to\infty} \int_{-R}^{R} \frac{f'(x)}{f(x) - w_0}\, dx.$$

Deduce that, since

$$\int_P \frac{dw}{w - w_0} = \lim_{R\to\infty} \int_{-R}^{R} \frac{f'(x)}{f(x) - w_0}\, dx,$$

$N = 1$ if w_0 is interior to P and that $N = 0$ if w_0 is exterior to P. Thus show that the mapping of the half plane Im $z > 0$ onto the interior of P is one to one.

96. Fluid Flow in a Channel through a Slit

We now present a further example of the idealized steady flow treated in Chap. 9, an example that will help to show how sources and sinks can be accounted for in problems of fluid flow.

Consider the two-dimensional steady flow of fluid between two parallel planes $v = 0$ and $v = \pi$ when the fluid is entering through a narrow slit along the line in the first plane which is perpendicular to the uv plane at the origin (Fig. 96). Let the rate of flow of fluid into the channel through the slit be Q

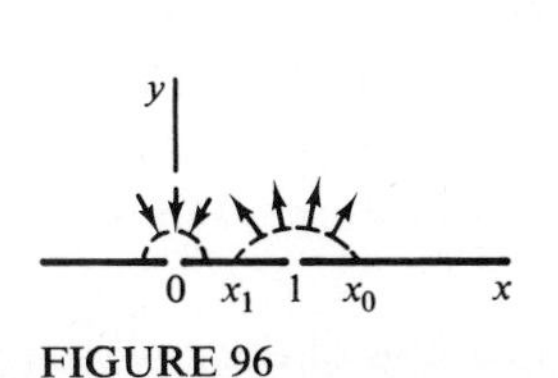

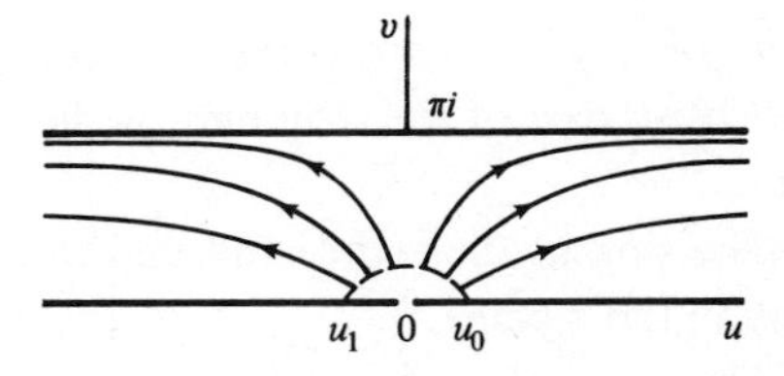

FIGURE 96

units of volume per unit time for each unit of depth of the channel, where the depth is measured perpendicular to the uv plane. The rate of flow out at either end is then $Q/2$.

The transformation $w = \text{Log } z$, derived in the preceding section, is a one to one mapping of the upper half of the z plane onto the strip in the w plane. The inverse transformation

$$z = e^w = e^u e^{iv} \tag{1}$$

thus maps the strip onto the half plane. Under transformation (1) the image of the u axis is the positive half of the x axis, and the image of the line $v = \pi$ is the negative half of the x axis. Hence the boundary of the strip is transformed into the boundary of the half plane.

The image of the point $w = 0$ is the point $z = 1$. The image of a point $w = u_0$ where $u_0 > 0$ is a point $z = x_0$ where $x_0 > 1$. The rate of flow of fluid across a curve joining the point $w = u_0$ to a point (u,v) within the strip is a stream function $\psi(u,v)$ for the flow (Sec. 88). If u_1 is a negative real number, then the rate of flow into the channel through the slit can be written

$$\psi(u_1, 0) = Q.$$

Now, under a conformal transformation, the function ψ is transformed into a function of x and y that represents the stream function for the flow in the corresponding region of the z plane; that is, the rate of flow is the same across corresponding curves in the two planes. As in Chap. 9, the same symbol ψ is used to represent the different stream functions in the two planes. Since the image of the point $w = u_1$ is a point $z = x_1$ where $0 < x_1 < 1$, the rate of flow across any curve connecting the points $z = x_0$ and $z = x_1$ and lying in the upper half of the z plane is also equal to Q. Hence there is a source at the point $z = 1$ equal to the source at $w = 0$.

The above argument applies in general to show that *under a conformal transformation a source or sink at a given point corresponds to an equal source or sink at the image of that point.*

As Re w tends to $-\infty$, the image of w approaches the point $z = 0$. A sink of strength $Q/2$ at the latter point corresponds to the sink infinitely far to the left in the strip. To apply the above argument in this case, we consider the rate of flow across a curve connecting the boundaries $v = 0$ and $v = \pi$ of the left-hand part of the strip and the flow across the image of that curve in the z plane.

The sink at the right-hand end of the strip is transformed into a sink at infinity in the z plane.

The stream function ψ for the flow in the upper half of the z plane in this case must be a function whose values are constant along each of the three parts of the x axis. Moreover, its value must increase by Q as the point z moves around the point $z = 1$ from the position $z = x_0$ to the position $z = x_1$, and its value must decrease by $Q/2$ as z moves about the origin in the corresponding manner. We see that the function

$$\psi(x,y) = \frac{Q}{\pi}\left[\operatorname{Arg}(z-1) - \tfrac{1}{2}\operatorname{Arg} z\right]$$

satisfies those requirements. Furthermore, this function is harmonic in the half plane $\operatorname{Im} z > 0$ because it is the imaginary part of the function

$$F(z) = \frac{Q}{\pi}\left[\operatorname{Log}(z-1) - \tfrac{1}{2}\operatorname{Log} z\right]$$

$$= \frac{Q}{\pi}\operatorname{Log}(z^{1/2} - z^{-1/2}).$$

The function F is a complex potential function for the flow in the upper half of the z plane. Since $z = e^w$, a complex potential function $F(w)$ for the flow in the channel is

$$F(w) = \frac{Q}{\pi}\operatorname{Log}(e^{w/2} - e^{-w/2}).$$

By dropping an additive constant, we can write

$$F(w) = \frac{Q}{\pi}\operatorname{Log}\left(\sinh\frac{w}{2}\right). \tag{2}$$

We have used the same symbol F to denote three distinct functions, once in the z plane and twice in the w plane.

The velocity vector $\overline{F'(w)}$ is given by the formula

$$V = \frac{Q}{2\pi}\coth\frac{\overline{w}}{2}. \tag{3}$$

From this it can be seen that

$$\lim_{|u|\to\infty} V = \frac{Q}{2\pi}.$$

Also, the point $w = \pi i$ is a *stagnation point*; that is, the velocity is zero there. Hence the fluid pressure along the wall $v = \pi$ of the channel is greatest at points opposite the slit.

The stream function $\psi(u,v)$ for the channel is the imaginary part of the function $F(w)$ given by equation (2). The streamlines $\psi(u,v) = c$ are therefore the curves

$$\frac{Q}{\pi}\operatorname{Arg}\left(\sinh\frac{w}{2}\right) = c.$$

This equation reduces to

$$\tan\frac{v}{2} = k\tanh\frac{u}{2} \tag{4}$$

where k is any real constant. Some of these streamlines are indicated in Fig. 96.

97. Flow in a Channel with an Offset

To illustrate further the use of the Schwarz-Christoffel transformation, let us find the complex potential for the flow of a fluid in a channel with an abrupt change in its breadth (Fig. 97). We take our unit of length such that the breadth of the wide part of the channel is π units; then $h\pi$, where $0 < h < 1$, represents the breadth of the narrow part. Let the real constant V_0 denote the velocity of the fluid far from the offset in the wide part; that is,

$$\lim_{u\to-\infty} V = V_0$$

where the complex variable V represents the velocity vector. The rate of flow per unit depth through the channel, or the strength of the source on the left and of the sink on the right, is then

$$Q = \pi V_0. \tag{1}$$

The cross section of the channel can be considered as the limiting case of the quadrilateral with the vertices w_1, w_2, w_3, and w_4, shown in the figure, as the first and last of these vertices are moved infinitely far to the left and to the right, respectively. In the limit the exterior angles become

$$k_1\pi = \pi, \qquad k_2\pi = \frac{\pi}{2}, \qquad k_3\pi = -\frac{\pi}{2}, \qquad k_4\pi = \pi.$$

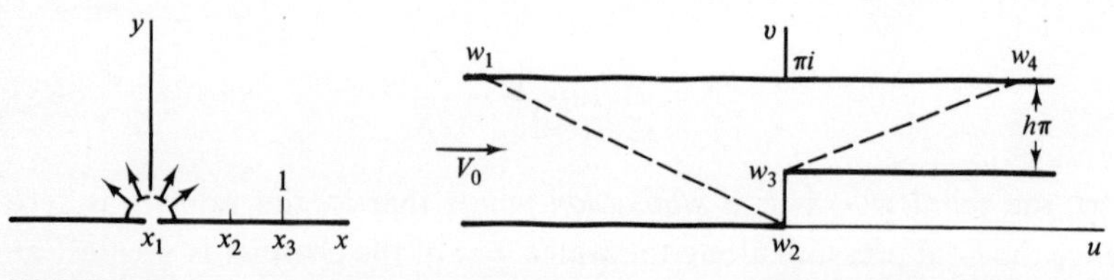

FIGURE 97

If we write $x_1 = 0$, $x_3 = 1$, $x_4 = \infty$ and leave x_2 to be determined, where $0 < x_2 < 1$, the derivative of the mapping function becomes

$$\frac{dw}{dz} = Az^{-1}(z - x_2)^{-1/2}(z - 1)^{1/2}. \tag{2}$$

In order to simplify the determination of the constants A and x_2 here, we proceed at once to the use of the complex potential of the flow. The source of the flow in the channel infinitely far to the left corresponds to an equal source at $z = 0$ (Sec. 96). The entire boundary of the cross section of the channel is the image of the x axis. In view of equation (1), then, the function

$$F(z) = V_0 \operatorname{Log} z = V_0 \operatorname{Log} r + iV_0 \theta \tag{3}$$

is the potential for the flow in the upper half of the z plane with the required source at the origin. Note that the sink on the right of the channel must correspond to a sink at infinity in the z plane.

The complex conjugate of the velocity V in the w plane can be written

$$\overline{V(w)} = \frac{dF}{dw} = \frac{dF}{dz}\frac{dz}{dw}.$$

Thus, by referring to equations (2) and (3), we can write

$$\overline{V(w)} = \frac{V_0}{A}\left(\frac{z - x_2}{z - 1}\right)^{1/2}. \tag{4}$$

At the limiting position of the point w_1, which corresponds to $z = 0$, the velocity is the real constant V_0. It therefore follows from equation (4) that

$$V_0 = \frac{V_0}{A}\sqrt{x_2}.$$

At the limiting position of w_4, which corresponds to $z = \infty$, let the real number V_4 denote the velocity. Now it seems plausible that, as a vertical line segment spanning the narrow part of the channel is moved infinitely far to the right, V approaches V_4 at each point on that segment. We could establish this conjecture as a fact by first finding w as a function of z from equation (2); but, to shorten our discussion, we assume that this is true. Then, since the flow is steady,

$$\pi h V_4 = \pi V_0 = Q,$$

or $V_4 = V_0/h$. Letting z tend to infinity in equation (4), we therefore find that

$$\frac{V_0}{h} = \frac{V_0}{A}.$$

Thus

$$A = h, \qquad x_2 = h^2, \tag{5}$$

and

$$\overline{V(w)} = \frac{V_0}{h}\left(\frac{z - h^2}{z - 1}\right)^{1/2}. \tag{6}$$

From equation (6) we can see that the magnitude $|V|$ of the velocity becomes infinite at the corner w_3 of the offset since it is the image of the point $z = 1$. Also, the corner w_2 is a stagnation point, a point where $v = 0$. Along the boundary of the channel, the fluid pressure is therefore greatest at w_2 and least at w_3.

In order to write the relation between the potential and the variable w, we must integrate equation (2), which can now be written

$$\frac{dw}{dz} = \frac{h}{z}\left(\frac{z - 1}{z - h^2}\right)^{1/2}. \tag{7}$$

By substituting a new variable s here, where

$$\frac{z - h^2}{z - 1} = s^2,$$

we can show that equation (7) reduces to

$$\frac{dw}{ds} = 2h\left(\frac{1}{1 - s^2} - \frac{1}{h^2 - s^2}\right).$$

Hence

$$w = h \operatorname{Log}\frac{1 + s}{1 - s} - \operatorname{Log}\frac{h + s}{h - s}. \tag{8}$$

The constant of integration here is zero because when $z = h^2$, s is zero and so therefore is w.

In terms of s, the potential F of equation (3) becomes

$$F = V_0 \operatorname{Log}\frac{h^2 - s^2}{1 - s^2};$$

consequently,

$$s^2 = \frac{\exp(F/V_0) - h^2}{\exp(F/V_0) - 1}. \tag{9}$$

By substituting s from this equation into equation (8), we get an implicit relation which defines the potential F as a function of w.

98. Electrostatic Potential about an Edge of a Conducting Plate

Two parallel conducting plates of infinite extent are kept at the electrostatic potential $V = 0$; and a parallel semi-infinite plate, placed midway between them, is kept at the potential $V = 1$. The coordinate system and the unit of length are chosen so that the plates lie in the planes $v = 0$, $v = \pi$, and $v = \pi/2$ (Fig. 98). Let us determine the potential function $V(u,v)$ in the region between those plates.

The cross section of that region in the uv plane has the limiting form of the quadrilateral bounded by the broken lines in the figure, as the points w_1 and w_3 move out to the right and w_4 to the left. In applying the Schwarz-Christoffel transformation here, we let the point x_4, corresponding to the vertex w_4, be the point at infinity. We choose the points $x_1 = -1$, $x_3 = 1$ and leave x_2 to be determined. The limiting values of the exterior angles of the quadrilateral are

$$k_1\pi = \pi, \qquad k_2\pi = -\pi, \qquad k_3\pi = k_4\pi = \pi.$$

Thus

$$\frac{dw}{dz} = A(z+1)^{-1}(z - x_2)(z-1)^{-1}$$

$$= A\frac{z - x_2}{z^2 - 1} = \frac{A}{2}\left(\frac{1 + x_2}{z+1} + \frac{1 - x_2}{z-1}\right),$$

and so the transformation of the upper half of the z plane into the divided strip in the w plane has the form

$$w = \frac{A}{2}\,[(1 + x_2)\,\mathrm{Log}\,(z+1) + (1 - x_2)\,\mathrm{Log}\,(z-1)] + B. \tag{1}$$

Let A_1, A_2 and B_1, B_2 denote the real and imaginary parts of the constants A and B. When $z = x$, the point w lies on the boundary of the divided strip; and, according to equation (1),

$$\begin{aligned} u + iv = \tfrac{1}{2}(A_1 + iA_2)\{(1 + x_2)[\mathrm{Log}\,|x+1| + i\arg(x+1)] \\ + (1 - x_2)[\mathrm{Log}\,|x-1| + i\arg(x-1)]\} + B_1 + iB_2. \end{aligned} \tag{2}$$

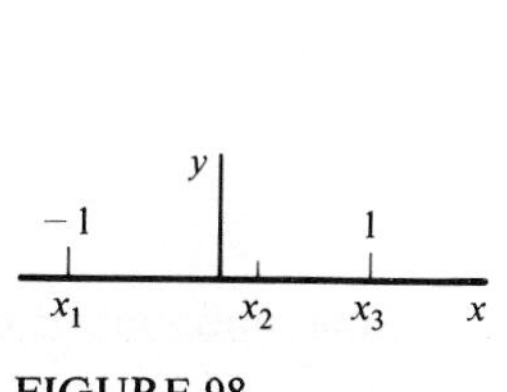

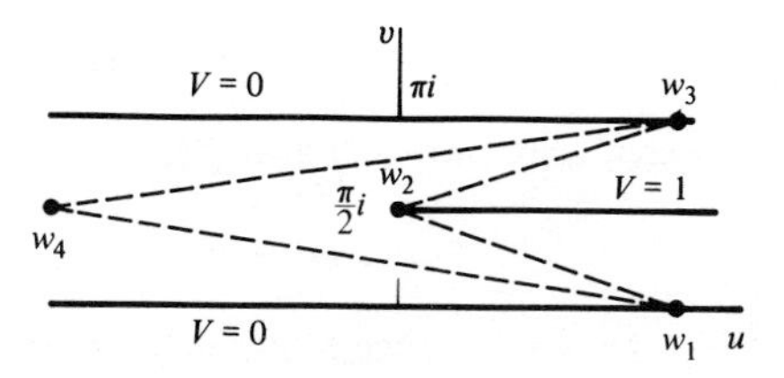

FIGURE 98

In order to determine the constants here, we first note that the limiting position of the line segment joining the points w_1 and w_4 is the u axis. That line is the image of the part of the x axis to the left of the point $x_1 = -1$; this is so because the line segment joining w_3 and w_4 is the image of the part of the x axis to the right of $x_3 = 1$, and the other two sides of the quadrilateral are the images of the remaining two segments of the x axis. Hence when $v = 0$ and u tends to infinity through positive values, the corresponding point x approaches the point $z = -1$ from the left. Thus

$$\arg(x + 1) = \pi, \qquad \arg(x - 1) = \pi,$$

and $\text{Log}|x + 1|$ tends to $-\infty$. Also, since $-1 < x_2 < 1$, the real part of the quantity inside the braces in equation (2) tends to $-\infty$. Since $v = 0$, it follows that $A_2 = 0$; otherwise, the imaginary part on the right would become infinite. By equating imaginary parts on the two sides, we now see that

$$0 = \tfrac{1}{2}A_1[(1 + x_2)\pi + (1 - x_2)\pi] + B_2.$$

Hence

$$-\pi A_1 = B_2, \qquad A_2 = 0. \tag{3}$$

The limiting position of the line segment joining the points w_1 and w_2 is the half line $v = \pi/2$, $u \geqq 0$. Points on that half line are images of the points $z = x$ where $-1 < x \leqq x_2$; consequently,

$$\arg(x + 1) = 0, \qquad \arg(x - 1) = \pi.$$

Identifying the imaginary parts on the two sides of equation (2) for those points, we see that

$$\frac{\pi}{2} = \frac{A_1}{2}(1 - x_2)\pi + B_2. \tag{4}$$

Finally, the limiting positions of the points on the line segment joining w_3 to w_4 are the points $u + \pi i$, which are the images of the points x where $x > 1$. By identifying the imaginary parts in equation (2) for those points, we find that

$$\pi = B_2.$$

Then, in view of equations (3) and (4),

$$A_1 = -1, \qquad x_2 = 0.$$

Thus $x = 0$ is the point whose image is the vertex $w = \pi i/2$; and, upon substituting these values into equation (2) and identifying real parts, we see that $B_1 = 0$.

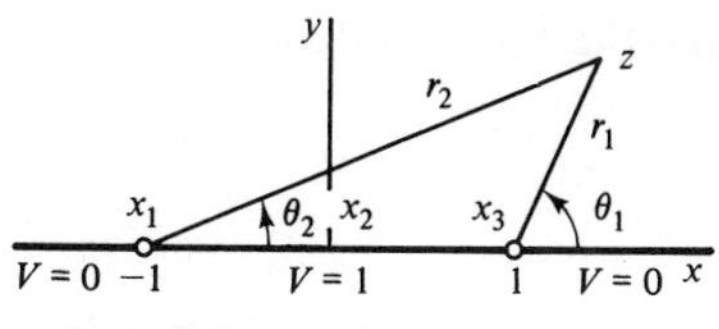

FIGURE 99

Transformation (1) now becomes

$$w = -\tfrac{1}{2}[\text{Log}\,(z+1) + \text{Log}\,(z-1)] + \pi i, \tag{5}$$

or

$$z^2 = 1 + e^{-2w}. \tag{6}$$

Under this transformation, the required harmonic function $V(u,v)$ becomes a harmonic function of x and y in the region $y > 0$ which satisfies the boundary conditions indicated in Fig. 99. Note that $x_2 = 0$ now. The harmonic function in that half plane assuming those values on the boundary is the imaginary part of the analytic function

$$F(z) = \frac{1}{\pi}\,\text{Log}\,\frac{z-1}{z+1} = \frac{1}{\pi}\,\text{Log}\,\frac{r_1}{r_2} + \frac{i}{\pi}(\theta_1 - \theta_2),$$

where θ_1 and θ_2 range from zero to π. Writing the tangents of these angles as functions of x and y and simplifying, we find that

$$\tan \pi V = \tan(\theta_1 - \theta_2) = \frac{2y}{x^2 + y^2 - 1}. \tag{7}$$

Equation (6) furnishes expressions for $x^2 + y^2$ and $x^2 - y^2$ in terms of u and v. Then from formula (7) we find that the relation between the potential V and the coordinates u and v can be written

$$\tan \pi V = \frac{1}{s}\sqrt{e^{-4u} - s^2} \tag{8}$$

where

$$s = -1 + \sqrt{1 + 2e^{-2u}\cos 2v + e^{-4u}}.$$

EXERCISES

1. Use the Schwarz-Christoffel transformation to obtain formally the mapping function given with Fig. 22, Appendix 2.
2. Explain why the solution of the problem of flow in a channel with a semi-infinite

rectangular obstruction (Fig. 100) is included in the solution of the problem treated in Sec. 97.

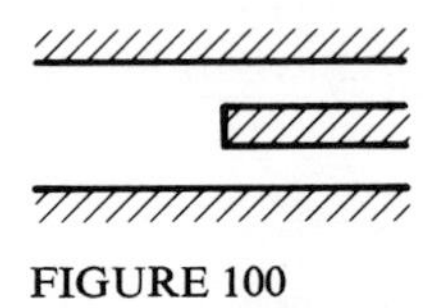

FIGURE 100

3. Refer to Fig. 29, Appendix 2. As the point z moves to the right along the negative part of the real axis where $x \leqq -1$, its image point w is to move to the right along the half line $u \leqq 0$, $v = h$. As the point z moves to the right along the segment $-1 \leqq x \leqq 1$, $y = 0$, its image point w is to move in the direction of decreasing v along the segment $u = 0$, $0 \leqq v \leqq h$. Finally, as z moves to the right along the positive part of the real axis where $x \geqq 1$, its image point w is to move to the right along the positive real axis. Note the changes in the direction of the motion of w at the images of the points $z = -1$ and $z = 1$. These changes indicate that the derivative of a mapping function might be

$$\frac{dw}{dz} = k\left(\frac{z+1}{z-1}\right)^{1/2}$$

where k is some constant. Thus obtain formally the transformation given there. Verify that the transformation, written in the form

$$w = \frac{h}{\pi}\{(z+1)^{1/2}(z-1)^{1/2} + \text{Log}\,[z + (z+1)^{1/2}(z-1)^{1/2}]\}$$

where $0 \leqq \arg(z \pm 1) \leqq \pi$, maps the boundary in the manner indicated in the figure.

4. Let $T(u,v)$ denote the bounded steady-state temperatures in the shaded region of the w plane in Fig. 29, Appendix 2, with the boundary conditions $T(u,h) = 1$ when $u < 0$ and $T = 0$ on the rest ($B'C'D'$) of the boundary. In terms of the real parameter α $(0 < \alpha < \pi/2)$, show that the image of each point $z = i\tan\alpha$ on the positive y axis is the point

$$w = \frac{h}{\pi}\left[\text{Log}\,(\tan\alpha + \sec\alpha) + i\left(\frac{\pi}{2} + \sec\alpha\right)\right]$$

(see Exercise 3) and that the temperature at that point w is

$$T(u,v) = \frac{\alpha}{\pi} \qquad \left(0 < \alpha < \frac{\pi}{2}\right).$$

5. Let $F(w)$ denote the complex potential function for the flow of a fluid over a step in the bed of a deep stream represented by the shaded region of the w plane in Fig. 29, Appendix 2, where the fluid velocity V approaches a real constant V_0 as

$|w|$ tends to infinity in that region. The transformation that maps the upper half of the z plane onto that region is noted in Exercise 3. Using the identity $dF/dw = (dF/dz)(dz/dw)$, show that

$$\overline{V(w)} = V_0(z-1)^{1/2}(z+1)^{-1/2};$$

and, in terms of the points $z = x$ whose images are the points along the bed of the stream, show that

$$|V| = |V_0| \sqrt{\left|\frac{x-1}{x+1}\right|}.$$

Thus note that the speed increases from $|V_0|$ along $A'B'$ until $|V| = \infty$ at B', then diminishes to zero at C', and increases toward $|V_0|$ from C' to D'; note also hat the speed is $|V_0|$ at the point $w = ih(\frac{1}{2} + 1/\pi)$ between B' and C'.

11

INTEGRAL FORMULAS OF POISSON TYPE

In this chapter we develop a theory which enables us to obtain solutions to a variety of boundary value problems where those solutions are expressed in terms of definite or improper integrals. Many of the integrals occurring are then readily evaluated.

99. The Poisson Integral Formula

Suppose that a function f is analytic within and on a simple closed contour C_0 which is described in the positive sense. The Cauchy integral formula

$$f(z) = \frac{1}{2\pi i}\int_{C_0} \frac{f(s)}{s - z}\, ds \tag{1}$$

expresses the value of f at any point z interior to C_0 in terms of the values of f at points s on C_0. When C_0 is a circle, we can obtain from formula (1) a corresponding formula for a harmonic function; that is, we can solve the Dirichlet problem for the circle.

Consider the case when C_0 is the circle $s = r_0 \exp(i\phi)$, $0 \leqq \phi \leqq 2\pi$, and write $z = r\exp(i\theta)$ where $0 < r < r_0$ (Fig. 101). The inverse of the nonzero

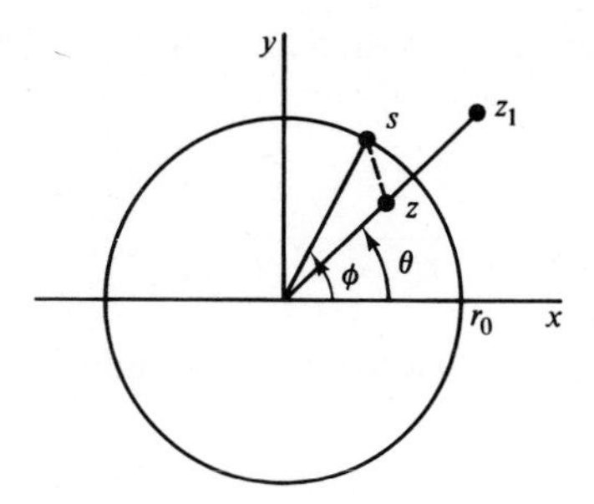

FIGURE 101

point z with respect to the circle is the point z_1 lying on the same ray as z and satisfying the condition $|z_1||z| = r_0^2$; that is,

$$z_1 = \frac{r_0^2}{r} \exp(i\theta) = \frac{r_0^2}{\bar{z}} = \frac{s\bar{s}}{\bar{z}}. \tag{2}$$

Since z_1 is exterior to the circle C_0, it follows from the Cauchy-Goursat theorem that the value of the integral in equation (1) is zero when z is replaced by z_1 in the integrand. Thus, using the stated parametric representation for C_0, we can write

$$f(z) = \frac{1}{2\pi} \int_0^{2\pi} \left(\frac{s}{s - z} - \frac{s}{s - z_1} \right) f(s)\, d\phi$$

where, for convenience, we retain the s instead of writing $r_0 \exp(i\phi)$.

Note that, in view of the last of expressions (2) for z_1, the factor inside the parentheses here can be written

$$\frac{s}{s - z} - \frac{1}{1 - \bar{s}/\bar{z}} = \frac{s}{s - z} + \frac{\bar{z}}{\bar{s} - \bar{z}} = \frac{r_0^2 - r^2}{|s - z|^2}. \tag{3}$$

An alternate form of the Cauchy integral formula (1) is therefore

$$f(re^{i\theta}) = \frac{r_0^2 - r^2}{2\pi} \int_0^{2\pi} \frac{f(r_0 e^{i\phi})}{|s - z|^2}\, d\phi \tag{4}$$

when $0 < r < r_0$. This form is also valid when $r = 0$; in that case it reduces directly to formula (1) with $z = 0$.

The quantity $|s - z|$ is the distance between the points s and z, and the law of cosines yields (see Fig. 101)

$$|s - z|^2 = r_0^2 - 2r_0 r \cos(\phi - \theta) + r^2 > 0. \tag{5}$$

Hence, if u is the real part of the analytic function f, it follows from formula (4) that

$$u(r, \theta) = \frac{1}{2\pi}\int_0^{2\pi} \frac{(r_0^2 - r^2)u(r_0,\phi)}{r_0^2 - 2r_0 r\cos(\phi - \theta) + r^2}\, d\phi \qquad (r < r_0). \tag{6}$$

This is the *Poisson integral formula* for the harmonic function u in the open disk bounded by the circle $r = r_0$.

Formula (6) defines a linear integral transformation of $u(r_0,\phi)$ into $u(r,\theta)$. The kernel of the transformation is, except for the factor $1/(2\pi)$, the real-valued function

$$P(r_0,r,\phi - \theta) = \frac{r_0^2 - r^2}{r_0^2 - 2r_0 r\cos(\phi - \theta) + r^2} \tag{7}$$

which is known as the *Poisson kernel.* The function $P(r_0,r,\phi - \theta)$ is also represented by expressions (3), and we see from the third of those expressions that it is always positive. Moreover, since $\bar{z}/(\bar{s} - \bar{z})$ and its complex conjugate $z/(s - z)$ have the same real parts, we find from the second of expressions (3) that

$$P(r_0,r,\phi - \theta) = \operatorname{Re}\left(\frac{s}{s - z} + \frac{z}{s - z}\right) = \operatorname{Re}\left(\frac{s + z}{s - z}\right). \tag{8}$$

Thus $P(r_0,r,\phi - \theta)$ is a harmonic function of r and θ interior to C_0 for each fixed s on C_0. From equation (7) we see that $P(r_0,r,\phi - \theta)$ is an even periodic function of $\phi - \theta$, with period 2π; and its value is 1 when $r = 0$.

The Poisson integral formula (6) can now be written

$$u(r,\theta) = \frac{1}{2\pi}\int_0^{2\pi} P(r_0,r,\phi - \theta)u(r_0,\phi)\, d\phi \qquad (r < r_0). \tag{9}$$

In the special case when $f(z) = u(x,y) = 1$, equation (9) shows that P has the property

$$\frac{1}{2\pi}\int_0^{2\pi} P(r_0,r,\phi - \theta)\, d\phi = 1 \qquad (r < r_0). \tag{10}$$

We have assumed that f is analytic not only interior to C_0 but on C_0 itself and that u is therefore harmonic in a domain which includes all points on that circle. In particular, u is continuous on C_0. The conditions will now be relaxed.

100. A Dirichlet Problem for a Disk

Let F be a given piecewise continuous function of θ $(0 \leqq \theta \leqq 2\pi)$. We shall prove that the function

$$U(r,\theta) = \frac{1}{2\pi}\int_0^{2\pi} P(r_0,r,\phi - \theta)F(\phi)\,d\phi \qquad (r < r_0), \tag{1}$$

which we may call the Poisson integral transform of F, satisfies the following conditions: *U is harmonic throughout the interior of the circle $r = r_0$ and*

$$\lim_{r\to r_0} U(r,\theta) = F(\theta) \qquad (r < r_0) \tag{2}$$

for each fixed θ where F is continuous. Thus U is a solution of the Dirichlet problem for the disk $r < r_0$ in the sense that the boundary value $F(\theta)$ is approached by $U(r,\theta)$ as the point (r,θ) approaches (r_0,θ) along a radius, except at the finite number of points (r_0,θ) where the discontinuities of F may occur.

Before proving the above statement, let us apply it to find the potential $V(r,\theta)$ inside a cylinder $r = 1$ where the boundary conditions are those illustrated in Fig. 72, the potential being zero on one half of the surface and unity on the other half. This problem was solved by conformal mapping in Sec. 86. In formula (1) write V for U, $r_0 = 1$, and $F(\phi) = 0$ when $0 < \phi < \pi$ and $F(\phi) = 1$ when $\pi < \phi < 2\pi$ to get

$$V(r,\theta) = \frac{1}{2\pi}\int_\pi^{2\pi} P(1,r,\phi - \theta)\,d\phi = \frac{1}{2\pi}\int_\pi^{2\pi} \frac{(1 - r^2)\,d\phi}{1 + r^2 - 2r\cos(\phi - \theta)}.$$

An indefinite integral of $P(1,r,\psi)$ is

$$\int P(1,r,\psi)\,d\psi = 2\arctan\left(\frac{1+r}{1-r}\tan\frac{\psi}{2}\right) \tag{3}$$

since the integrand here is the derivative with respect to ψ of the function on the right. That function is assigned the value $-\pi$ when $\psi/2 = -\pi/2$ and the value π when $\psi/2 = \pi/2$ so that it can be a continuous function whose values increase from $-\pi$ to 2π as $\psi/2$ increases from $-\pi/2$ to π. This is the required range of values for $\psi/2$ since $\psi = \phi - \theta$ and the angles ϕ and θ range from π to 2π and 0 to 2π, respectively. Thus

$$\pi V(r,\theta) = \arctan\left(\frac{1+r}{1-r}\tan\frac{2\pi - \theta}{2}\right) - \arctan\left(\frac{1+r}{1-r}\tan\frac{\pi - \theta}{2}\right),$$

where it is physically evident that the values of $\pi V(r,\theta)$ must range from 0 to π.

By simplifying the expression for $\tan(\pi V)$ obtained from this last equation, we find that

$$V(r,\theta) = \frac{1}{\pi} \arctan\left(\frac{1 - r^2}{2r \sin \theta}\right) \qquad (0 \leq \arctan t \leq \pi). \tag{4}$$

This is the solution obtained earlier in terms of cartesian coordinates.

The function U defined by formula (1) is harmonic interior to the circle $r = r_0$ because P is a harmonic function of r and θ there. To be precise, we note that since F is piecewise continuous, integral (1) can be written as the sum of a finite number of definite integrals each of which has an integrand that is continuous in r, θ, and ϕ. The partial derivatives of those integrands with respect to r and θ are likewise continuous. Since the order of integration and differentiation with respect to r and θ may then be interchanged and since P satisfies Laplace's equation in the polar coordinates r and θ, it follows that U satisfies that equation too.

To establish condition (2), we need to show that if F is continuous at θ, there is corresponding to each positive number ε a positive number δ such that

$$|U(r,\theta) - F(\theta)| < \varepsilon \qquad \text{when} \qquad 0 < r_0 - r < \delta. \tag{5}$$

In view of property (10), Sec. 99, the first inequality here can be written

$$\left| \frac{1}{2\pi} \int_0^{2\pi} P(r_0,r,\phi - \theta)[F(\phi) - F(\theta)]\, d\phi \right| < \varepsilon. \tag{6}$$

We shall find it convenient to let F be extended periodically with period 2π so that the integrand is periodic in ϕ with that same period.

Since F is continuous at θ, there is a small positive number α corresponding to the given positive number ε such that

$$|F(\phi) - F(\theta)| < \frac{\varepsilon}{2} \qquad \text{when} \qquad |\phi - \theta| \leq \alpha.$$

Let us write

$$I_1(r) = \frac{1}{2\pi} \int_{\theta-\alpha}^{\theta+\alpha} P(r_0,r,\phi - \theta)[F(\phi) - F(\theta)]\, d\phi,$$

$$I_2(r) = \frac{1}{2\pi} \int_{\theta+\alpha}^{\theta-\alpha+2\pi} P(r_0,r,\phi - \theta)[F(\phi) - F(\theta)]\, d\phi.$$

Then inequality (6) can be written

$$|I_1(r) + I_2(r)| < \varepsilon. \tag{7}$$

Since P is a positive-valued function and because of property (10), Sec. 99,

$$|I_1(r)| \leqq \frac{1}{2\pi}\int_{\theta-\alpha}^{\theta+\alpha} P(r_0,r,\phi-\theta)|F(\phi)-F(\theta)|\,d\phi$$

$$< \frac{\varepsilon}{4\pi}\int_0^{2\pi} P(r_0,r,\phi-\theta)\,d\phi = \frac{\varepsilon}{2}$$

whenever $r < r_0$.

Next, recall that $P(r_0,r,\phi-\theta) = (r_0^2 - r^2)/|s-z|^2$ and observe in Fig. 101 that when $r \leqq r_0$, the quantity $|s-z|^2$ has a positive minimum value $m(\alpha)$ as ϕ, the argument of s, varies between $\theta+\alpha$ and $\theta-\alpha+2\pi$. If M denotes an upper bound of $|F(\phi)-F(\theta)|$ for all values of θ and ϕ, it follows that

$$|I_2(r)| \leqq \frac{(r_0^2-r^2)M}{2\pi m(\alpha)}\,2\pi < \frac{2Mr_0}{m(\alpha)}(r_0-r) < \frac{\varepsilon}{2}$$

when $r_0 - r < m(\alpha)\varepsilon/(4Mr_0)$. Consequently, if α is selected corresponding to the given ε, the number $m(\alpha)$ is determined; and $|I_1(r)| + |I_2(r)| < \varepsilon$ when $r_0 - r < \delta$ if

$$\delta = \frac{m(\alpha)\varepsilon}{4Mr_0}.$$

This is then a value of δ such that inequality (7), or (6), is satisfied. Statement (5) is therefore true when δ has this value.

According to formula (1), the value of U at $r = 0$ is

$$\frac{1}{2\pi}\int_0^{2\pi} F(\phi)\,d\phi.$$

Thus *the value of a harmonic function at the center of the circle is the average of the boundary values on the circle.*

It is left to the exercises below to prove that P and U can be represented by series involving the elementary harmonic functions $r^n \cos n\theta$ and $r^n \sin n\theta$ as follows:

$$P(r_0,r,\phi-\theta) = 1 + 2\sum_{n=1}^{\infty}\left(\frac{r}{r_0}\right)^n \cos n(\phi-\theta) \qquad (r < r_0), \tag{8}$$

$$U(r,\theta) = \frac{1}{2}a_0 + \sum_{n=1}^{\infty}\left(\frac{r}{r_0}\right)^n (a_n \cos n\theta + b_n \sin n\theta) \qquad (r < r_0), \tag{9}$$

where

$$a_n = \frac{1}{\pi}\int_0^{2\pi} F(\phi)\cos n\phi\,d\phi, \qquad b_n = \frac{1}{\pi}\int_0^{2\pi} F(\phi)\sin n\phi\,d\phi. \tag{10}$$

101. Related Boundary Value Problems

Details of proofs of results given below are left to the exercises. The function F representing boundary values on the circle $r = r_0$ is assumed to be piecewise continuous.

Suppose that $F(2\pi - \theta) = -F(\theta)$. The Poisson integral formula (1) of Sec. 100 then becomes

$$U(r,\theta) = \frac{1}{2\pi}\int_0^{\pi} [P(r_0,r,\phi - \theta) - P(r_0,r,\phi + \theta)]F(\phi)\, d\phi. \tag{1}$$

This function U has zero values on the horizontal radii $\theta = 0$ and $\theta = \pi$ of the circle, as we expect when we interpret U as a steady temperature. Formula (1) therefore solves *the Dirichlet problem for the semicircular region* $r < r_0$, $0 < \theta < \pi$ (Fig. 102) *where* $U = 0$ *on the diameter* AB *and*

$$\lim_{r \to r_0} U(r,\theta) = F(\theta) \qquad (r < r_0,\ 0 < \theta < \pi) \tag{2}$$

for each fixed θ *where* F *is continuous.*

If $F(2\pi - \theta) = F(\theta)$, then

$$U(r,\theta) = \frac{1}{2\pi}\int_0^{\pi} [P(r_0,r,\phi - \theta) + P(r_0,r,\phi + \theta)]F(\phi)\, d\phi; \tag{3}$$

and $U_\theta(r,\theta) = 0$ when $\theta = 0$ or $\theta = \pi$. Formula (3) thus furnishes a function U that is *harmonic in the semicircular region* $r < r_0$, $0 < \theta < \pi$ (Fig. 102) *and satisfies condition* (2) *as well as the condition that its normal derivative is zero on the diameter* AB.

The analytic function $z = r_0^2/Z$ maps the circle $|Z| = r_0$ in the Z plane onto the circle $|z| = r_0$ in the z plane, and it maps the exterior of the first circle onto the interior of the second. Writing $z = r \exp(i\theta)$ and $Z = R \exp(i\psi)$, we note that $r = r_0^2/R$ and $\theta = 2\pi - \psi$. The harmonic function $U(r,\theta)$ represented by formula (1), Sec. 100, is then transformed into the function

$$U\left(\frac{r_0^2}{R}, 2\pi - \psi\right) = -\frac{1}{2\pi}\int_0^{2\pi} \frac{r_0^2 - R^2}{r_0^2 - 2r_0 R\cos(\phi + \psi) + R^2} F(\phi)\, d\phi$$

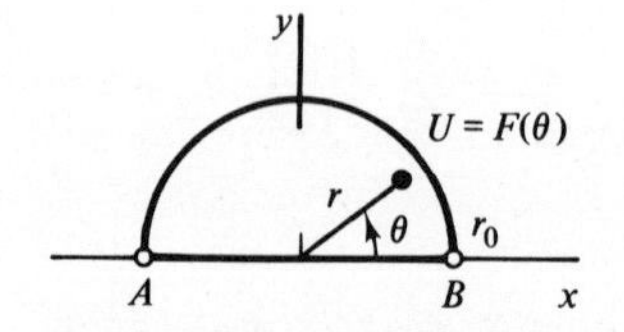

FIGURE 102

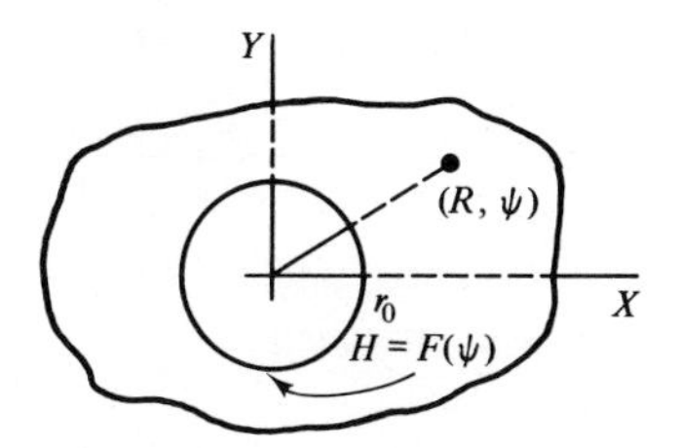

FIGURE 103

which is harmonic in the domain $R > r_0$. Now, in general, if $u(r, \theta)$ is harmonic, then so is $u(r, -\theta)$. (See Exercise 10 of this section.) Hence the function $H(R,\psi) = U(r_0{}^2/R,\psi - 2\pi)$, or

$$H(R,\psi) = -\frac{1}{2\pi}\int_0^{2\pi} P(r_0,R,\phi - \psi)F(\phi)\,d\phi \qquad (R > r_0), \tag{4}$$

is also harmonic. For each fixed ψ where $F(\psi)$ is continuous we find from condition (2), Sec. 100, that

$$\lim_{R\to r_0} H(R,\psi) = F(\psi) \qquad (R > r_0). \tag{5}$$

Thus formula (4) solves *the Dirichlet problem for the region exterior to the circle* $R = r_0$ in the Z plane (Fig. 103). We note that the Poisson kernel is negative there; also,

$$\frac{1}{2\pi}\int_0^{2\pi} P(r_0,R,\phi - \psi)\,d\phi = -1 \qquad (R > r_0), \tag{6}$$

$$\lim_{R\to\infty} H(R,\psi) = \frac{1}{2\pi}\int_0^{2\pi} F(\phi)\,d\phi. \tag{7}$$

EXERCISES

1. Use the Poisson integral formula (1), Sec. 100, to derive the formula

$$V(x,y) = \frac{1}{\pi}\arctan\frac{1 - x^2 - y^2}{(x-1)^2 + (y-1)^2 - 1} \qquad (0 \leqq \arctan t \leqq \pi)$$

for the electrostatic potential interior to a cylinder $x^2 + y^2 = 1$ if $V = 1$ on the first quadrant ($x > 0$, $y > 0$) of the cylindrical surface and $V = 0$ on the rest of that surface. Also note that $1 - V$ is the solution to Exercise 8, Sec. 86.

2. Let T denote the steady temperatures in a disk $r \leqq 1$ with insulated faces when $T = 1$ on the arc $0 < \theta < 2\theta_0$ of the edge $r = 1$ and $T = 0$ on the rest of the edge, where $0 < \theta_0 < \pi/2$. Use the Poisson integral formula to show that

$$T(x,y) = \frac{1}{\pi} \arctan \frac{(1 - x^2 - y^2)y_0}{(x-1)^2 + (y - y_0)^2 - y_0{}^2} \qquad (0 \leqq \arctan t \leqq \pi)$$

where $y_0 = \tan \theta_0$. Verify that this function T satisfies the boundary conditions.

3. Let I denote this *finite unit impulse function:*

$$I(h,\theta - \theta_0) = \begin{cases} \dfrac{1}{h} & \text{when} \quad \theta_0 < \theta < \theta_0 + h, \\ 0 & \text{when} \quad 0 \leqq \theta < \theta_0 \text{ or } \theta_0 + h < \theta < 2\pi \end{cases}$$

where h is a positive constant and $0 \leqq \theta_0 < 2\pi$. Note that

$$\int_0^{2\pi} I(h,\theta - \theta_0)\, d\theta = 1.$$

With the aid of a mean-value theorem for integrals, show that

$$\lim_{h \to 0} \int_0^{2\pi} P(r_0,r,\phi - \theta) I(h,\phi - \theta_0)\, d\phi = P(r_0,r,\theta - \theta_0) \qquad (r < r_0,\ h > 0).$$

Thus the Poisson kernel $P(r_0,r,\theta - \theta_0)$ is the limit, as h approaches 0 through positive values, of the harmonic function inside the circle $r = r_0$ whose boundary values are represented by the impulse function $2\pi I(h,\theta - \theta_0)$.

4. Show that the formula in Exercise 11, Sec. 66, for the sum of a cosine series can be written

$$1 + 2\sum_{n=1}^{\infty} k^n \cos n\theta = \frac{1 - k^2}{1 - 2k \cos \theta + k^2} \qquad (-1 < k < 1).$$

Then show that the Poisson kernel has the series representation (8), Sec. 100.

5. Show that the series in formula (8), Sec. 100, converges uniformly with respect to ϕ. Then obtain from formula (1) of that section the series representation (9) there.

6. Use formulas (9) and (10) of Sec. 100 to find the steady temperatures $T(r,\theta)$ in a solid cylinder $r \leqq r_0$ of infinite length if $T(r_0,\theta) = A \cos \theta$. Show that no heat flows across the plane $y = 0$.

Ans. $T = A(r/r_0) \cos \theta = Ax/r_0$.

7. Obtain the special cases

(*a*) $$H(R,\psi) = \frac{1}{2\pi} \int_0^{\pi} [P(r_0,R,\phi + \psi) - P(r_0,R,\phi - \psi)]F(\phi)\, d\phi,$$

(*b*) $$H(R,\psi) = -\frac{1}{2\pi} \int_0^{\pi} [P(r_0,R,\phi + \psi) + P(r_0,R,\phi - \psi)]F(\phi)\, d\phi$$

of formula (4), Sec. 101, for the harmonic function H in the unbounded region $R > r_0$, $0 < \psi < \pi$, shown in Fig. 104, if that function satisfies the boundary condition

$$\lim_{R \to r_0} H(R,\psi) = F(\psi) \qquad (R > r_0,\ 0 < \psi < \pi)$$

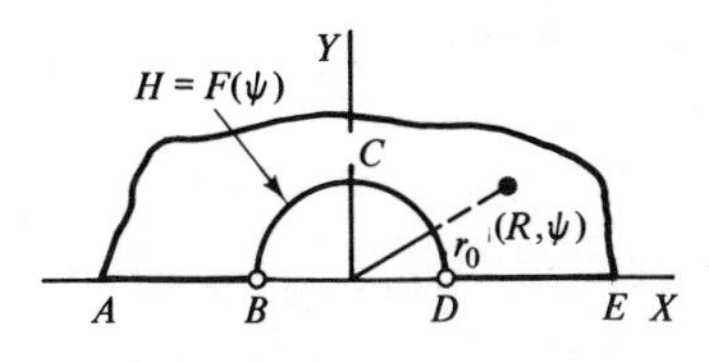

FIGURE 104

on the semicircle and (*a*) it is zero on the rays *BA* and *DE*; (*b*) its normal derivative is zero on the rays *BA* and *DE*.

8. Give the details needed in establishing formula (1) of Sec. 101 as a solution of the Dirichlet problem stated there for the region shown in Fig. 102.

9. Give the details needed in establishing formula (3) of Sec. 101 as a solution of the boundary value problem stated there.

10. Obtain formula (4), Sec. 101, as a solution of the Dirichlet problem for the region exterior to a circle (Fig. 103). To show that $u(r, -\theta)$ is harmonic when $u(r, \theta)$ is harmonic, refer to the polar form of Laplace's equation in Exercise 11 of Sec. 20.

11. State why formula (6), Sec. 101, is valid.

12. Establish equation (7), Sec. 101.

102. Integral Formulas for a Half Plane

Let f be an analytic function of z throughout the half plane $\text{Im } z \geqq 0$ such that, for some positive constants k and M, f satisfies the order property

$$|z^k f(z)| < M \qquad (\text{Im } z \geqq 0). \tag{1}$$

For a fixed point z above the real axis let C_R denote the upper half of a positively oriented circle of radius R centered at the origin, where $R > |z|$ (Fig. 105). Then, according to the Cauchy integral formula,

$$f(z) = \frac{1}{2\pi i}\int_{C_R} \frac{f(s)\,ds}{s - z} + \frac{1}{2\pi i}\int_{-R}^{R} \frac{f(t)\,dt}{t - z}. \tag{2}$$

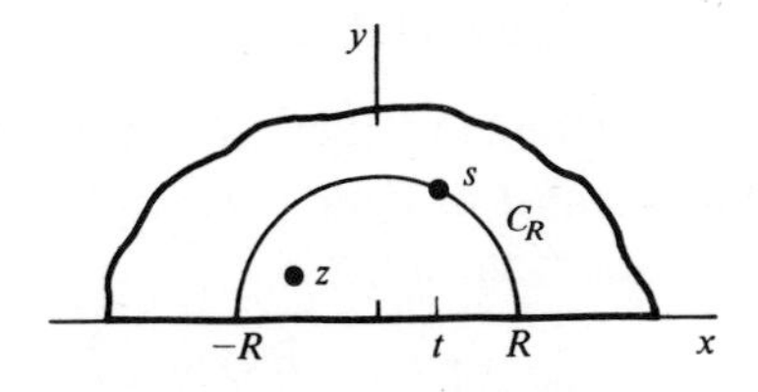

FIGURE 105

We find that the first of these integrals approaches 0 as R tends to ∞ because $|f(s)| < M/R^k$; consequently,

$$f(z) = \frac{1}{2\pi i}\int_{-\infty}^{\infty} \frac{f(t)}{t - z}\,dt \qquad (\text{Im } z > 0). \tag{3}$$

Because of condition (1), the improper integral here converges; and the number to which it converges is the same as its Cauchy principal value. (See Sec. 72.) Representation (3) is a *Cauchy integral formula for the half plane* Im $z > 0$.

When the point z lies below the real axis, the right-hand side of equation (2) is zero; hence integral (3) is zero for such a point. Consequently, when z is above the real axis, we have the following formula where c is an arbitrary constant:

$$f(z) = \frac{1}{2\pi i}\int_{-\infty}^{\infty} \left(\frac{1}{t - z} + \frac{c}{t - \bar{z}}\right) f(t)\,dt \qquad (\text{Im } z > 0). \tag{4}$$

In the two cases $c = -1$ and $c = 1$ this formula reduces, respectively, to

$$f(z) = \frac{1}{\pi}\int_{-\infty}^{\infty} \frac{yf(t)}{|t - z|^2}\,dt \qquad (y > 0), \tag{5}$$

$$f(z) = \frac{1}{\pi i}\int_{-\infty}^{\infty} \frac{(t - x)f(t)}{|t - z|^2}\,dt \qquad (y > 0). \tag{6}$$

If $f(z) = u(x,y) + iv(x,y)$, it follows from formulas (5) and (6) that the harmonic functions u and v are represented in the half plane $y > 0$ in terms of the boundary values of u by the formulas

$$u(x,y) = \frac{1}{\pi}\int_{-\infty}^{\infty} \frac{yu(t,0)}{|t - z|^2}\,dt = \frac{1}{\pi}\int_{-\infty}^{\infty} \frac{yu(t,0)}{(t - x)^2 + y^2}\,dt \qquad (y > 0), \tag{7}$$

$$v(x,y) = \frac{1}{\pi}\int_{-\infty}^{\infty} \frac{(x - t)u(t,0)}{(t - x)^2 + y^2}\,dt \qquad (y > 0). \tag{8}$$

Formula (7) is known as the *Poisson integral formula for the half plane*, or the *Schwarz integral formula*. In the next section we shall relax the conditions for the validity of formulas (7) and (8).

103. A Dirichlet Problem for a Half Plane

Let F denote a real-valued function of x that is bounded for all x and continuous except for at most a finite number of finite jumps. When $y \geqq \varepsilon$ and $|x| \leqq 1/\varepsilon$, where ε is any positive constant, the integral

$$I(x,y) = \int_{-\infty}^{\infty} \frac{F(t)\,dt}{(t - x)^2 + y^2}$$

converges uniformly with respect to x and y, as do the integrals of the partial derivatives of the integrand with respect to x and y. Each of these integrals is the sum of a finite number of improper or definite integrals over intervals where F is continuous; hence the integrand of each component integral is a continuous function of t, x, and y when $y \geqq \varepsilon$. Consequently, each partial derivative of $I(x,y)$ is represented by the integral of the corresponding derivative of the integrand whenever $y > 0$.

We write $U(x,y) = yI(x,y)/\pi$. Thus U is the Schwarz integral transform of F, suggested by equation (7), Sec. 102:

$$U(x,y) = \frac{1}{\pi}\int_{-\infty}^{\infty} \frac{yF(t)}{(t-x)^2 + y^2}\,dt \qquad (y > 0). \tag{1}$$

Except for the factor $1/\pi$, the kernel here is $y/|t - z|^2$. It is the imaginary part of the function $1/(t - z)$ which is analytic in z when $y > 0$. It follows that the kernel is harmonic, and so it satisfies Laplace's equation in x and y. Because the order of differentiation and integration can be interchanged, function (1) then satisfies that equation. Consequently, *U is harmonic when $y > 0$.*

To prove that

$$\lim_{y \to 0} U(x,y) = F(x) \qquad (y > 0) \tag{2}$$

for each fixed x at which F is continuous, we substitute $t = x + y \tan r$ in formula (1) and write

$$U(x,y) = \frac{1}{\pi}\int_{-\pi/2}^{\pi/2} F(x + y \tan r)\,dr \qquad (y > 0). \tag{3}$$

If

$$G(x,y,r) = F(x + y \tan r) - F(x)$$

and α is some small positive constant, then

$$\pi[U(x,y) - F(x)] = \int_{-\pi/2}^{\pi/2} G(x,y,r)\,dr = I_1(y) + I_2(y) + I_3(y) \tag{4}$$

where

$$I_1(y) = \int_{-\pi/2}^{-\pi/2+\alpha} G(x,y,r)\,dr, \qquad I_2(y) = \int_{-\pi/2+\alpha}^{\pi/2-\alpha} G(x,y,r)\,dr,$$

$$I_3(y) = \int_{\pi/2-\alpha}^{\pi/2} G(x,y,r)\,dr.$$

If M denotes an upper bound for $|F(x)|$, then $|G(x,y,r)| \leqq 2M$. For a given positive number ε we select α so that $6M\alpha < \varepsilon$; then

$$|I_1(y)| \leqq 2M\alpha < \frac{\varepsilon}{3} \qquad \text{and} \qquad |I_3(y)| \leqq 2M\alpha < \frac{\varepsilon}{3}.$$

We next show that corresponding to ε there is a positive number δ such that

$$|I_2(y)| < \frac{\varepsilon}{3} \qquad \text{when} \qquad 0 < y < \delta.$$

Since F is continuous at x, there is a positive number γ such that

$$|G(x,y,r)| < \frac{\varepsilon}{3\pi} \qquad \text{when} \qquad 0 < y|\tan r| < \gamma.$$

Note that the maximum value of $|\tan r|$ as r ranges between $-\pi/2 + \alpha$ and $\pi/2 - \alpha$ is $\tan(\pi/2 - \alpha) = \cot \alpha$. Hence if we write $\delta = \gamma \tan \alpha$, it follows that

$$|I_2(y)| < \frac{\varepsilon}{3\pi}(\pi - 2\alpha) < \frac{\varepsilon}{3} \qquad \text{when} \qquad 0 < y < \delta.$$

We have thus shown that

$$|I_1(y)| + |I_2(y)| + |I_3(y)| < \varepsilon \qquad \text{when} \qquad 0 < y < \delta.$$

Condition (2) now follows from this result and equation (4).

Formula (1) therefore solves *the Dirichlet problem for the half plane* $y > 0$ with the boundary condition (2). It is evident from form (3) of formula (1) that $|U(x,y)| \leqq M$ in the half plane where M is an upper bound of $|F(x)|$; that is, U is bounded. We note that $U(x,y) = F_0$ when $F(x) = F_0$, where F_0 is a constant.

According to formula (8) of the preceding section, under certain conditions on F the function

$$V(x,y) = \frac{1}{\pi}\int_{-\infty}^{\infty} \frac{(x-t)F(t)}{(t-x)^2 + y^2}\,dt \qquad (y > 0) \tag{5}$$

is a harmonic conjugate of the function U given by formula (1). Actually, *formula* (5) *furnishes a harmonic conjugate of* U *if* F *is everywhere continuous, except for at most a finite number of finite jumps, and if* F *satisfies an order property* $|x^k F(x)| < M$, *where* $k > 0$. For under those conditions we find that U and V satisfy the Cauchy-Riemann equations when $y > 0$.

Special cases of formula (1) when F is an odd or an even function are left to the exercises.

104. A Neumann Problem for a Disk

As in Sec. 99 and Fig. 101, we write $s = r_0 \exp(i\phi)$ and $z = r \exp(i\theta)$, where $r < r_0$. When s is fixed, the function

$$\begin{aligned} Q(r_0,r,\phi - \theta) &= -2r_0 \operatorname{Log}|s - z| \\ &= -r_0 \operatorname{Log}[r_0{}^2 - 2r_0 r\cos(\phi - \theta) + r^2] \end{aligned} \tag{1}$$

is harmonic interior to the circle $|z| = r_0$ because it is the real part of $-2r_0 \log(z - s)$ where the branch cut of $\log(z - s)$ is an outward ray from the point s. If, moreover, $r \neq 0$,

$$Q_r(r_0,r,\phi - \theta) = -\frac{r_0}{r}\frac{2r^2 - 2r_0 r\cos(\phi - \theta)}{r_0^2 - 2r_0 r\cos(\phi - \theta) + r^2} \tag{2}$$

$$= \frac{r_0}{r}[P(r_0,r,\phi - \theta) - 1]$$

where P is the Poisson kernel (7) of Sec. 99.

These observations suggest that the function Q may be used to write an integral representation for a harmonic function U whose normal derivative U_r on the circle $r = r_0$ assumes prescribed values $G(\theta)$.

If G is piecewise continuous and U_0 is an arbitrary constant, the function

$$U(r,\theta) = \frac{1}{2\pi}\int_0^{2\pi} Q(r_0,r,\phi - \theta)G(\phi)\,d\phi + U_0 \qquad (r < r_0) \tag{3}$$

is harmonic because the integrand is a harmonic function of r and θ. If the mean value of G over the circle is zero,

$$\int_0^{2\pi} G(\phi)\,d\phi = 0, \tag{4}$$

then, in view of equation (2),

$$U_r(r,\theta) = \frac{1}{2\pi}\int_0^{2\pi}\frac{r_0}{r}[P(r_0,r,\phi - \theta) - 1]G(\phi)\,d\phi$$

$$= \frac{r_0}{r}\frac{1}{2\pi}\int_0^{2\pi} P(r_0,r,\phi - \theta)G(\phi)\,d\phi.$$

Now according to equations (1) and (2) of Sec. 100,

$$\lim_{r\to r_0}\frac{1}{2\pi}\int_0^{2\pi} P(r_0,r,\phi - \theta)G(\phi)\,d\phi = G(\theta) \qquad (r < r_0).$$

Hence

$$\lim_{r\to r_0} U_r(r,\theta) = G(\theta) \qquad (r < r_0) \tag{5}$$

for each value of θ where G is continuous.

Since the value of Q is constant when $r = 0$, it follows from equations (3) and (4) that U_0 is the value of U at the center of the circle.

When G is piecewise continuous and satisfies condition (4), the formula

$$U(r,\theta) = -\frac{r_0}{2\pi}\int_0^{2\pi}\operatorname{Log}[r_0^2 - 2r_0 r\cos(\phi - \theta) + r^2]G(\phi)\,d\phi + U_0 \qquad (r < r_0) \tag{6}$$

therefore solves *the Neumann problem for the region interior to the circle* $r = r_0$, where $G(\theta)$ is the normal derivative of the harmonic function $U(r,\theta)$ at the boundary in the sense of condition (5).

The values $U(r,\theta)$ may represent steady temperatures in a disk $r < r_0$ with insulated faces. In that case condition (5) states that the flux of heat into the disk through its edge is proportional to $G(\theta)$. Condition (4) is the natural physical requirement that the total rate of flow of heat into the disk must be zero since temperatures do not vary with time.

A corresponding formula for a harmonic function H in the domain *exterior* to the circle $r = r_0$ can be written in terms of Q as

$$H(R,\psi) = -\frac{1}{2\pi}\int_0^{2\pi} Q(r_0,R,\phi - \psi)G(\phi)\, d\phi + H_0 \qquad (R > r_0), \tag{7}$$

where H_0 is a constant. As before, we assume that G is piecewise continuous and that

$$\int_0^{2\pi} G(\phi)\, d\phi = 0. \tag{8}$$

Then

$$H_0 = \lim_{R\to\infty} H(R,\psi)$$

and

$$\lim_{R\to r_0} H_R(R,\psi) = G(\psi) \qquad (R > r_0) \tag{9}$$

for each ψ where G is continuous.

The verification of formula (7), as well as special cases of formula (3) that apply to semicircular regions, is left to the exercises.

105. A Neumann Problem for a Half Plane

Let $G(x)$ be continuous for all real x, except for at most a finite number of finite jumps, and let it satisfy an order property

$$|x^k G(x)| < M \qquad (-\infty < x < \infty) \tag{1}$$

where $k > 1$. For each fixed real number t the function $\operatorname{Log}|z - t|$ is harmonic in the half plane $\operatorname{Im} z > 0$. Consequently, the function

$$\begin{aligned} U(x,y) &= \frac{1}{\pi}\int_{-\infty}^{\infty} \operatorname{Log}|z - t|G(t)\, dt + U_0 \\ &= \frac{1}{2\pi}\int_{-\infty}^{\infty} \operatorname{Log}[(t - x)^2 + y^2]G(t)\, dt + U_0 \qquad (y > 0), \end{aligned} \tag{2}$$

where U_0 is a real constant, is harmonic in that half plane.

Formula (2) was written with the Schwarz formula (1) of Sec. 103 in mind; for it follows from formula (2) that

$$U_y(x,y) = \frac{1}{\pi}\int_{-\infty}^{\infty} \frac{yG(t)}{(t-x)^2+y^2}\,dt \qquad (y>0). \tag{3}$$

In view of equations (1) and (2) of Sec. 103, then,

$$\lim_{y\to 0} U_y(x,y) = G(x) \qquad (y>0) \tag{4}$$

at each point x where G is continuous.

Integral formula (2) therefore solves *the Neumann problem for the half plane* $y > 0$ with boundary condition (4). But we have not presented conditions on G that are sufficient to ensure that the harmonic function U is bounded as $|z|$ increases.

When G is an odd function, formula (2) can be written

$$U(x,y) = \frac{1}{2\pi}\int_0^{\infty} \operatorname{Log}\frac{(t-x)^2+y^2}{(t+x)^2+y^2}\,G(t)\,dt \qquad (x>0,\ y>0). \tag{5}$$

This represents a function which is harmonic in the *first quadrant* $x>0$, $y>0$ and which satisfies the boundary conditions

$$U(0,y) = 0 \qquad (y>0), \tag{6}$$

$$\lim_{y\to 0} U_y(x,y) = G(x) \qquad (x>0,\ y>0). \tag{7}$$

The kernels of all the integral formulas for harmonic functions presented in this chapter can be described in terms of a single real-valued function of the complex variables $z = x + iy$ and $w = u + iv$:

$$K(z,w) = \operatorname{Log}|z-w| \qquad (z \neq w). \tag{8}$$

This is *Green's function* for the *logarithmic potential* in the z plane. The function is symmetric; that is, $K(w,z) = K(z,w)$. Expressions for kernels used earlier, in terms of K and its derivatives, are given in the exercises.

EXERCISES

1. Obtain the special case of formula (1), Sec. 103,

$$U(x,y) = \frac{y}{\pi}\int_0^{\infty}\left[\frac{1}{(t-x)^2+y^2} - \frac{1}{(t+x)^2+y^2}\right]F(t)\,dt \qquad (x>0,\ y>0)$$

for a bounded function U which is harmonic in the *first quadrant* and which satisfies the boundary conditions

$$U(0,y) = 0 \qquad (y > 0),$$
$$\lim_{y \to 0} U(x,y) = F(x) \qquad (x > 0,\ x \neq x_j,\ y > 0),$$

where F is bounded for all positive x and continuous except for at most a finite number of finite jumps at the points x_j $(j = 1, 2, \ldots, n)$.

2. Obtain the special case of formula (1), Sec. 103,

$$U(x,y) = \frac{y}{\pi}\int_0^\infty \left[\frac{1}{(t-x)^2+y^2} + \frac{1}{(t+x)^2+y^2}\right] F(t)\,dt \qquad (x > 0,\ y > 0)$$

for a bounded function U which is harmonic in the *first quadrant* and which satisfies the boundary conditions

$$U_x(0,y) = 0 \qquad (y > 0),$$
$$\lim_{y \to 0} U(x,y) = F(x) \qquad (x > 0,\ x \neq x_j,\ y > 0),$$

where F is bounded for all positive x and continuous except possibly for finite jumps at a finite number of points $x = x_j$ $(j = 1, 2, \ldots, n)$.

3. Interchange the x and y axes in Sec. 103 to write the solution

$$U(x,y) = \frac{1}{\pi}\int_{-\infty}^{\infty} \frac{xF(t)}{(t-y)^2+x^2}\,dt \qquad (x > 0)$$

of the Dirichlet problem for the half plane $x > 0$. Write

$$\lim_{x \to 0} U(x,y) = \begin{cases} 1 & \text{when} \quad x > 0,\ -1 < y < 1, \\ 0 & \text{when} \quad x > 0,\ |y| > 1. \end{cases}$$

Then obtain these formulas for U and its harmonic conjugate $-V$:

$$U(x,y) = \frac{1}{\pi}\left(\arctan\frac{y+1}{x} - \arctan\frac{y-1}{x}\right),$$

$$V(x,y) = \frac{1}{2\pi}\,\mathrm{Log}\,\frac{x^2+(y+1)^2}{x^2+(y-1)^2}$$

where $-\pi/2 \leqq \arctan t \leqq \pi/2$. Also, show that $\pi[V(x,y) + iU(x,y)] = \mathrm{Log}\,(z+i) - \mathrm{Log}\,(z-i)$, where $z = x + iy$.

4. Let $T(x,y)$ denote the bounded steady temperatures in a plate $x > 0$, $y > 0$ with insulated faces when

$$\lim_{y \to 0} T(x,y) = F_1(x) \quad \text{and} \quad \lim_{x \to 0} T(x,y) = F_2(y) \qquad (x > 0,\ y > 0)$$

(Fig. 106). Here F_1 and F_2 are bounded and continuous except for at most a finite number of finite jumps. Write $x + iy = z$ and show with the aid of the formula in Exercise 1 that

$$T(x,y) = T_1(x,y) + T_2(x,y) \qquad (x > 0,\ y > 0)$$

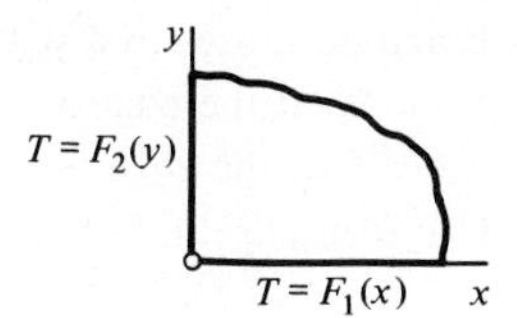

FIGURE 106

where

$$T_1(x,y) = \frac{y}{\pi}\int_0^\infty \left(\frac{1}{|t-z|^2} - \frac{1}{|t+z|^2}\right)F_1(t)\,dt,$$

$$T_2(x,y) = \frac{x}{\pi}\int_0^\infty \left(\frac{1}{|it-z|^2} - \frac{1}{|it+z|^2}\right)F_2(t)\,dt.$$

5. Establish formula (7), Sec. 104, as a solution of the Neumann problem for the region exterior to a circle, using earlier results found in that section.

6. Obtain the special case of formula (3), Sec. 104,

$$U(r,\theta) = \frac{1}{2\pi}\int_0^\pi [Q(r_0,r,\phi-\theta) - Q(r_0,r,\phi+\theta)]G(\phi)\,d\phi$$

for a function U which is harmonic in the *semicircular region* $r < r_0$, $0 < \theta < \pi$ and which satisfies the boundary conditions

$$U(r,0) = U(r,\pi) = 0 \qquad (r < r_0)$$

and

$$\lim_{r\to r_0} U_r(r,\theta) = G(\theta) \qquad (r < r_0,\ 0 < \theta < \pi)$$

for each θ where G is continuous.

7. Obtain the special case of formula (3), Sec. 104,

$$U(r,\theta) = \frac{1}{2\pi}\int_0^\pi [Q(r_0,r,\phi-\theta) + Q(r_0,r,\phi+\theta)]G(\phi)\,d\phi + U_0$$

for a function U which is harmonic in the *semicircular region* $r < r_0$, $0 < \theta < \pi$ and which satisfies the boundary conditions

$$U_\theta(r,0) = U_\theta(r,\pi) = 0 \qquad (r < r_0)$$

and

$$\lim_{r\to r_0} U_r(r,\theta) = G(\theta) \qquad (r < r_0,\ 0 < \theta < \pi)$$

for each θ where G is continuous, provided that

$$\int_0^\pi G(\phi)\,d\phi = 0.$$

8. Let $T(x,y)$ denote the steady temperatures in a plate $x \geqq 0$, $y \geqq 0$. The faces of the plate are insulated, and $T = 0$ on the edge $x = 0$. The flux of heat into the plate along the segment $0 < x < 1$ of the edge $y = 0$ is a constant A, and the rest of that edge is insulated. Use formula (5), Sec. 105, to show that the flux out of the plate along the edge $x = 0$ is

$$\frac{A}{\pi} \operatorname{Log}\left(1 + \frac{1}{y^2}\right).$$

9. Show that the Poisson kernel is given in terms of Green's function

$$K(z,w) = \operatorname{Log}|z - w| = \tfrac{1}{2} \operatorname{Log}[\rho^2 - 2\rho r \cos(\phi - \theta) + r^2],$$

where $z = r \exp(i\theta)$ and $w = \rho \exp(i\phi)$, by the equation

$$P(\rho,r,\phi - \theta) = 2\rho \frac{\partial K}{\partial \rho} - 1.$$

10. Show that the kernel used in the Schwarz integral transform (Sec. 103) can be written in terms of Green's function

$$K(z,w) = \operatorname{Log}|z - w| = \tfrac{1}{2} \operatorname{Log}[(x - u)^2 + (y - v)^2],$$

where $z = x + iy$ and $w = u + iv$, as

$$\frac{y}{|u - z|^2} = \left.\frac{\partial K}{\partial y}\right]_{v=0} = -\left.\frac{\partial K}{\partial v}\right]_{v=0}.$$

Here K is to be interpreted as a function of the four real variables x, y, u, and v.

12

FURTHER THEORY OF FUNCTIONS

Many topics in the theory of functions which were not essential to the continuity of the presentation in the earlier chapters were omitted there. Several such topics do, however, deserve a place in an introductory course because of their general interest, and we include them in this chapter.

A. ANALYTIC CONTINUATION

We consider first how the behavior of an analytic function in a domain is determined by its behavior in a smaller set contained in that domain. We then consider the problem of extending the domain of definition of an analytic function.

106. Conditions under which $f(z) \equiv 0$

In Sec. 66 we proved that the zeros of an analytic function are isolated unless the function is identically zero. That is, when a function f is analytic at a point z_0, there is a neighborhood $|z - z_0| < \varepsilon$ such that either $f(z) \equiv 0$ throughout that neighborhood or else f has no zeros in it except possibly at the point z_0 itself.

Suppose now that z_0 is an accumulation point of an infinite set and that $f(z) = 0$ at each point z of that set. Then every neighborhood of z_0 contains a zero of f other than the point z_0 itself; and if f is analytic at z_0, it follows that there is some neighborhood of z_0 throughout which $f(z) \equiv 0$. All the coefficients $f(z_0), f^{(n)}(z_0)/n!$ $(n = 1, 2, 3, \ldots)$ in the Taylor series of $f(z)$ about z_0 are therefore zero. Hence if this function f is analytic interior to a circle $|z - z_0| = r_0$, it follows that $f(z) \equiv 0$ in the open disk $|z - z_0| < r_0$.

In particular, if $f(z) = 0$ at each point z in some *domain* containing z_0, or at each point on some *arc* containing z_0, and if f is analytic in an open disk $|z - z_0| < r_0$, then $f(z)$ is identically zero throughout that disk.

We now present the main result of this section.

Theorem. *If a function f is analytic throughout a domain D and $f(z) = 0$ at each point z of a domain or an arc interior to D, then $f(z) = 0$ at each point of D.*

We first prove the theorem for the case when $f(z) = 0$ at each point z of a domain D_0 interior to D. Let s_0 be any point in D_0 and let σ be any point which lies in D but not in D_0. Since a domain is connected, there exists a polygonal path C, consisting of a finite number of line segments joined end to end, which lies in D and joins s_0 to σ (Fig. 107).

Now the analytic function f has a Taylor series expansion about each point s of C, the radius of the circle of convergence being some positive number $R(s)$. We agree, however, to write $R(s) = 1$ whenever that radius is greater

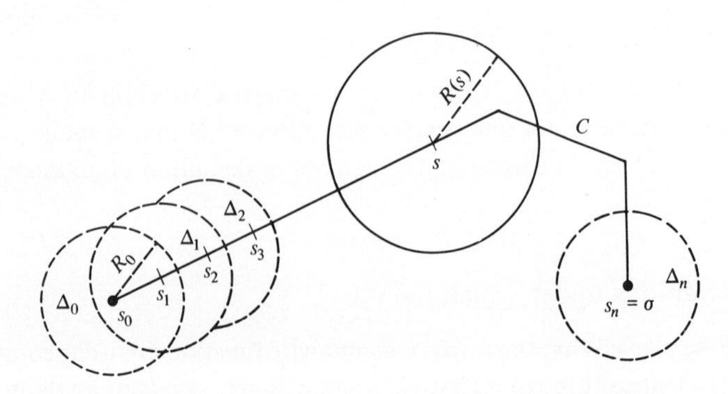

FIGURE 107

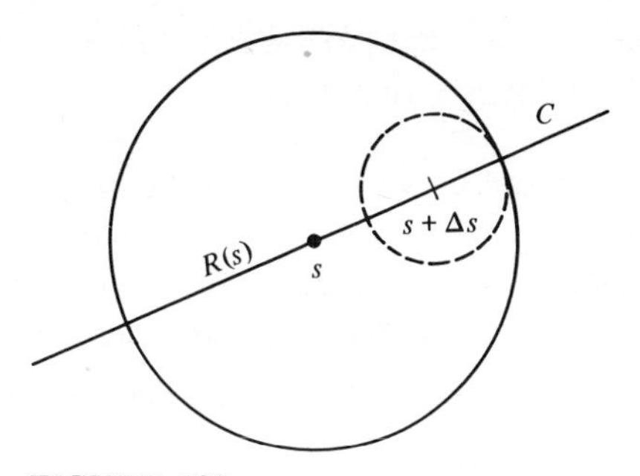

FIGURE 108

than unity; hence, $0 < R(s) \leqq 1$. A circle $|z - s| = R(s)$ may, of course, extend beyond D.

We shall need the fact that R is a continuous function of s. To show this, let s be any point on C and let $s + \Delta s$ be a point on C close enough to s such that $|\Delta s| < R(s)$. It follows that f is analytic in the open disk

$$|z - (s + \Delta s)| < R(s) - |\Delta s|$$

centered at $s + \Delta s$ (Fig. 108). But f may actually be analytic in a larger open disk centered there. Hence $R(s + \Delta s) \geqq R(s) - |\Delta s|$, or

$$-[R(s + \Delta s) - R(s)] \leqq |\Delta s|. \tag{1}$$

If $R(s + \Delta s) \leqq R(s)$, inequality (1) can be written

$$|R(s + \Delta s) - R(s)| \leqq |\Delta s|. \tag{2}$$

Suppose, on the other hand, that $R(s + \Delta s) > R(s)$. Observe that if z is a point lying in the disk

$$|z - s| < R(s + \Delta s) - |\Delta s|, \tag{3}$$

then

$$|z - (s + \Delta s)| \leqq |z - s| + |\Delta s| < R(s + \Delta s).$$

The function f is therefore analytic at z because that point lies within the circle of convergence about $s + \Delta s$. Consequently, disk (3) is contained in the disk $|z - s| < R(s)$. That is, $R(s + \Delta s) - |\Delta s| \leqq R(s)$; and again inequality (2) is satisfied.

With inequality (2), we see that $|R(s + \Delta s) - R(s)|$ is less than any positive number ε when $|\Delta s|$ is less than both ε and $R(s)$. That is,

$$\lim_{\Delta s \to 0} R(s + \Delta s) = R(s);$$

and the continuity of R at s is established.

When the contour C is given a parametric representation $z = z(t)$, $a \leqq t \leqq b$, the function R can be considered as a real-valued function $R[z(t)]$ of a real variable which is continuous and positive on a closed bounded interval. The function R has, therefore, a positive minimum value R_0. Thus the function f is analytic in the disk $|z - s_0| < R_0$, denoted by Δ_0. Since $f(z) = 0$ at each point in the domain D_0 which contains s_0, we conclude that $f(z) = 0$ at each point z in the disk Δ_0. This follows from the remarks that precede the statement of the theorem.

Let $s_0, s_1, s_2, \ldots, s_n = \sigma$ be a sequence of points on C such that

$$\tfrac{1}{2}R_0 \leqq |s_j - s_{j-1}| < R_0 \qquad (j = 1, 2, \ldots, n).$$

As illustrated in Fig. 107, there is about each point s_j an open disk Δ_j of radius R_0 throughout which f is analytic. Because the center s_1 of Δ_1 lies in the domain Δ_0 where $f(z)$ is always zero, $f(z) = 0$ throughout Δ_1. Likewise, the center of Δ_2 lies in the domain Δ_1, and so $f(z) = 0$ throughout Δ_2. Continuing in this manner, we eventually reach Δ_n and find that $f(\sigma) = 0$. The proof of the theorem is thus complete for the case when $f(z) = 0$ at each point of a domain D_0 interior to D.

Suppose now that $f(z) = 0$ along an arc in D. Then about any point on the arc there is an open disk, or domain, interior to D; and, by the remarks preceding the statement of the theorem, $f(z) = 0$ throughout that disk. Hence, in view of the case just completed, we can conclude that $f(z) = 0$ at each point of D.

107. Permanence of Forms of Functional Identities

Suppose that two functions f and g are analytic in the same domain D and that $f(z) = g(z)$ at each point z of some domain or arc contained in D. The function h defined by $h(z) = f(z) - g(z)$ is also analytic in D, and $h(z) = 0$ throughout the subdomain or along the arc. Thus $h(z) = 0$ throughout D; that is, $f(z) = g(z)$ there. We arrive at the following result.

Theorem 1. *A function that is analytic in a domain D is uniquely determined over D by its values over a domain, or along an arc, interior to D.*

As an illustration, the function e^z is the only entire function that can assume the values e^x along a segment of the real axis. Moreover, since e^{-z} is entire and since $e^x e^{-x} = 1$ whenever x is real, the function

$$e^z e^{-z} - 1$$

is entire and takes on zero values all along the real axis. Consequently,

$$e^z e^{-z} - 1 = 0$$

everywhere; and the identity $e^{-z} = 1/e^z$ is valid for all complex numbers z.

Such permanence of forms of other identities between functions, in passing from a real to a complex variable, is established in the same way. We limit our attention in the theorem below to the important class of identities that involve only polynomials in the functions.

Theorem 2. *Let $P(w_1, w_2, \ldots, w_n)$ be a polynomial in the n variables w_j, and let f_j $(j = 1, 2, \ldots, n)$ be analytic functions of z in a domain D that contains some interval $a < x < b$ of the x axis. If on that interval the functions f_j satisfy the identity*

$$P[f_1(x), f_2(x), \ldots, f_n(x)] = 0, \tag{1}$$

then

$$P[f_1(z), f_2(z), \ldots, f_n(z)] = 0 \tag{2}$$

throughout the domain D.

The left-hand side of equation (2) represents an analytic function of z in the given domain; and it is zero along an arc in that domain, according to identity (1). Hence identity (2) is valid throughout the domain.

To illustrate this theorem, we consider the polynomial $P(w_1, w_2) = w_1^2 + w_2^2 - 1$ and the entire functions $f_1(z) = \sin z$ and $f_2(z) = \cos z$. On the real axis $P[f_1(x), f_2(x)] = \sin^2 x + \cos^2 x - 1 = 0$. Thus, $P[f_1(z), f_2(z)] = \sin^2 z + \cos^2 z - 1 = 0$, or $\sin^2 z + \cos^2 z = 1$, throughout the entire z plane.

108. Uniqueness of Analytic Continuation

The *intersection* of two domains D_1 and D_2 is the domain $D_1 \cap D_2$ consisting of all points that are common to both D_1 and D_2. If the two domains have points in common, their *union* $D_1 \cup D_2$, consisting of the totality of points that lie in either D_1 or D_2, is also a domain.

If we have two domains D_1 and D_2 with points in common (Fig. 109) and a function f_1 that is analytic in D_1, there *may* exist a function f_2 which is

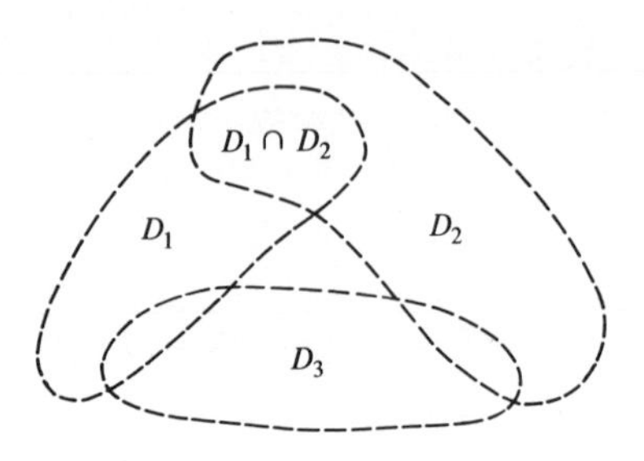

FIGURE 109

analytic in D_2 such that $f_2(z) = f_1(z)$ for each z in the intersection $D_1 \cap D_2$. If so, we call f_2 *the analytic continuation of f_1 into the domain D_2*.

Whenever that analytic continuation f_2 exists, it is unique, according to Theorem 1 of the preceding section; for not more than one function can be analytic in D_2 and also assume the value $f_1(z)$ at each point z of the domain $D_1 \cap D_2$ interior to D_2. However, if there is an analytic continuation f_3 of f_2 from D_2 into a domain D_3 which intersects D_1 as indicated in Fig. 109, it is not necessarily true that $f_3(z) = f_1(z)$ for each z in $D_1 \cap D_3$. In the following section we illustrate the fact that such a chain of continuations of a given function from a domain D_1 may lead to a different function defined on D_1.

If f_2 is the analytic continuation of f_1 from a domain D_1 into a domain D_2, then the function F defined by

$$F(z) = \begin{cases} f_1(z) & \text{when} \quad z \text{ is in } D_1, \\ f_2(z) & \text{when} \quad z \text{ is in } D_2 \end{cases}$$

is analytic in the union $D_1 \cup D_2$. The function F is the analytic continuation into $D_1 \cup D_2$ of either f_1 or f_2; and f_1 and f_2 are called *elements* of F.

109. Examples

Consider first the function f_1 defined by the equation

$$f_1(z) = \sum_{n=0}^{\infty} z^n. \tag{1}$$

The power series here converges if and only if $|z| < 1$. It is the Maclaurin series expansion of the function $1/(1 - z)$. Hence

$$f_1(z) = \frac{1}{1 - z} \qquad \text{when} \qquad |z| < 1,$$

and f_1 is not defined when $|z| \geqq 1$.

Now the function

$$f_2(z) = \frac{1}{1 - z} \qquad (z \neq 1) \tag{2}$$

is defined and analytic everywhere except at the point $z = 1$. Since $f_2(z) = f_1(z)$ inside the circle $|z| = 1$, the function f_2 is the analytic continuation of f_1 into the domain consisting of all points in the z plane except for $z = 1$. It is the only possible analytic continuation of f_1 into that domain, according to the results established in the preceding section. In this example f_1 is also an element of f_2.

It is of interest to note that if we begin with the information that the power series

$$\sum_{n=0}^{\infty} z^n$$

converges and represents an analytic function of z when $|z| < 1$ and that its sum is $1/(1 - x)$ when $z = x$, then we can conclude that its sum is $1/(1 - z)$ whenever $|z| < 1$. This follows from the fact that the function $1/(1 - z)$ is the analytic function interior to the circle $|z| = 1$ that assumes the values $1/(1 - x)$ along the segment of the x axis inside the circle.

For another illustration of analytic continuation, consider the function

$$g_1(z) = \int_0^{\infty} e^{-zt}\, dt. \tag{3}$$

Straightforward integration reveals that integral (3) exists only when $\operatorname{Re} z > 0$ and that

$$g_1(z) = \frac{1}{z}. \tag{4}$$

The domain of definition $\operatorname{Re} z > 0$ is denoted by D_1 in Fig. 110; the function g_1 is analytic there. Let g_2 be defined in terms of a geometric series by the equation

$$g_2(z) = i \sum_{n=0}^{\infty} \left(\frac{z + i}{i}\right)^n \qquad (|z + i| < 1). \tag{5}$$

Within its circle of convergence, the unit circle about the point $z = -i$, the series is convergent. Hence

$$g_2(z) = i \frac{1}{1 - (z + i)/i} = \frac{1}{z} \tag{6}$$

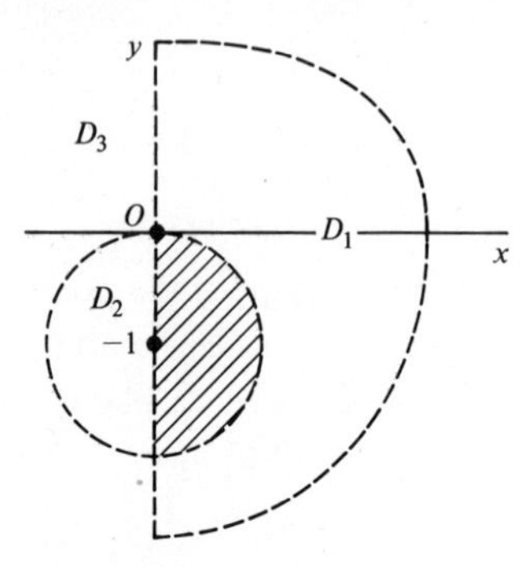

FIGURE 110

when z is in the domain $|z + i| < 1$, denoted by D_2. Evidently then, $g_2(z) = g_1(z)$ for each z in the intersection $D_1 \cap D_2$; and g_2 is the analytic continuation of g_1 into D_2.

The function $G(z) = 1/z$ $(z \neq 0)$ is the analytic continuation of both g_1 and g_2 into the domain D_3 consisting of all points in the z plane except the origin. The functions g_1 and g_2 are then elements of G.

Finally, consider this branch of $z^{1/2}$:

$$h_1(z) = \sqrt{r} \exp \frac{i\theta}{2} \qquad (r > 0, 0 < \theta < \pi).$$

The analytic continuation h_2 across the negative real axis into the lower half plane is

$$h_2(z) = \sqrt{r} \exp \frac{i\theta}{2} \qquad \left(r > 0, \frac{\pi}{2} < \theta < 2\pi\right).$$

The analytic continuation of h_2 across the positive real axis into the first quadrant is then

$$h_3(z) = \sqrt{r} \exp \frac{i\theta}{2} \qquad \left(r > 0, \pi < \theta < \frac{5\pi}{2}\right).$$

Note that $h_3(z) \neq h_1(z)$ in the first quadrant; in fact, $h_3(z) = -h_1(z)$ there.

110. The Principle of Reflection

In Chap. 3 we found that some elementary functions $f(z)$ possess the property $f(\bar{z}) = \overline{f(z)}$, and others do not. For examples of those that do, we can cite the functions

$$z, \qquad z^2 + 1, \qquad e^z, \qquad \sin z;$$

for when z is replaced by its conjugate, the value of each of these functions changes to the conjugate of the original value. On the other hand, the functions

$$iz, \qquad z^2 + i, \qquad e^{iz}, \qquad (1 + i) \sin z$$

do not satisfy the property that the reflection of z in the real axis corresponds to the reflection of $f(z)$ in the real axis.

The following theorem, known as the *reflection principle*, explains these observations.

Theorem. *Let a function f be analytic in some domain D that includes a segment of the x axis and is symmetric to the x axis. If $f(x)$ is real whenever x is a point on that segment, then*

$$f(\bar{z}) = \overline{f(z)} \tag{1}$$

whenever z is a point in D. Conversely, if condition (1) *is satisfied, then $f(x)$ is real.*

Equation (1) represents the same condition on f as the equation

$$\overline{f(\bar{z})} = f(z), \tag{2}$$

where $f(z) = u(x,y) + iv(x,y)$ and

$$\overline{f(\bar{z})} = u(x,-y) - iv(x,-y). \tag{3}$$

When condition (2) is satisfied at a point on the real axis, then

$$f(x) = u(x,0) + iv(x,0) = u(x,0) - iv(x,0);$$

hence $v(x,0) = 0$, and $f(x)$ is real. The converse statement in the theorem is therefore true.

In order to prove the direct statement in the theorem, we first show that the function $\overline{f(\bar{z})}$ is analytic throughout the domain D. We write

$$F(z) = \overline{f(\bar{z})} = U(x,y) + iV(x,y).$$

Then, according to equation (3),

$$U(x,y) = u(r,t), \qquad V(x,y) = -v(r,t) \tag{4}$$

where $r = x$ and $t = -y$. Since $f(r + it)$ is an analytic function of $r + it$, the functions $u(r,t)$ and $v(r,t)$, together with their partial derivatives, are continuous throughout D; and the Cauchy-Riemann equations

$$u_r = v_t, \qquad u_t = -v_r$$

are satisfied there. Now, in view of equations (4), we see that

$$U_x = u_r, \qquad V_y = -v_t \frac{dt}{dy} = v_t$$

and therefore $U_x = U_y$. Similarly, we find that

$$U_y = -V_x.$$

These partial derivatives of U and V are continuous, and the function F is then analytic in the domain D.

Since $f(x)$ is real, $v(x,0) = 0$. Hence

$$F(x) = U(x,0) + iV(x,0) = u(x,0);$$

that is, $F(z) = f(z)$ when the point z is on the segment of the x axis within the domain. It follows from Theorem 1 of Sec. 107 that $F(z) = f(z)$ at each point z of D since both functions are analytic there. Thus condition (2) is established, and the proof of the theorem is complete.

EXERCISES

1. Given that the hyperbolic sine and hyperbolic cosine functions, the exponential function, and the sine and cosine functions are all entire, use Theorem 2 of Sec. 107 to obtain each of these identities for all complex z from the corresponding identities when z is real:
(*a*) $\sinh z + \cosh z = e^z$; (*b*) $\sin 2z = 2 \sin z \cos z$;
(*c*) $\cosh^2 z - \sinh^2 z = 1$; (*d*) $\sin (\pi/2 - z) = \cos z$.

2. Show that the function

$$f_2(z) = \frac{1}{z^2 + 1} \qquad (z \neq \pm i)$$

is the analytic continuation of the function

$$f_1(z) = \sum_{n=0}^{\infty} (-1)^n z^{2n} \qquad (|z| < 1)$$

into the domain consisting of all points in the z plane except $z = \pm i$.

3. Show that the function $1/z^2$ represents the analytic continuation of the function defined by the series

$$\sum_{n=0}^{\infty} (n + 1)(z + 1)^n \qquad (|z + 1| < 1)$$

into the domain consisting of all points in the z plane except $z = 0$.

4. State why the function

$$h_4(z) = \sqrt{r} \exp \frac{i\theta}{2} \qquad (r > 0, -\pi < \theta < \pi)$$

is the analytic continuation of the function (Sec. 109)

$$h_1(z) = \sqrt{r} \exp \frac{i\theta}{2} \qquad (r > 0, 0 < \theta < \pi)$$

across the positive real axis into the lower half plane.

5. Find the analytic continuation of Log z from the upper half plane Im $z > 0$ into the lower half plane across the negative real axis. Note that this analytic continuation is different from Log z in the lower half plane.

Ans. $\text{Log}\, r + i\theta \quad (r > 0, 0 < \theta < 2\pi)$.

6. Find the analytic continuation of the function

$$f(z) = \int_0^\infty te^{-zt}\, dt \qquad (\text{Re}\, z > 0)$$

into the domain consisting of all points in the z plane except the origin.

Ans. $1/z^2$.

7. Show that the function $1/(z^2 + 1)$ is the analytic continuation of the function

$$f(z) = \int_0^\infty e^{-zt} \sin t\, dt \qquad (\text{Re}\, z > 0)$$

into the domain consisting of all points in the z plane except $z = \pm i$.

8. Show that if the condition that $f(x)$ be real in the Theorem of Sec. 110 is replaced by the condition that $f(x)$ be pure imaginary, the conclusion is changed to $f(\bar{z}) = -\overline{f(z)}$.

9. Let S denote a set of points in a domain D such that S has an accumulation point in D. Generalize Theorem 1 of Sec. 107 by proving that a function which is analytic in D is uniquely determined by its values on the set S.

B. SINGULAR POINTS AND ZEROS

We now examine further the behavior of functions near their singular points.

111. Poles and Zeros

It was pointed out in Sec. 70 that if z_0 is a pole of any order of a function f, then

$$\lim_{z \to z_0} f(z) = \infty; \tag{1}$$

that is, for each positive number ε there exists a positive number δ such that

$$|f(z)| > \frac{1}{\varepsilon} \qquad \text{whenever} \qquad 0 < |z - z_0| < \delta. \tag{2}$$

As a consequence, there is always some neighborhood of a pole containing no zeros of the function f.

Since poles are isolated singular points, it follows that *if z_0 is a pole of a function f, there is a neighborhood of z_0 which contains neither a zero of f nor a singular point of f other than z_0 itself.*

According to Exercise 12, Sec. 71, if z_0 is a zero of order m of a function $f(z)$, then z_0 is a pole of order m of the reciprocal function $1/f(z)$. A converse is easily established. For if z_0 is a pole of order m of a function $g(z)$, the function $(z - z_0)^m g(z)$ has a removable singular point at z_0. The value assigned to the latter function at z_0 so that the resulting function is analytic in some disk $|z - z_0| < r_0$ about z_0 is a number other than zero. If ϕ denotes that analytic function, then

$$\phi(z) = (z - z_0)^m g(z) \qquad \text{when} \qquad 0 < |z - z_0| < r_0 \tag{3}$$

and

$$\phi(z_0) \neq 0.$$

Now the function $1/\phi(z)$ is analytic at z_0; and, for some positive number r_1, it is represented by a Taylor series

$$\frac{1}{\phi(z)} = \sum_{n=0}^{\infty} a_n (z - z_0)^n \qquad (|z - z_0| < r_1),$$

where $r_1 \leqq r_0$ and $a_0 = 1/\phi(z_0) \neq 0$. It follows from equation (3) that

$$\frac{1}{g(z)} = (z - z_0)^m \sum_{n=0}^{\infty} a_n (z - z_0)^n \qquad (|z - z_0| < r_1). \tag{4}$$

Therefore, *if z_0 is a pole of order m of a function $g(z)$, then it is a zero of order m of the reciprocal function $1/g(z)$.*

In contrast to condition (2), suppose f is bounded and analytic in a domain $0 < |z - z_0| < \delta$. Then the following theorem due to Riemann applies.

Theorem. *If a function f is bounded and analytic throughout a domain $0 < |z - z_0| < \delta$, then either f is analytic at z_0 or else z_0 is a removable singular point of that function.*

To prove this, observe that $f(z)$ is represented by its Laurent series in the given domain about z_0. If C denotes a circle $|z - z_0| = r$, where $r < \delta$, the coefficients b_n of $1/(z - z_0)^n$ in that series are (Sec. 59)

$$b_n = \frac{1}{2\pi i}\int_C \frac{f(z)\,dz}{(z - z_0)^{-n+1}} = \frac{r^n}{2\pi}\int_0^{2\pi} f(z_0 + re^{i\theta})e^{in\theta}\,d\theta,$$

where $n = 1, 2, \ldots$. Since f is bounded, there is a positive real number M such that

$$|f(z)| < M \qquad (0 < |z - z_0| < \delta);$$

hence

$$|b_n| < Mr^n.$$

But the coefficients are constants; and, since r can be chosen arbitrarily small, $b_n = 0$ $(n = 1, 2, \ldots)$. Thus the Laurent series for $f(z)$ reduces to a power series

$$f(z) = \sum_{n=0}^{\infty} a_n(z - z_0)^n \qquad (0 < |z - z_0| < \delta).$$

If $f(z_0)$ is defined to be the number a_0, it follows that f is analytic at z_0. This proves the theorem.

112. Essential Singular Points

The behavior of a function near an essential singular point is quite irregular. This was pointed out in Sec. 69 where Picard's theorem was stated; namely, *in any neighborhood of an essential singular point the function assumes every finite value, with one possible exception, an infinite number of times.* We also illustrated Picard's theorem by showing that the function exp $(1/z)$, which has an essential singular point at the origin, assumes the value -1 an infinite number of times in any neighborhood of that singular point. We do not prove Picard's theorem, but do prove a related theorem due to Weierstrass. This theorem shows that the value of a function is arbitrarily close to any prescribed number c at points arbitrarily near an essential singular point of that function.

Theorem. *Let z_0 be an essential singular point of a function f and let c be any given complex number. Then for each positive number ε, however small, the inequality*

$$(1) \qquad |f(z) - c| < \varepsilon$$

is satisfied at some point z different from z_0 in each neighborhood of z_0.

To prove the theorem, suppose that condition (1) is not satisfied at any point in a neighborhood $|z - z_0| < \delta$, where δ is small enough so that f is analytic in the domain $0 < |z - z_0| < \delta$. Then $|f(z) - c| \geqq \varepsilon$ for all points in that domain, and the function

$$g(z) = \frac{1}{f(z) - c} \qquad (0 < |z - z_0| < \delta) \tag{2}$$

is analytic and bounded there. According to Riemann's theorem (Sec. 111), z_0 is a removable singular point of g. Let $g(z_0)$ be defined so that g is analytic at z_0. Since f cannot be a constant function, neither can g; and, in view of the Taylor series for g at z_0, either $g(z_0) \neq 0$ or else g has a zero of some finite order at z_0. Consequently, its reciprocal,

$$\frac{1}{g(z)} = f(z) - c,$$

is either analytic at z_0 or else it has a pole there. But this contradicts the hypothesis that z_0 is an essential singular point of f. Hence condition (1) must be satisfied at some point in the given neighborhood.

113. The number of Zeros and Poles

The properties of the logarithmic derivative found in Exercises 13 and 14, Sec. 71, can be generalized.

Let a function f be analytic inside and on a simple closed contour C except for at most a finite number of poles interior to C. Also, let f have no zeros on C and at most a finite number of zeros interior to C. Then, if C is described in the positive sense,

$$\frac{1}{2\pi i} \int_C \frac{f'(z)}{f(z)}\, dz = N - P \tag{1}$$

where N is the total number of zeros of f inside C and P is the total number of poles of f there. A zero of order m_0 is to be counted m_0 times, and a pole of order m_p is to be counted m_p times.

To prove statement (1), we show that the integer $N - P$ is the sum of the residues of the function $f'(z)/f(z)$ at its singular points inside the simple closed contour C. Those singular points are, of course, the zeros and poles of f interior to C.

Let z_0 be a zero of f of order m_0. In some neighborhood of z_0 we can write

$$f(z) = (z - z_0)^{m_0} g(z) \tag{2}$$

where g is analytic in that neighborhood and $g(z_0) \neq 0$. Hence

$$f'(z) = m_0(z - z_0)^{m_0 - 1} g(z) + (z - z_0)^{m} g'(z),$$

or

$$\frac{f'(z)}{f(z)} = \frac{m_0}{z - z_0} + \frac{g'(z)}{g(z)}.$$

Since $g'(z)/g(z)$ is analytic at z_0, the function $f'(z)/f(z)$ has a simple pole at z_0 with residue m_0. The sum of the residues of $f'(z)/f(z)$ at all the zeros of f inside C is therefore the integer N.

If z_p is a pole of f of order m_p, the function

$$h(z) = (z - z_p)^{m_p} f(z) \qquad (z \neq z_p) \tag{3}$$

can be defined at z_p so that h is analytic there; moreover, $h(z_p) \neq 0$. Thus in some neighborhood of z_p, except at the point $z = z_p$ itself,

$$f(z) = (z - z_p)^{-m_p} h(z), \tag{4}$$

and

$$f'(z) = -m_p(z - z_p)^{-m_p-1} h(z) + (z - z_p)^{-m} h'(z).$$

Therefore

$$\frac{f'(z)}{f(z)} = -\frac{m_p}{z - z_p} + \frac{h'(z)}{h(z)},$$

and we see that $f'(z)/f(z)$ has a simple pole at z_p with residue $-m_p$. Hence the sum of the residues of $f'(z)/f(z)$ at all the poles of f inside C is the integer $-P$; and formula (1) is established.

One form of what is known as the Bolzano-Weierstrass theorem can be stated as follows.[1] *A set of infinitely many points each of which lies in a closed bounded region has at least one accumulation point in that region.* The proof can be made by selecting any infinite sequence $z_1, z_2, \ldots$ of points from the set and applying to that sequence a process for subdividing squares, a procedure used in Exercise 13, Sec. 50.

In view of that theorem, the condition that the number of zeros and poles inside C be finite which was used in proving formula (1) can be removed. For the number of zeros and poles within the simple closed contour C must necessarily be finite if the function f is to be analytic within and on C, except possibly for poles within C, because zeros and poles are isolated. The full argument is left as an exercise.

[1] See, for example, A. E. Taylor and W. R Mann, "Advanced, Calculus" 2d ed., pp. 545 and 549, 1972.

114. The Argument Principle

Let C be a simple closed contour having positive orientation in the z plane and let f be a function which is analytic within and on C, except possibly for poles interior to C. Also, let f have no zeros on C. The image Γ of C under the transformation $w = f(z)$ is a closed contour in the w plane (Fig. 111). As a point z traverses C in the positive direction, its image w traverses Γ in a particular direction which determines the orientation of Γ.

Since f has no zeros on C, the contour Γ does not pass through the origin in the w plane. Let w_0 be a fixed point on Γ and let ϕ_0 be a value of arg w_0. Then let arg w vary continuously, starting with the value ϕ_0, as the point w begins at w_0 and traverses Γ once in the direction of orientation assigned to it by the mapping $w = f(z)$. When w returns to the starting point w_0, arg w assumes a particular value of arg w_0 which we denote by ϕ_1. Thus, the change in arg w as w describes Γ once in its direction of orientation is $\phi_1 - \phi_0$. Note that this change is independent of the particular point w_0 chosen to determine it.

The number $\phi_1 - \phi_0$ is also the change in the argument of $f(z)$ as z describes C once in the positive direction, and we write

$$\Delta_C \arg f(z) = \phi_1 - \phi_0. \tag{1}$$

The value of $\Delta_C \arg f(z)$ is a multiple of 2π, and the integer

$$\frac{1}{2\pi}\Delta_C \arg f(z)$$

represents the number of times the point w winds around the origin in the w plane as z describes C once in the positive direction. If, for instance, this integer is -1, then Γ winds around the origin once in the clockwise direction. In Fig. 111 the value of $\Delta_C \arg f(z)$ is zero. The value of $\Delta_C \arg f(z)$ is also

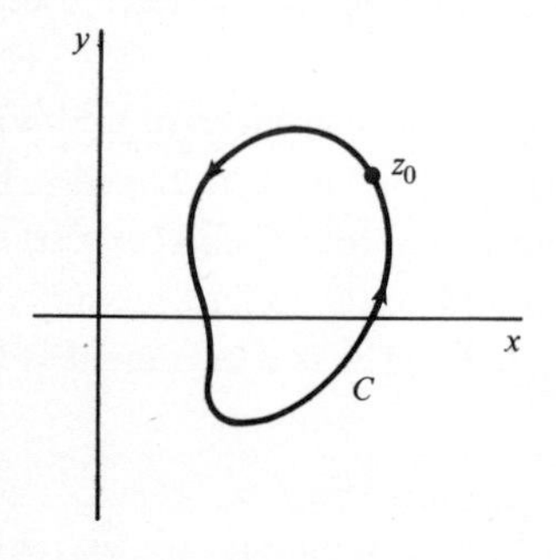

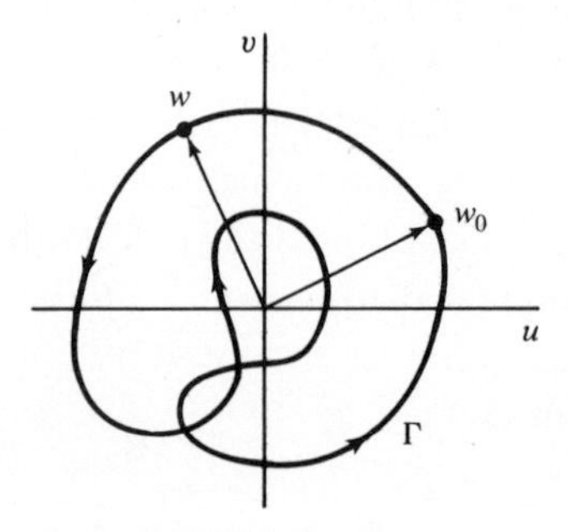

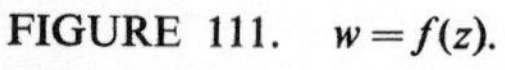
FIGURE 111. $w = f(z)$.

zero when the contour Γ does not enclose the origin; the verification of this fact for a special case is left to the exercises.

The value of $\Delta_C \arg f(z)$ can be determined from the number of zeros and poles of f interior to C.

Theorem 1. *Let C be a simple closed contour described in the positive sense and let f be a function which is analytic inside and on C, except possibly for poles interior to C. Also, let f have no zeros on C. Then*

$$\frac{1}{2\pi}\Delta_C \arg f(z) = N - P \tag{2}$$

where N and P are the number of zeros and the number of poles of f, counting multiplicities, interior to C.

Our proof of this result, known as the *argument principle*, is based on the formula

$$\frac{1}{2\pi i}\int_C \frac{f'(z)}{f(z)}\,dz = N - P \tag{3}$$

obtained in the previous section. If C is expressed parametrically as $z = z(t)$, $a \leqq t \leqq b$, a parametric representation for its image Γ under the transformation $w = f(z)$ is then

$$w = w(t) = f[z(t)] \qquad (a \leqq t \leqq b).$$

Now according to Exercise 7, Sec. 43,

$$w'(t) = f'[z(t)]z'(t)$$

along each of the smooth arcs making up the contour Γ. Since $z'(t)$ and $w'(t)$ are piecewise continuous on the interval $a \leqq t \leqq b$, we can write

$$\int_a^b \frac{f'[z(t)]}{f[z(t)]}\,z'(t)\,dt = \int_a^b \frac{w'(t)}{w(t)}\,dt.$$

That is,

$$\int_C \frac{f'(z)}{f(z)}\,dz = \int_\Gamma \frac{dw}{w}.$$

Formula (3) therefore becomes

$$\frac{1}{2\pi i}\int_\Gamma \frac{dw}{w} = N - P. \tag{4}$$

Since Γ never passes through the origin in the w plane, we can express each point on that contour in polar form as $w = \rho \exp(i\phi)$. If we then represent Γ in terms of some parameter τ as

$$w = w(\tau) = \rho(\tau) \exp[i\phi(\tau)] \qquad (c \leqq \tau \leqq d),$$

we obtain the equation

$$w'(\tau) = \rho'(\tau) \exp[i\phi(\tau)] + \rho(\tau) \exp[i\phi(\tau)]i\phi'(\tau),$$

where $\rho'(\tau)$ and $\phi'(\tau)$ are piecewise continuous on the interval $c \leqq \tau \leqq d$. Thus we can write

$$\int_\Gamma \frac{dw}{w} = \int_c^d \frac{w'(\tau)}{w(\tau)}\, d\tau = \int_c^d \frac{\rho'(\tau)}{\rho(\tau)}\, d\tau + i\int_c^d \phi'(\tau)\, d\tau,$$

or

$$\int_\Gamma \frac{dw}{w} = \text{Log}\, \rho(\tau)\Big]_c^d + i\phi(\tau)\Big]_c^d.$$

But $\rho(d) = \rho(c)$ and

$$\phi(d) - \phi(c) = \phi_1 - \phi_0 = \Delta_C \arg f(z).$$

Hence

$$\int_\Gamma \frac{dw}{w} = i\Delta_C \arg f(z). \tag{5}$$

Formula (2) now follows immediately from equations (4) and (5).

We next present a useful consequence of the argument principle known as *Rouché's theorem.*

Theorem 2. *Let f and g be functions which are analytic inside and on a positively oriented simple closed contour C. If $|f(z)| > |g(z)|$ at each point z on C, the functions $f(z)$ and $f(z) + g(z)$ have the same number of zeros, counting multiplicities, inside C.*

To prove this, observe first that since $|f(z)| > |g(z)| \geqq 0$ on C, $f(z)$ has no zeros on C. Moreover,

$$|f(z) + g(z)| \geqq |f(z)| - |g(z)| > 0$$

on C, and so the function $f(z) + g(z)$ has no zeros on C either. Now

$$\frac{1}{2\pi}\Delta_C \arg f(z) = N_f \tag{6}$$

and

$$\frac{1}{2\pi} \Delta_C \arg [f(z) + g(z)] = N_{f+g} \tag{7}$$

where N_f is the number of zeros of $f(z)$ interior to C and N_{f+g} is the number of zeros of $f(z) + g(z)$ there. Equations (6) and (7) follow from the argument principle and the fact that the functions $f(z)$ and $f(z) + g(z)$ have no poles inside C. Observe that

$$\begin{aligned} \Delta_C \arg [f(z) + g(z)] &= \Delta_C \arg \left\{ f(z) \left[1 + \frac{g(z)}{f(z)} \right] \right\} \\ &= \Delta_C \arg f(z) + \Delta_C \arg \left[1 + \frac{g(z)}{f(z)} \right]. \end{aligned}$$

Under the transformation $w = 1 + g(z)/f(z)$, the image Γ of C lies inside the circle $|w - 1| = 1$ because $|w - 1| = |g(z)/f(z)| < 1$ on C. Hence the point $w = 0$ is not enclosed by the curve Γ, and the value of $\Delta_C \arg [1 + g(z)/f(z)]$ is zero. Thus $\Delta_C \arg [f(z) + g(z)] = \Delta_C \arg f(z)$; and $f(z) + g(z)$ has the same number of zeros as $f(z)$ interior to C, according to equations (6) and (7).

For an application of Rouché's theorem, let us determine the number of roots of the equation $z^7 - 4z^3 + z - 1 = 0$ interior to the circle $|z| = 1$. Write $f(z) = -4z^3$ and $g(z) = z^7 + z - 1$, and observe that $|f(z)| = 4$ and $|g(z)| \leqq 3$ when $|z| = 1$. The conditions in Rouché's theorem are thus satisfied. Consequently, since $f(z)$ has three zeros, counting multiplicities, interior to the circle $|z| = 1$, so does $f(z) + g(z)$. That is, the equation $z^7 - 4z^3 + z - 1 = 0$ has three roots interior to the circle $|z| = 1$.

EXERCISES

1. Let c be a fixed complex number different from zero. Show that the function $\exp(1/z)$, which has an essential singular point at $z = 0$, assumes the value c an infinite number of times in any neighborhood of the origin.

Suggestion: Write $c = c_0 \exp(i\gamma)$, where $c_0 > 0$, and show that $\exp(1/z)$ assumes the value c at the points $z = r \exp(i\theta)$ where r and θ satisfy the equations

$$r^2 = \frac{1}{\gamma^2 + (\text{Log } c_0)^2},$$

$$\sin\theta = \frac{-\gamma}{\sqrt{\gamma^2 + (\text{Log } c_0)^2}}, \qquad \cos\theta = \frac{\text{Log } c_0}{\sqrt{\gamma^2 + (\text{Log } c_0)^2}}.$$

Note that r can be made arbitrarily small by adding integral multiples of 2π to the angle γ, leaving c unaltered.

2. *If a function f is analytic in a domain $0 < |z - z_0| < r_0$ and if z_0 is an accumulation point of zeros of the function, then z_0 is an essential singular point of f or else $f(z)$ is identically zero.* Prove this theorem with the aid of results found in Secs. 66 and 111.
3. Examine the set of zeros of the function $z^2 \sin(1/z)$ and apply the theorem stated in Exercise 2 to show that the origin is an essential singular point of the function. Note that this conclusion also follows from the nature of the Laurent series that represents this function in the domain $|z| > 0$.
4. Let C be a simple closed contour in the z plane, described in the positive sense, and let w_0 be any given complex number. Let a function g be analytic within and on C, and suppose that $g'(z) \neq 0$ at any point z interior to C. If $g(z) \neq w_0$ at any point z on C, then

$$\frac{1}{2\pi i}\int_C \frac{g'(z)}{g(z) - w_0}\,dz = N$$

where the integer N is the number of points z interior to C at which $g(z) = w_0$. Show how this result follows from those given in Sec. 113. (Compare the result with that found in Exercise 12, Sec. 95.)
5. Complete the argument (Sec. 113), based on the Bolzano-Weierstrass theorem, that if a function f is analytic within and on a simple closed contour C, except possibly for poles inside C, and if $f(z) \neq 0$ at any point z on C, then the zeros and poles of f inside C are finite in number and formula (1) of Sec. 113 is valid.
6. Let f be a function which is analytic inside and on a simple closed contour C, and suppose that $f(z)$ is never zero on C. Let the image of C under the transformation $w = f(z)$ be the closed contour Γ shown in Fig. 112. Using the contour Γ, determine the value of $\Delta_C \arg f(z)$. Also, determine the number of zeros of f interior to C.

Ans. 6π; 3.

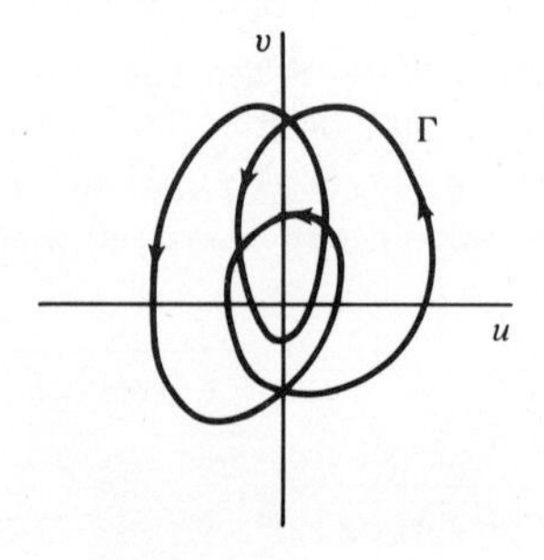

FIGURE 112

7. Let C denote the unit circle $|z| = 1$ described in the positive sense. Determine the value of $\Delta_C \arg f(z)$ for the function

$$(a) \quad f(z) = z^2; \qquad (b) \quad f(z) = \frac{z^3 + 2}{z}.$$

Also, for each of the transformations $w = f(z)$ defined by these functions, state the number of times the image point w winds around the origin in the w plane as the point z describes C once in the positive direction.

Ans. (*a*) $4\pi, 2$; (*b*) $-2\pi, -1$.

8. Using the notation of Sec. 114, show that when Γ does not enclose the point $w = 0$ and there is a ray from that point which does not intersect Γ, then $\Delta_C \arg f(z) = 0$.

Suggestion: Note that the change in the value of $\arg f(z)$ must be less than 2π in absolute value when a point z makes one cycle around C. Then use the fact that $\Delta_C \arg f(z)$ is an integral multiple of 2π.

9. Determine the number of zeros of the polynomial (*a*) $z^6 - 5z^4 + z^3 - 2z$; (*b*) $2z^4 - 2z^3 + 2z^2 - 2z + 9$ inside the circle $|z| = 1$.

Ans. (*a*) 4; (*b*) 0.

10. Determine the number of roots of the equation $2z^5 - 6z^2 + z + 1 = 0$ in the region $1 \leqq |z| < 2$.

Ans. 3.

11. Show that if c is a complex number such that $|c| > e$, the equation $cz^n = e^z$ has n roots inside the circle $|z| = 1$.

12. Using Rouché's theorem, prove that any polynomial

$$P(z) = a_0 + a_1 z + \cdots + a_{n-1} z^{n-1} + a_n z^n \qquad (a_n \neq 0),$$

where $n \geqq 1$, has precisely n zeros. Thus give an alternative proof of the fundamental theorem of algebra (Sec. 55).

Suggestion: Note that it is sufficient to let a_n be unity. Then, writing

$$f(z) = z^n, \qquad g(z) = a_0 + a_1 z + \cdots + a_{n-1} z^{n-1},$$

show that $P(z)$ has n zeros inside a circle $|z| = R$ where R is larger than either of the two numbers 1 and $|a_0| + |a_1| + \cdots + |a_{n-1}|$. To show that $P(z)$ has no other zeros, show that

$$|z^n + g(z)| \geqq |z|^n - |g(z)| > 0$$

when $|z| \geqq R$.

C. RIEMANN SURFACES

A Riemann surface is a generalization of the complex plane to a surface of more than one sheet such that a multiple-valued function has only one value corresponding to each point on that surface. Once such a surface is devised for a given function, the function is single-valued on the surface and the theory of single-valued functions applies there. Complexities arising because the function is multiple-valued are thus relieved by a geometrical device. However, the description of those surfaces and the arrangement of proper connections between the sheets can become quite involved. We limit our attention to fairly simple examples.

115. A Surface for $\log z$

Corresponding to each nonzero number z, the multiple-valued function

$$\log z = \operatorname{Log} r + i\theta$$

has infinitely many values. In order to describe $\log z$ as a single-valued function, we replace the z plane, with the origin deleted, by a surface on which a new point is located whenever the argument of the number z is increased or decreased by 2π, or an integral multiple of 2π.

Consider the z plane with the origin deleted as a thin sheet R_0 which is cut along the positive half of the real axis. On that sheet let θ range from zero to 2π. Let a second sheet R_1 be cut in the same way and placed in front of the sheet R_0. The lower edge of the slit in R_0 is then joined to the upper edge of the slit in R_1. On R_1 the angle θ ranges from 2π to 4π; so when z is represented by a point on R_1, the imaginary part of $\log z$ ranges from 2π to 4π.

A sheet R_2 is then cut in the same way and placed in front of R_1, and the lower edge of the slit in R_1 is joined to the upper edge of the slit in this new sheet, and similarly for sheets $R_3, R_4, \ldots$. A sheet R_{-1} on which θ varies from zero to -2π is cut and placed behind R_0, with the lower edge of its slit connected to the upper edge of the slit in R_0; the sheets $R_{-2}, R_{-3}, \ldots$ are constructed in like manner. The coordinates r and θ of a point on any sheet can be considered as polar coordinates of the projection of the point onto the original z plane, the angular coordinate θ being restricted to a definite range of 2π radians on each sheet.

Consider any continuous curve on this connected surface of infinitely many sheets. As a point z describes that curve, the values of $\log z$ vary continuously since θ, in addition to r, now varies continuously; and $\log z$ assumes just one value corresponding to each point on the curve. For example, as the point makes a complete cycle around the origin on the sheet R_0 over the path indicated in Fig. 113, the angle changes from zero to 2π. As it moves across the line $\theta = 2\pi$, the point passes to the sheet R_1 of the surface. As the point completes a cycle in R_1, the angle θ varies from 2π to 4π; and as it crosses the line $\theta = 4\pi$, the point passes to the sheet R_2.

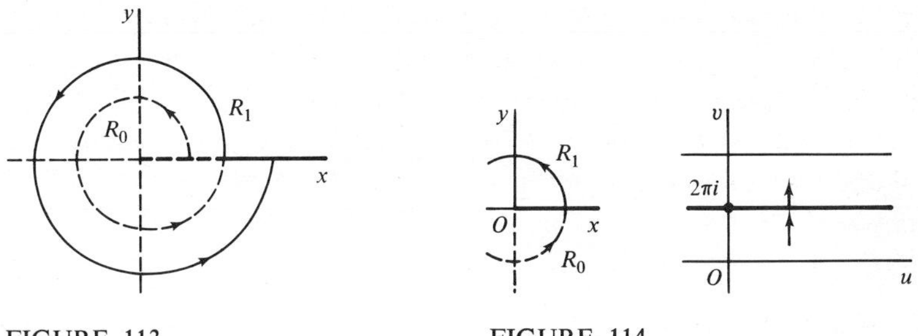

FIGURE 113

FIGURE 114

The surface described here is a Riemann surface for log z. It is a connected surface of infinitely many sheets arranged so that log z is a single-valued function of points on it.

The transformation $w = \log z$ maps the whole Riemann surface in a one to one manner onto the entire w plane. The image of the sheet R_0 is the strip $0 \leqq v \leqq 2\pi$. As a point z moves onto the sheet R_1 over the arc shown in Fig. 114, its image w moves upward across the line $v = 2\pi$, as indicated in the figure.

Note that log z defined on the sheet R_1 represents the analytic continuation of the single-valued analytic function

$$\text{Log}\, r + i\theta \qquad (0 < \theta < 2\pi)$$

upward across the positive real axis. In this sense, log z is not only a single-valued function of all points z on the Riemann surface but also an *analytic* function at all points there.

The sheets could, of course, be cut along the negative real axis, or along any other half line starting at the origin, and properly joined along the slits to form other Riemann surfaces for log z.

116. A Surface for $z^{1/2}$

Corresponding to each point in the z plane other than the origin,

$$z^{1/2} = \sqrt{r}\left(\cos\frac{\theta}{2} + i\sin\frac{\theta}{2}\right)$$

has two values. A Riemann surface for $z^{1/2}$ is obtained by replacing the z plane with a surface made up of two sheets R_0 and R_1, each cut along the positive real axis with R_1 placed in front of R_0. The lower edge of the slit in R_0 is

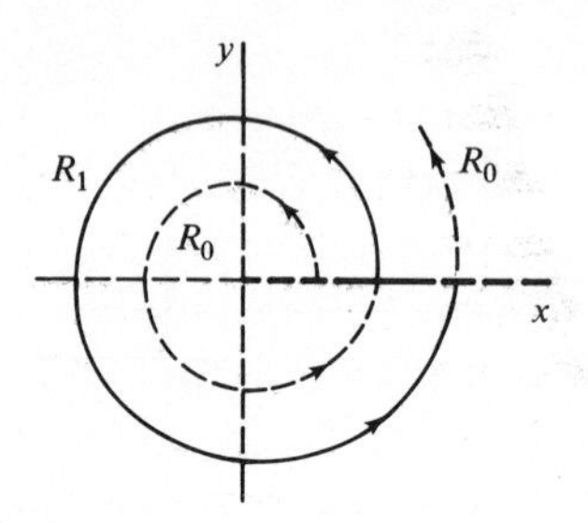

FIGURE 115

joined to the upper edge of the slit in R_1, and the lower edge of the slit in R_1 is joined to the upper edge of the slit in R_0.

As a point z starts from the upper edge of the slit in R_0 and describes a continuous circuit around the origin in the counterclockwise direction (Fig. 115), the angle θ increases from 0 to 2π. The point then passes from the sheet R_0 to the sheet R_1 where θ increases from 2π to 4π. As the point moves still further, it passes back to the sheet R_0 where the values of θ can vary from 4π to 6π or from 0 to 2π, a choice that does not affect the value of $z^{1/2}$, etc. Note that the value of $z^{1/2}$ at a point where the circuit passes from the sheet R_0 to the sheet R_1 is different from the value of $z^{1/2}$ at a point where the circuit passes from the sheet R_1 to the sheet R_0.

We have thus constructed a Riemann surface on which $z^{1/2}$ is single-valued for each nonzero z. In that construction the edges of the sheets R_0 and R_1 are joined in pairs in such a way that the resulting surface is closed and connected. The points where two of the edges are joined are distinct from the points where the other two edges are joined. Thus, it is physically impossible to build a model of that Riemann surface. In visualizing a Riemann surface, it is important to understand how we are to proceed when we arrive at an edge of a slit.

The origin is a special point on this Riemann surface. It is common to both sheets, and a curve around the origin on the surface must wind around it twice in order to be a closed curve. A point of this kind on a Riemann surface is called a branch point.

The image of the sheet R_0 under the transformation $w = z^{1/2}$ is the upper half of the w plane since the argument of w is $\theta/2$ and $0 \leqq \theta/2 \leqq \pi$ on R_0. Likewise, the image of the sheet R_1 is the lower half of the w plane. As defined on either sheet, the function is the analytic continuation, across the cut, of the function defined on the other sheet. In this respect, the single-valued function $z^{1/2}$ of points on the Riemann surface is analytic at all points except the origin.

117. Surfaces for Other Irrational Functions

Let us describe a Riemann surface for the double-valued function

$$f(z) = (z^2 - 1)^{1/2} = \sqrt{r_1 r_2} \exp\left(i\frac{\theta_1 + \theta_2}{2}\right) \tag{1}$$

where $z - 1 = r_1 \exp(i\theta_1)$ and $z + 1 = r_2 \exp(i\theta_2)$, as shown in Fig. 116. A branch of this function, with the line segment P_1P_2 between the branch points $z = \pm 1$ as a branch cut, was described in Sec. 37. That branch is given by formula (1) with the restrictions $r_1 > 0$, $r_2 > 0$, $r_1 + r_2 > 2$, and $0 \leqq \theta_k < 2\pi$ $(k = 1, 2)$. The branch is not defined on the segment P_1P_2.

A Riemann surface for the double-valued function (1) must consist of two sheets R_0 and R_1. Let both sheets be cut along the segment P_1P_2. The lower edge of the slit in R_0 is then joined to the upper edge of the slit in R_1, and the lower edge in R_1 is joined to the upper edge in R_0.

On the sheet R_0 let the angles θ_1 and θ_2 range from 0 to 2π. If a point on the sheet R_0 describes a simple closed curve which encloses the segment P_1P_2 once in the counterclockwise direction, then both θ_1 and θ_2 change by the amount 2π upon the return of the point to its original position. The change in $(\theta_1 + \theta_2)/2$ is also 2π, and the value of f is unchanged. If a point starting on the sheet R_0 describes a path which passes twice around just the branch point $z = 1$, it crosses from the sheet R_0 onto the sheet R_1 and then back onto the sheet R_0 before it returns to its original position. In this case, the value of θ_1 changes by the amount 4π while the value of θ_2 does not change at all. Similarly, for a circuit twice around the point $z = -1$, the value of θ_2 changes by 4π while the value of θ_1 remains unchanged. Again, the change in $(\theta_1 + \theta_2)/2$ is 2π; and the value of f is unchanged. Thus on the sheet R_0 the range of the angles θ_1 and θ_2 may be extended by changing both θ_1 and θ_2 by the same integral multiple of 2π or by changing just one of the angles by a multiple of 4π. In either case the total change in both angles is an even integral multiple of 2π.

To obtain the range of values for θ_1 and θ_2 on the sheet R_1, we note that if a point starts on the sheet R_0 and describes a path around just one of the

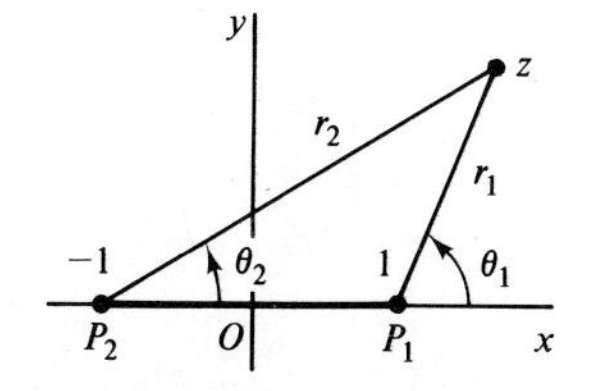

FIGURE 116

branch points once, it crosses onto the sheet R_1 and does not return to the sheet R_0. In this case the value of one of the angles is changed by 2π while the value of the other remains unchanged. Hence, on the sheet R_1 one angle can range from 2π to 4π while the other angle ranges from 0 to 2π. Their sum then ranges from 2π to 4π, and the value of $(\theta_1 + \theta_2)/2$, the argument of $f(z)$, ranges from π to 2π. Again the range of the angles is extended by changing the value of just one of the angles by an integral multiple of 4π or by changing the value of both of the angles by the same integral multiple of 2π.

The double-valued function given in equation (1) may now be considered as a single-valued function of the points on the Riemann surface just constructed. The transformation $w = f(z)$ maps each of the sheets used in the construction of that surface onto the entire w plane.

For another example, consider the double-valued function

$$g(z) = [z(z^2 - 1)]^{1/2} = \sqrt{rr_1r_2}\, \exp\left(i\,\frac{\theta + \theta_1 + \theta_2}{2}\right) \tag{2}$$

(Fig. 117). The points $z = 0, \pm 1$ are branch points of this function. We note that if the point z describes a circuit that includes all three of those points, the argument of $g(z)$ changes by the angle 3π and the value of the function thus changes. Consequently, a branch cut must run from one of those branch points to the point at infinity in order to describe a single-valued branch of g. Hence the point at infinity is also a branch point, as we can show by noting that the function $g(1/z)$ has a branch point at $z = 0$.

Let two sheets be cut along the line segment L_2 from $z = -1$ to $z = 0$ and along the part L_1 of the real axis to the right of the point $z = 1$. We specify that each of the three angles θ, θ_1, and θ_2 may range from 0 to 2π on the sheet R_0 and from 2π to 4π on the sheet R_1. We also specify that the angles corresponding to a point on either sheet may be changed by integral multiples of 2π in such a way that the sum of the three angles changes by an integral multiple of 4π; the value of the function g is therefore unaltered.

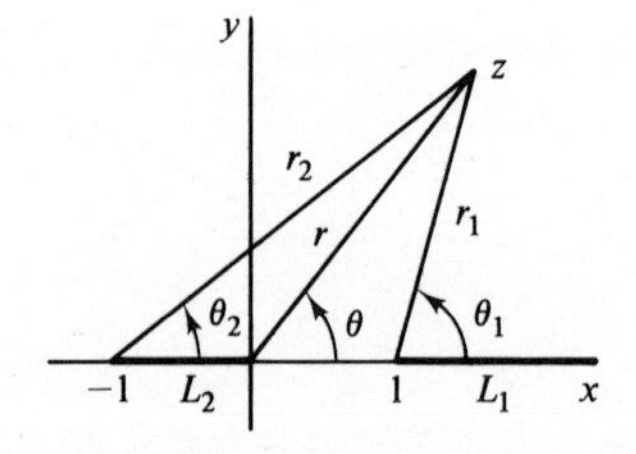

FIGURE 117

A Riemann surface for the double-valued function (2) is obtained by joining the lower edges in R_0 of the slits along L_1 and L_2 to the upper edges in R_1 of the slits along L_1 and L_2, respectively. The lower edges in R_1 of the slits along L_1 and L_2 are then joined to the upper edges in R_0 of the slits along L_1 and L_2, respectively. It is readily verified with the aid of Fig. 117 that one branch of the function is represented by its values at points on R_0 and the other branch at points on R_1.

EXERCISES

1. Describe a Riemann surface for the triple-valued function $w = (z - 1)^{1/3}$, and point out which third of the w plane represents the image of each sheet of that surface.

2. Describe the Riemann surface for $\log z$ obtained by cutting the z plane along the negative real axis. Compare this Riemann surface for $\log z$ with the one obtained in Sec. 115.

3. Determine the image under the transformation $w = \log z$ of the sheet R_n, where n is an arbitrary integer, of the Riemann surface for $\log z$ given in Sec. 115.

4. Verify that under the transformation $w = z^{1/2}$ the sheet R_1 of the Riemann surface for $z^{1/2}$ given in Sec. 116 is mapped onto the lower half of the w plane.

5. Describe the curve, on a Riemann surface for $z^{1/2}$, whose image is the entire circle $|w| = 1$ under the transformation $w = z^{1/2}$.

6. Corresponding to each point on the Riemann surface described in Sec. 117 for the function $w = g(z)$ there is just one value of w. Show that corresponding to each value of w there are in general three points on the surface.

7. Describe a Riemann surface for the multiple-valued function

$$f(z) = \left(\frac{z-1}{z}\right)^{1/2}.$$

8. Let C denote the circle $|z - 2| = 1$ on the Riemann surface described in Sec. 116 for $z^{1/2}$, where the upper half of that circle lies on the sheet R_0 and the lower half on R_1. Note that for each point z on C we can write

$$z^{1/2} = \sqrt{r} \exp \frac{i\theta}{2} \qquad \text{where} \qquad 4\pi - \frac{\pi}{2} < \theta < 4\pi + \frac{\pi}{2}.$$

State why it follows that

$$\int_C z^{1/2}\, dz = 0.$$

Generalize this result to fit the case of other simple closed curves that cross from one sheet to another without enclosing the branch points. Generalize to other functions, thus extending the Cauchy-Goursat theorem to integrals of multiple-valued functions.

9. Note that the Riemann surface described in Sec. 117 for $(z^2 - 1)^{1/2}$ is also a Riemann surface for the function

$$h(z) = z + (z^2 - 1)^{1/2}.$$

Let f_0 denote the branch of $(z^2 - 1)^{1/2}$ defined on the sheet R_0, and show that the branches h_0 and h_1 of h on the two sheets are given by the equations

$$h_0(z) = \frac{1}{h_1(z)} = z + f_0(z).$$

10. In Exercise 9 the branch f_0 of $(z^2 - 1)^{1/2}$ can be described by the equation

$$f_0(z) = \sqrt{r_1 r_2} \exp \frac{i\theta_1}{2} \exp \frac{i\theta_2}{2},$$

where θ_1 and θ_2 range from 0 to 2π and

$$z - 1 = r_1 \exp (i\theta_1), \qquad z + 1 = r_2 \exp (i\theta_2).$$

Note that $2z = r_1 \exp (i\theta_1) + r_2 \exp (i\theta_2)$ and show that the branch h_0 of the function $h(z) = z + (z^2 - 1)^{1/2}$ can be written in the form

$$h_0(z) = \frac{1}{2}\left(\sqrt{r_1} \exp \frac{i\theta_1}{2} + \sqrt{r_2} \exp \frac{i\theta_2}{2}\right)^2.$$

Find $h_0(z)\overline{h_0(z)}$ and note that $r_1 + r_2 \geqq 2$ and $\cos [(\theta_1 - \theta_2)/2] \geqq 0$, for all z, to prove that $|h_0(z)| \geqq 1$. Then show that the transformation $w = z + (z^2 - 1)^{1/2}$ maps the sheet R_0 of the Riemann surface onto the region $|w| \geqq 1$, the sheet R_1 onto the region $|w| \leqq 1$, and the branch cut between the points $z = \pm 1$ onto the circle $|w| = 1$. Note that the transformation used here is an inverse of the transformation

$$z = \frac{1}{2}\left(w + \frac{1}{w}\right),$$

and compare the result obtained with the result of Exercise 18, Sec. 41.

APPENDIX 1
BIBLIOGRAPHY

The following list of supplementary books is far from exhaustive. Further references can be found in many of the books listed here.

THEORY

AHLFORS, L. V.: "Complex Analysis," 2d ed., McGraw-Hill Book Company, Inc., New York, 1966.

BIEBERBACH, L.: "Conformal Mapping," Chelsea Publishing Company, New York, 1953.

———: "Lehrbuch der Funktionentheorie," vols. 1 and 2, B. G. Teubner, Berlin, 1934.

CARATHEODORY, C.: "Conformal Representation," Cambridge University Press, London, 1952.

———: "Theory of Functions of a Complex Variable," vols. 1 and 2, Chelsea Publishing Company, New York, 1954.

COPSON, E. T.: "Theory of Functions of a Complex Variable," Oxford University Press, London, 1957.

DETTMAN, J. W.: "Applied Complex Variables," Macmillan Company, New York, 1965.

DIENES, P.: "The Taylor Series: An Introduction to the Theory of Functions of a Complex Variable," Dover Publications, New York, 1957.

EVANS, G. C.: "The Logarithmic Potential," American Mathematical Society, Providence, R.I., 1927.

FORSYTH, A. R.: "Theory of Functions of a Complex Variable," Cambridge University Press, London, 1918.

HILLE, E.: "Analytic Function Theory," vols. 1 and 2, Ginn & Company, Boston, 1959, 1962.

HURWITZ, A., and R. COURANT: "Vorlesungen über allgemeine Funktionentheorie und elliptische Funktionen," Interscience Publishers, Inc., New York, 1944.

KAPLAN, W.: "Advanced Calculus," 2d ed., Addison-Wesley Publishing Company, Inc., Reading, Mass., 1973.

KELLOGG, O. D.: "Foundations of Potential Theory," Dover Publications, New York, 1953.

KNOPP, K.: "Elements of the Theory of Functions," Dover Publications, New York, 1952.

LEVINSON, N., and R. REDHEFFER: "Complex Variables," Holden-Day, Inc., San Francisco, 1970.

MACROBERT, T. M.: "Functions of a Complex Variable," Macmillan & Co., Ltd., London, 1954.

MARKUSHEVICH, A. I.: "Theory of Functions of a Complex Variable," vols. 1, 2, and 3, Prentice-Hall, Inc., Englewood Cliffs, N.J., 1965, 1967.

MITRINOVIĆ, D. S.: "Calculus of Residues," P. Noordhoff, Ltd., Groningen, 1966.

NEHARI, Z.; "Introduction to Complex Analysis," rev. ed., Allyn and Bacon, Inc., Boston, 1962.

PENNISI, L. L.: "Elements of Complex Variables," Holt, Rinehart and Winston, Inc., New York, 1963.

SPRINGER, G.: "Introduction to Riemann Surfaces," Addison-Wesley Publishing Company, Reading, Mass., 1957.

STERNBERG, W. J., and T. L. SMITH: "Theory of Potential and Spherical Harmonics," University of Toronto Press, Toronto, 1944.

TAYLOR, A. E., and W. R. MANN: "Advanced Calculus," 2d ed., Xerox Publishing Company, Lexington, Mass., 1972.

THRON, W. J.: "Introduction to the Theory of Functions of a Complex Variable," John Wiley and Sons, Inc., New York, 1953.

TITCHMARSH, E. C.: "Theory of Functions," Oxford University Press, London, 1939.

WHITTAKER, E. T., and G. N. WATSON: "Modern Analysis," Cambridge University Press, London, 1950.

APPLICATIONS

BOWMAN, F.: "Introduction to Elliptic Functions, with Applications," English Universities Press, London, 1953.

CHURCHILL, R. V.: "Operational Mathematics," 3d ed., McGraw-Hill Book Co., Inc., New York, 1972.

———: "Fourier Series and Boundary Value Problems," 2d ed., McGraw-Hill Book Co., Inc., New York, 1963.

GLAUERT, H.: "The Elements of Aerofoil and Airscrew Theory," Cambridge University Press, London, 1948.

GUILLEMIN, E. A.: "The Mathematics of Circuit Analysis," John Wiley & Sons, Inc., New York, 1951.

JEANS, J. H.: "Mathematical Theory of Electricity and Magnetism," Cambridge University Press, London, 1925.

KOBER, H.: "Dictionary of Conformal Representations," Dover Publications, New York, 1952.

LAMB, H.: "Hydrodynamics," Dover Publications, New York, 1945.

LEBEDEV, N. N.: "Special Functions, and their Applications," Prentice-Hall, Inc., Englewood Cliffs, N.J., 1965.

LOVE, A. E. H.: "Elasticity," Dover Publications, New York, 1944.

MILNE-THOMSON, L. M.: "Theoretical Hydrodynamics," Macmillan & Co., Ltd., London, 1955.

MUSKHELISHVILI, N. I.: "Some Basic Problems of the Mathematical Theory of Elasticity," P. Noordhoff, N. V., Groningen, Netherlands, 1953.

OBERHETTINGER, F., and W. MAGNUS: "Anwendung der elliptischen Funktionen in Physik und Technik," Springer-Verlag OHG, Berlin, 1949.

ROTHE, R., F. OLLENDORFF, and K. POHLHAUSEN: "Theory of Functions as Applied to Engineering Problems," Technology Press, Massachusetts Institute of Technology, Cambridge, Mass., 1948.

SMYTHE, W. R.: "Static and Dynamic Electricity," 2d ed., McGraw-Hill Book Company, Inc., New York, 1950.

SOKOLNIKOFF, I. S.: "Mathematical Theory of Elasticity," 2d ed., McGraw-Hill Book Company, Inc., New York, 1956.

WALKER, M.: "Conjugate Functions for Engineers," Oxford University Press, London, 1933.

APPENDIX 2

TABLE OF TRANSFORMATIONS OF REGIONS (See Sec. 41)

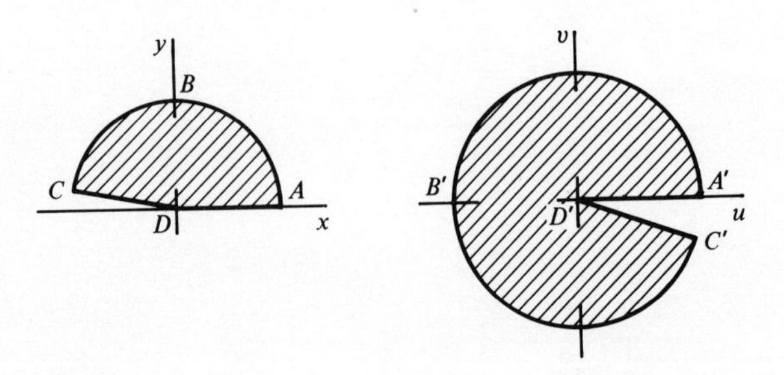

FIGURE 1.
$w = z^2$.

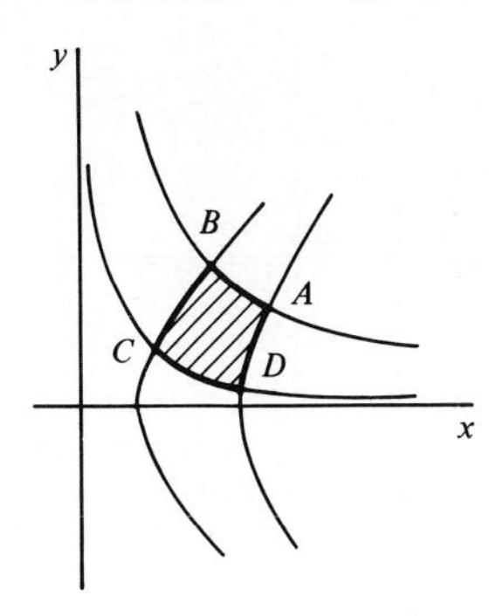

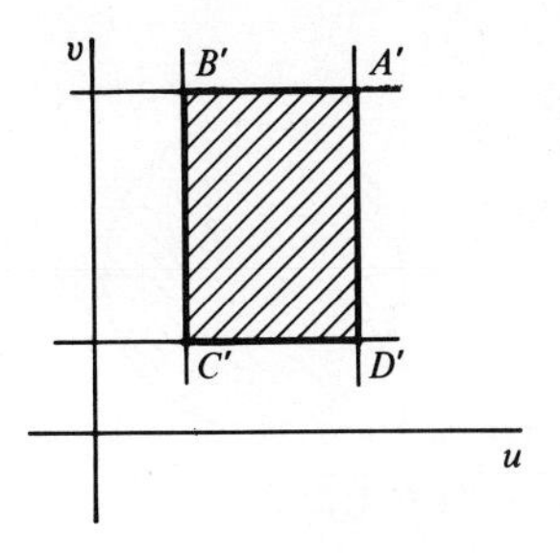

FIGURE 2.
$w = z^2$.

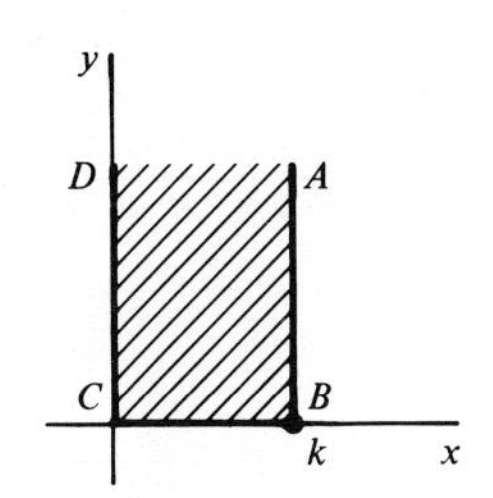

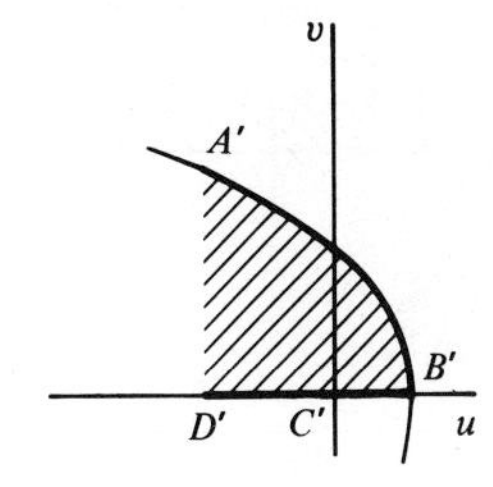

FIGURE 3.

$w = z^2$; $A'B'$ on parabola $\rho = \dfrac{2k^2}{1 + \cos \phi}$.

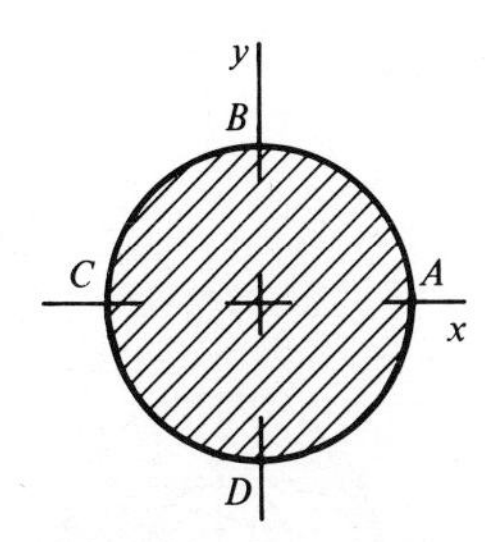

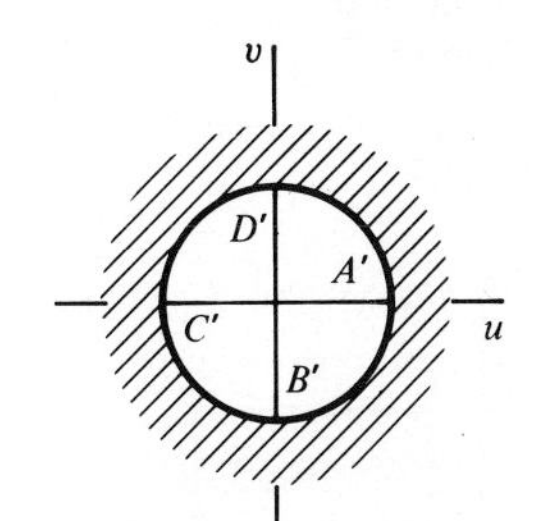

FIGURE 4.

$w = \dfrac{1}{z}$.

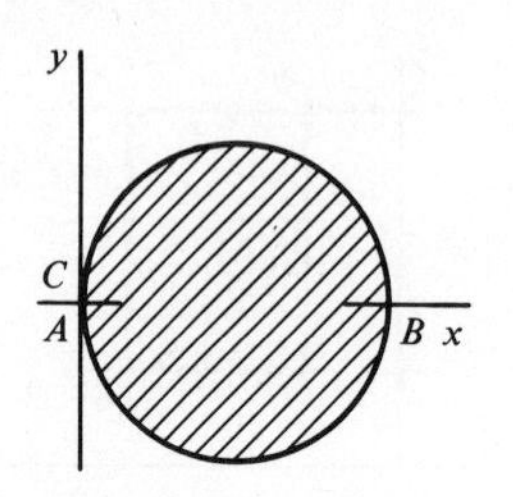

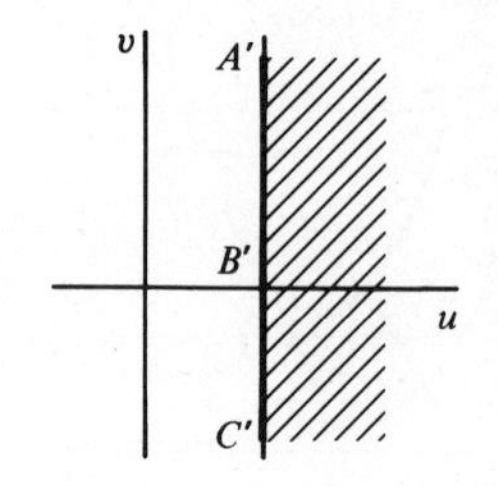

FIGURE 5.

$w = \dfrac{1}{z}$.

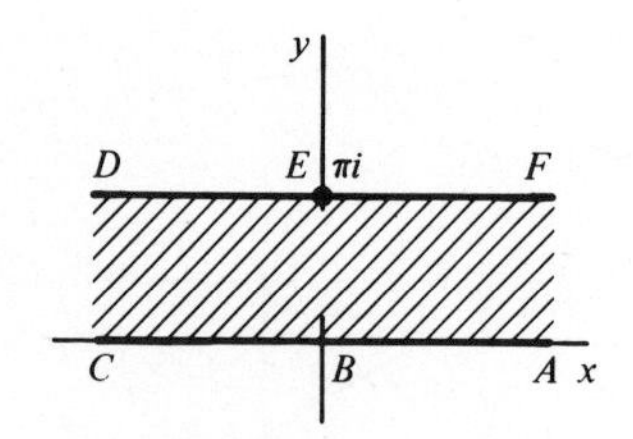

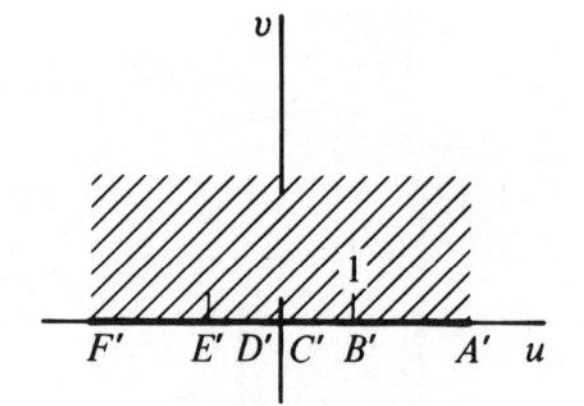

FIGURE 6.
$w = e^z$.

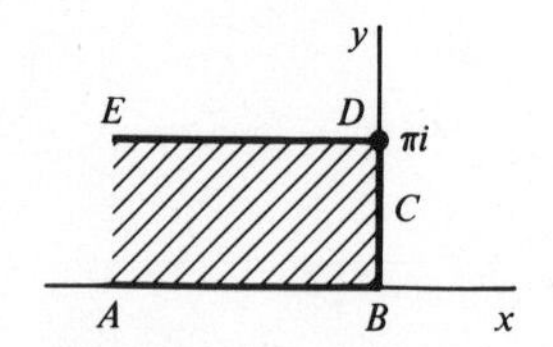

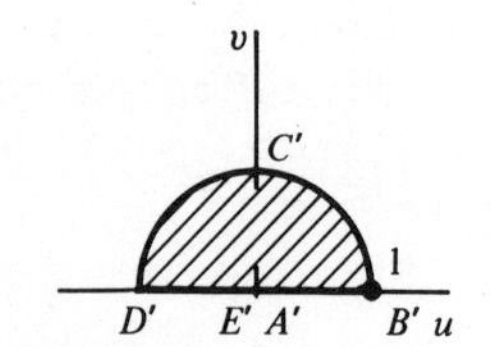

FIGURE 7.
$w = e^z$.

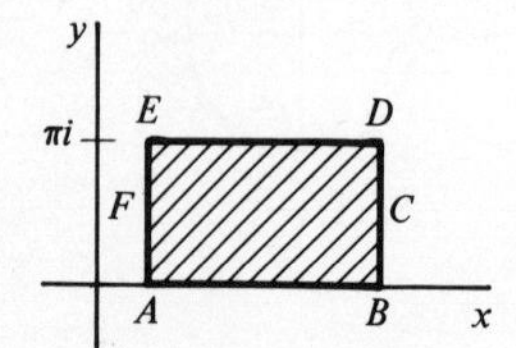

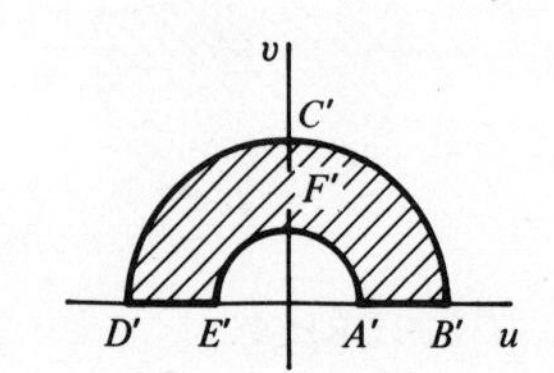

FIGURE 8.
$w = e^z$.

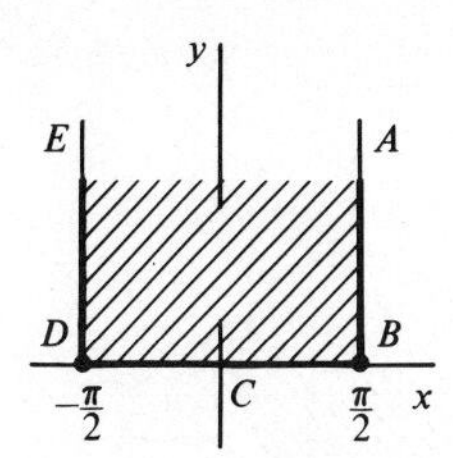

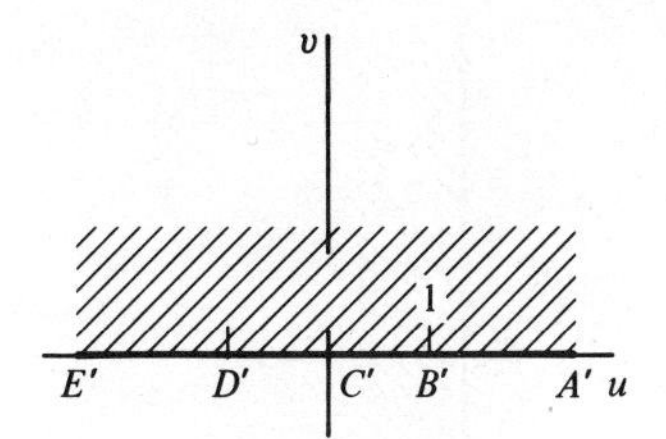

FIGURE 9.
$w = \sin z$.

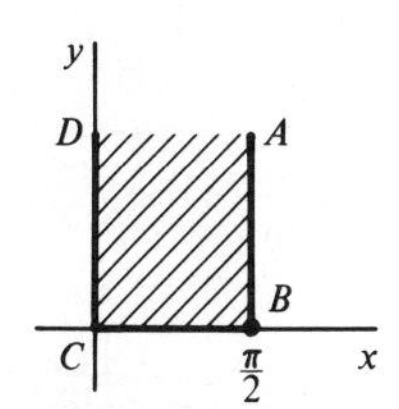

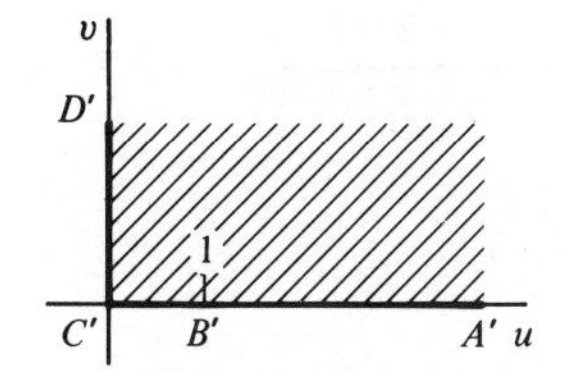

FIGURE 10.
$w = \sin z$.

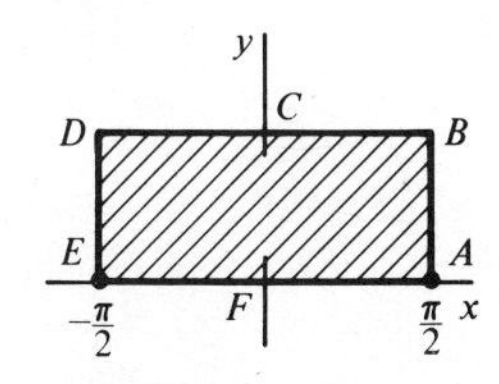

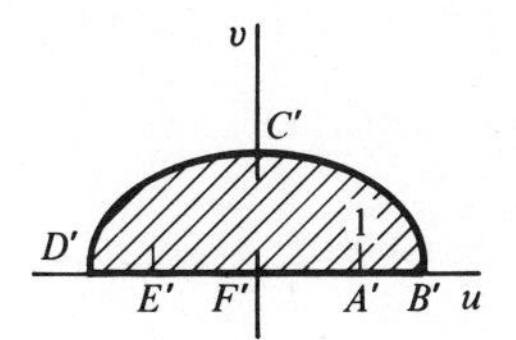

FIGURE 11.

$w = \sin z$; BCD on line $y = k$, $B'C'D'$ on ellipse

$$\left(\frac{u}{\cosh k}\right)^2 + \left(\frac{v}{\sinh k}\right)^2 = 1.$$

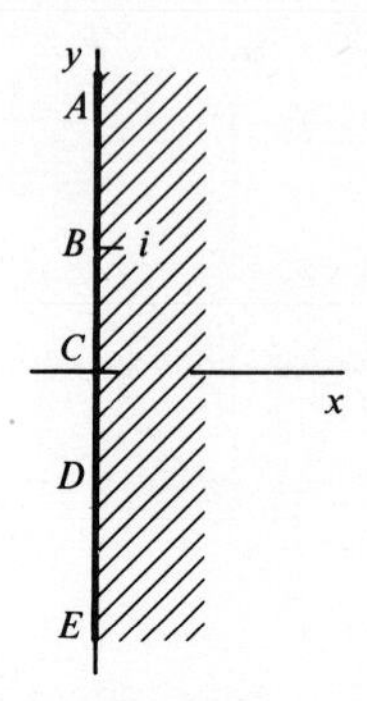

FIGURE 12.
$$w = \frac{z-1}{z+1}.$$

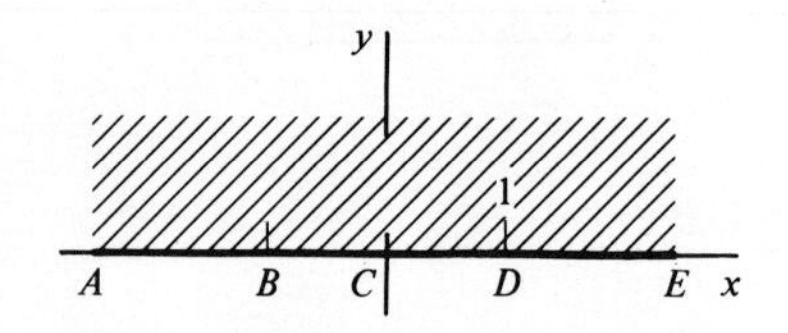

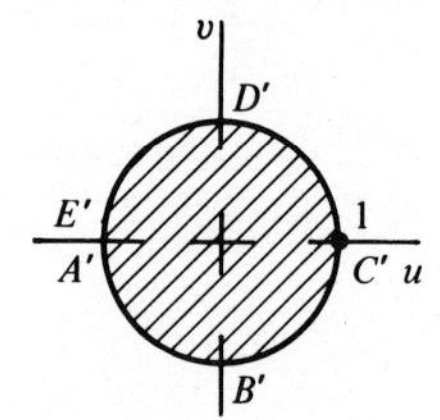

FIGURE 13.
$$w = \frac{i-z}{i+z}.$$

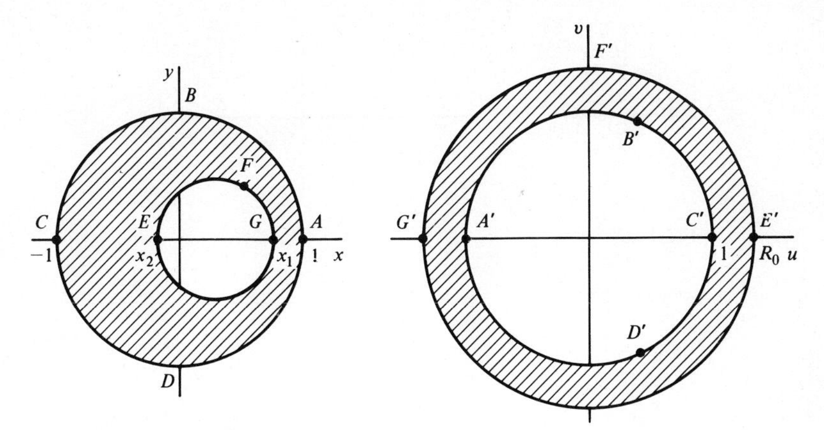

FIGURE 14.

$$w = \frac{z-a}{az-1};\ a = \frac{1 + x_1x_2 + \sqrt{(1-x_1^2)(1-x_2^2)}}{x_1 + x_2};$$

$$R_0 = \frac{1 - x_1x_2 + \sqrt{(1-x_1^2)(1-x_2^2)}}{x_1 - x_2} \quad (a > 1 \text{ and } R_0 > 1 \text{ when } -1 < x_2 < x_1 < 1).$$

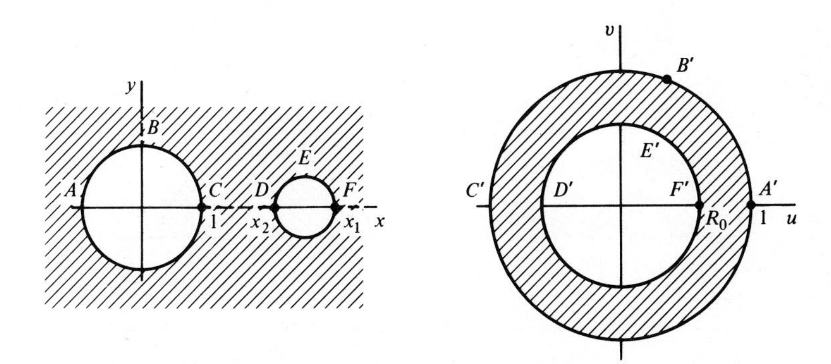

FIGURE 15.

$$w = \frac{z-a}{az-1};\ a = \frac{1 + x_1x_2 + \sqrt{(x_1^2-1)(x_2^2-1)}}{x_1 + x_2};$$

$$R_0 = \frac{x_1x_2 - 1 - \sqrt{(x_1^2-1)(x_2^2-1)}}{x_1 - x_2}$$

$(x_2 < a < x_1$ and $0 < R_0 < 1$ when $1 < x_2 < x_1)$.

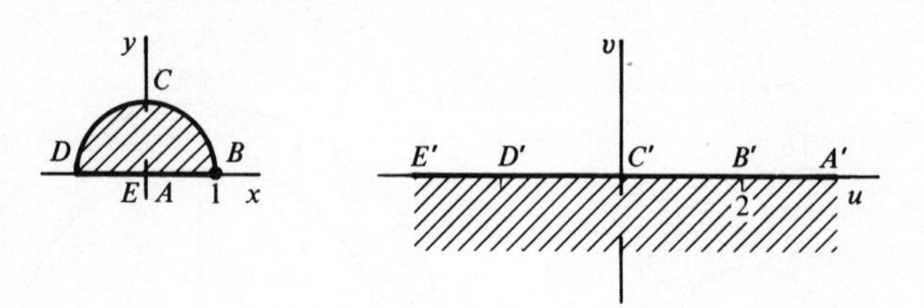

FIGURE 16.

$$w = z + \frac{1}{z}.$$

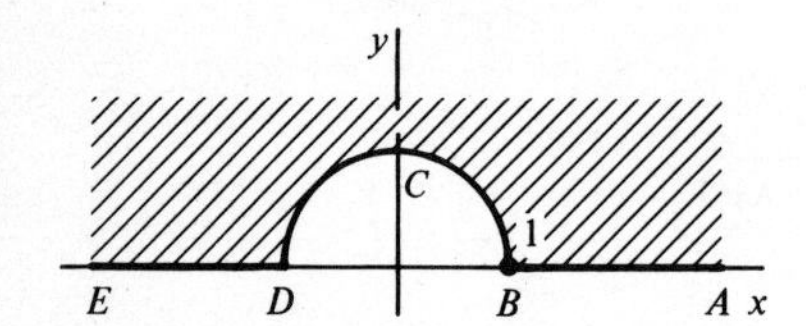

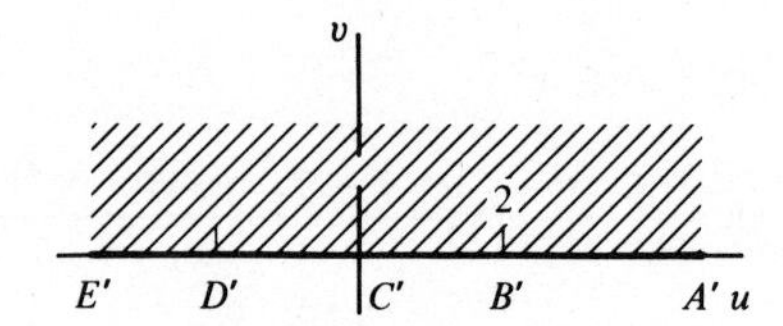

FIGURE 17.

$$w = z + \frac{1}{z}.$$

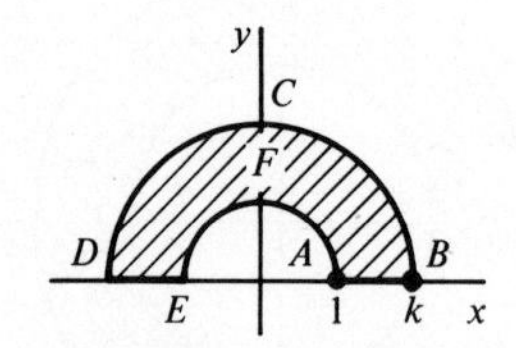

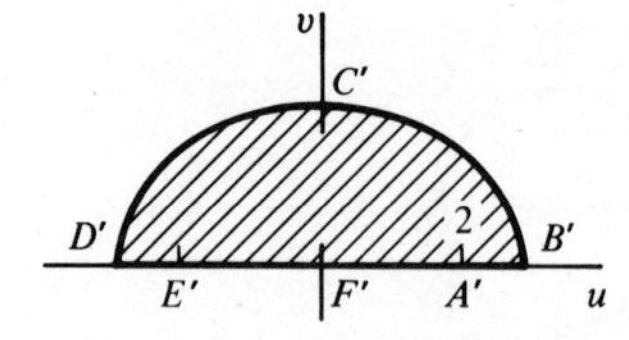

FIGURE 18.

$$w = z + \frac{1}{z};\ B'C'D' \text{ on ellipse } \left(\frac{ku}{k^2+1}\right)^2 + \left(\frac{kv}{k^2-1}\right)^2 = 1.$$

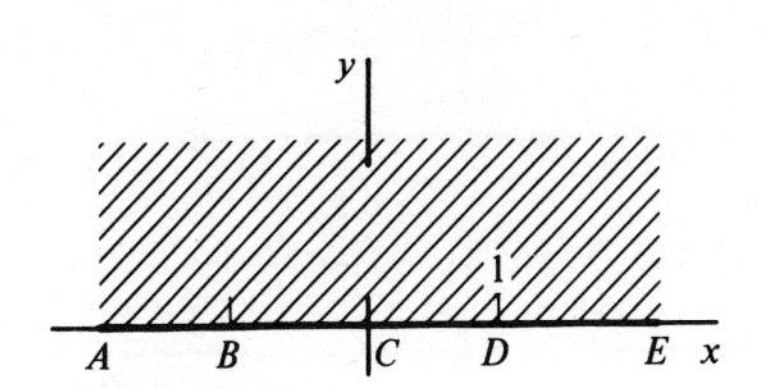

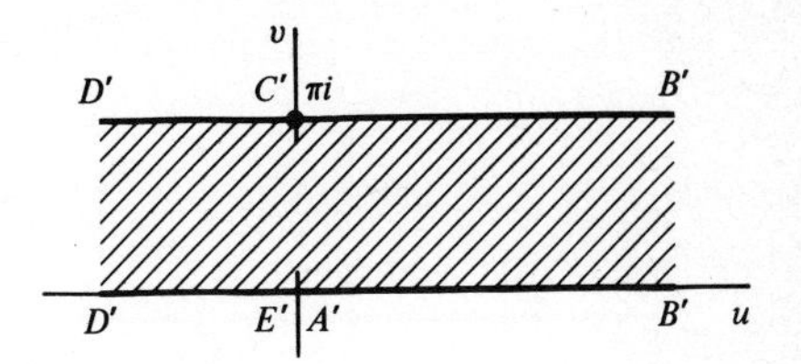

FIGURE 19.

$$w = \text{Log}\,\frac{z-1}{z+1};\ z = -\coth\frac{w}{2}.$$

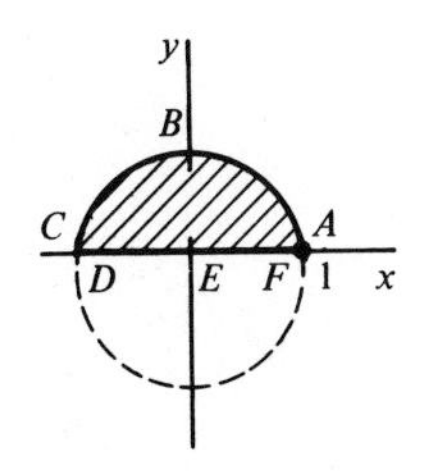

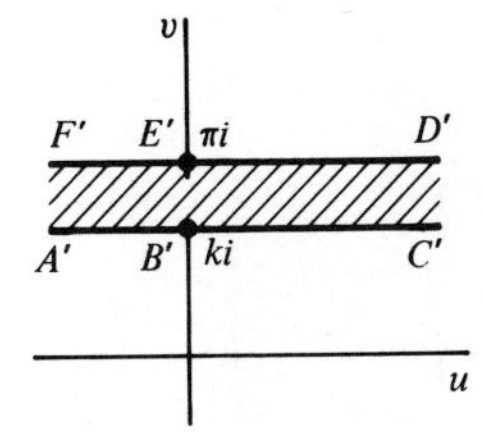

FIGURE 20.

$$w = \text{Log}\,\frac{z-1}{z+1};\ ABC \text{ on circle } x^2 + y^2 - 2y\cot k = 1.$$

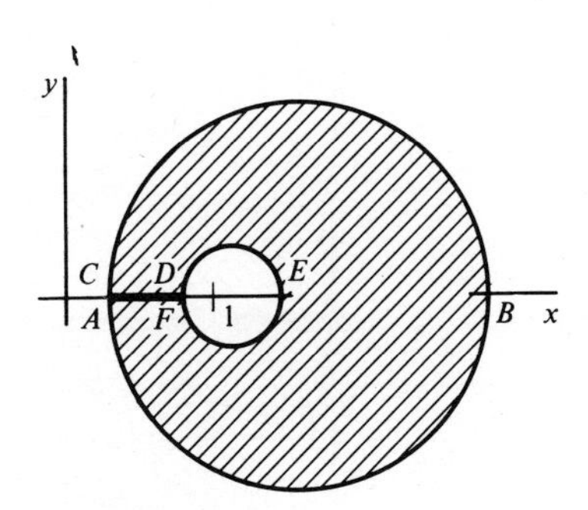

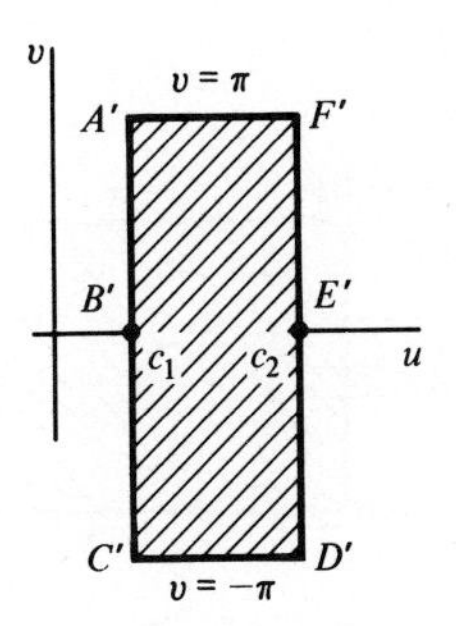

FIGURE 21.

$$w = \text{Log}\,\frac{z+1}{z-1};\ \text{centers of circles at } z = \coth c_n,$$

radii: $\operatorname{csch} c_n (n = 1, 2)$.

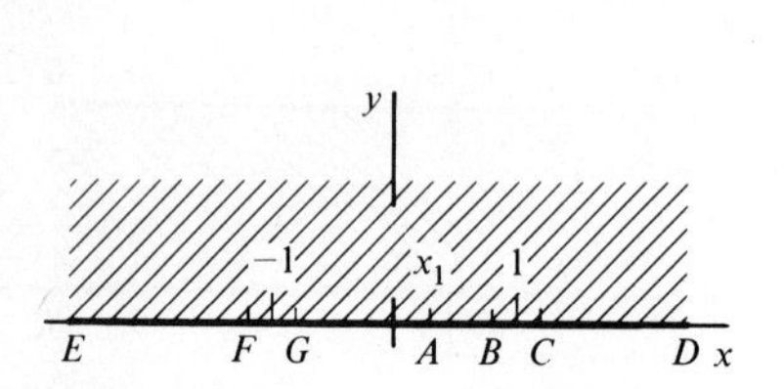

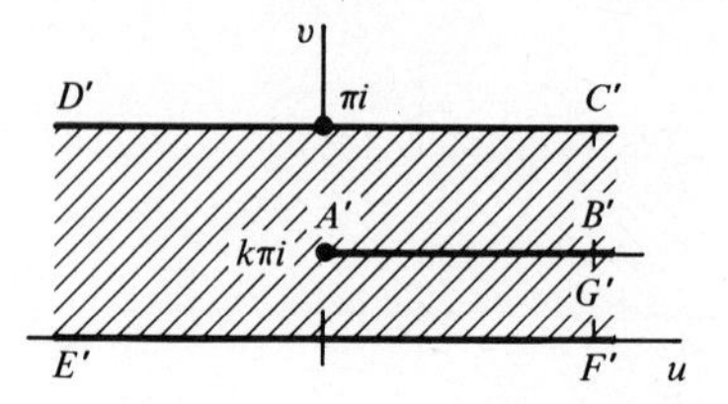

FIGURE 22.

$$w = k \operatorname{Log} \frac{k}{1-k} + \operatorname{Log} 2(1-k) + i\pi - k \operatorname{Log}(z+1) - (1-k) \operatorname{Log}(z-1),$$

$x_1 = 2k - 1.$

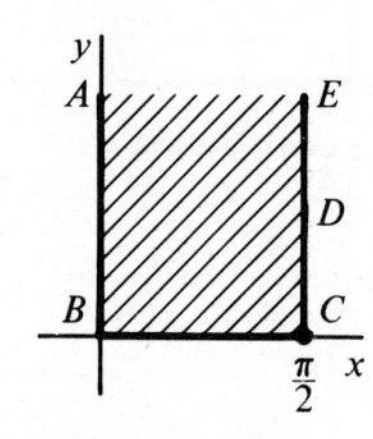

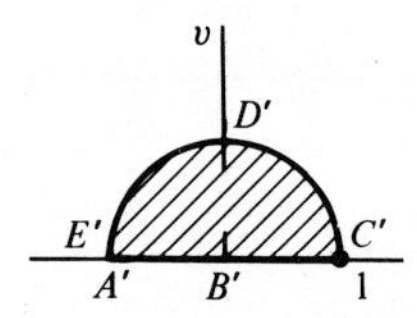

FIGURE 23.

$$w = \left(\tan \frac{z}{2}\right)^2 = \frac{1-\cos z}{1+\cos z}.$$

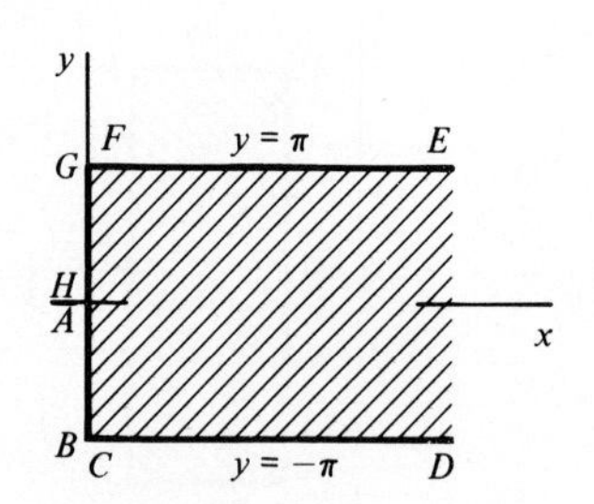

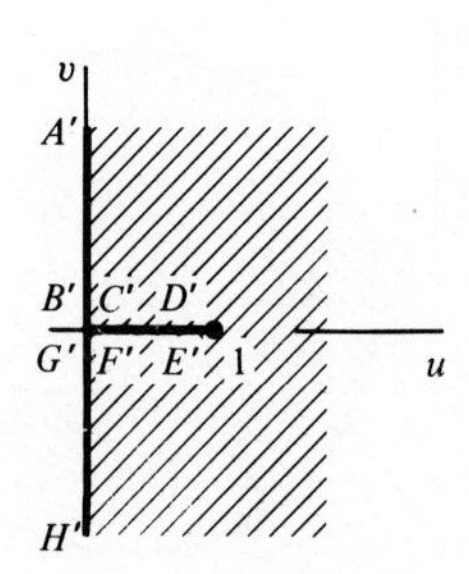

FIGURE 24.

$$w = \coth \frac{z}{2} = \frac{e^z + 1}{e^z - 1}.$$

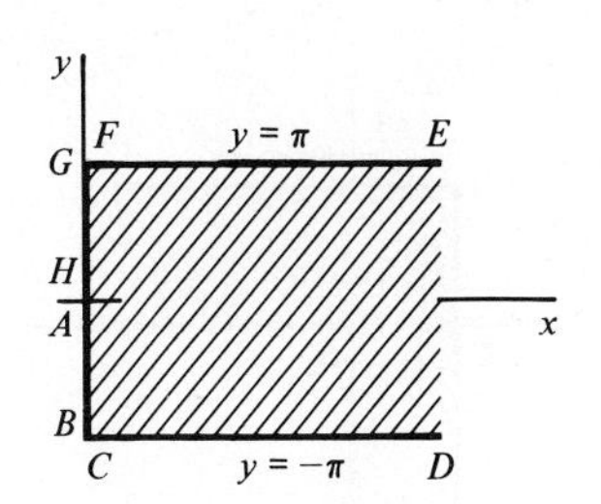

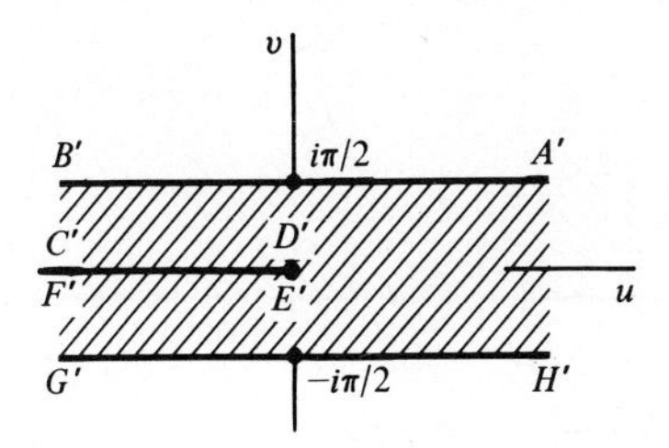

FIGURE 25.

$$w = \text{Log coth}\,\frac{z}{2}.$$

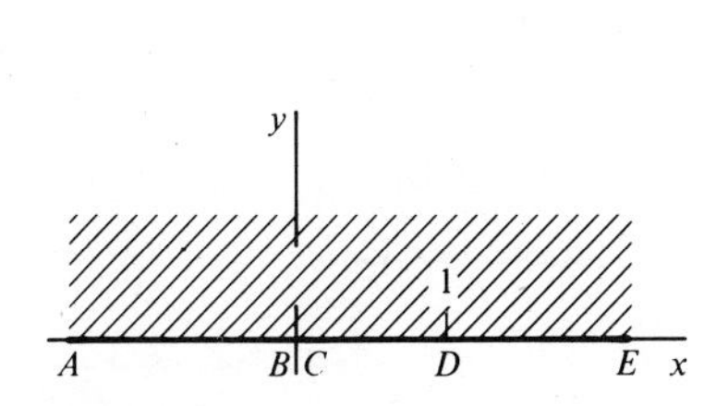

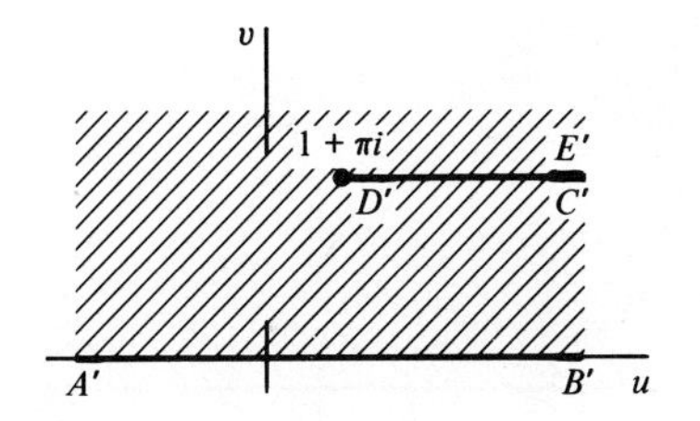

FIGURE 26.
$w = \pi i + z - \text{Log}\, z.$

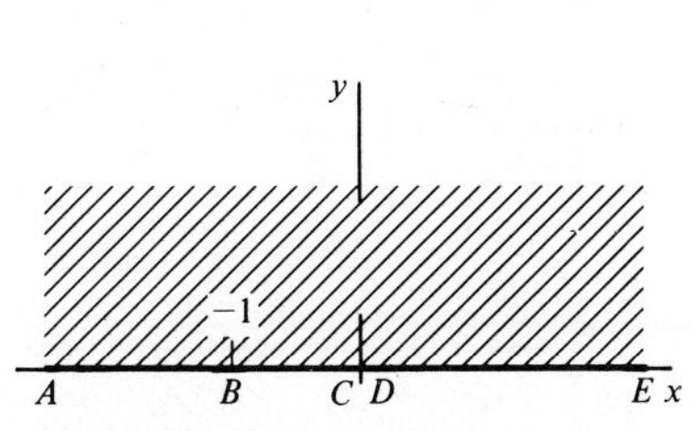

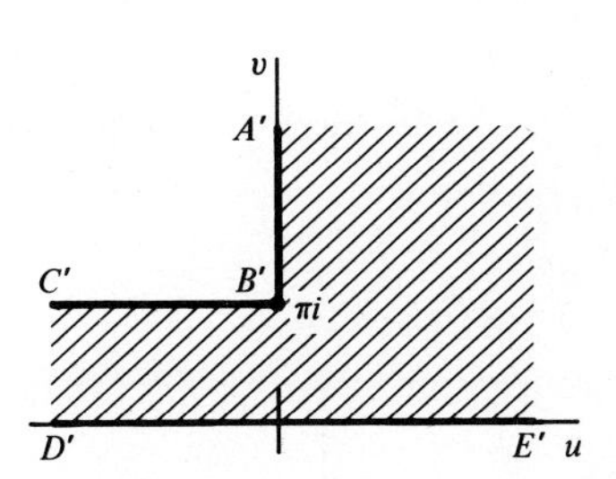

FIGURE 27.

$$w = 2(z+1)^{1/2} + \text{Log}\,\frac{(z+1)^{1/2} - 1}{(z+1)^{1/2} + 1}.$$

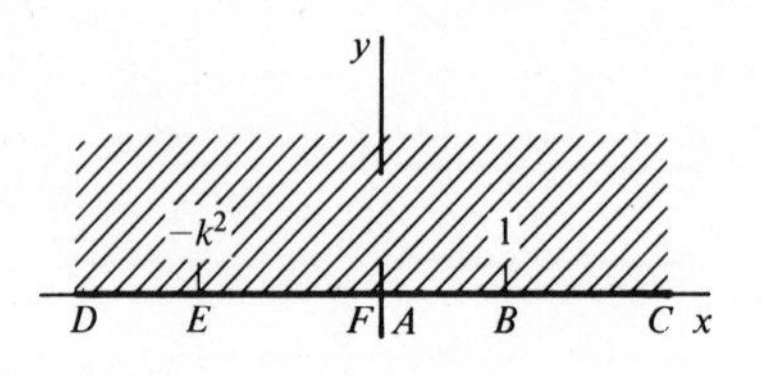

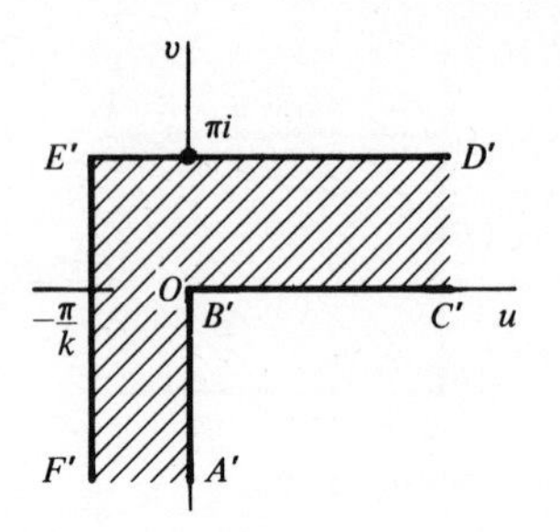

FIGURE 28.

$$w = \frac{i}{k} \text{Log} \frac{1+ikt}{1-ikt} + \text{Log} \frac{1+t}{1-t}; \ t = \left(\frac{z-1}{z+k^2}\right)^{1/2}.$$

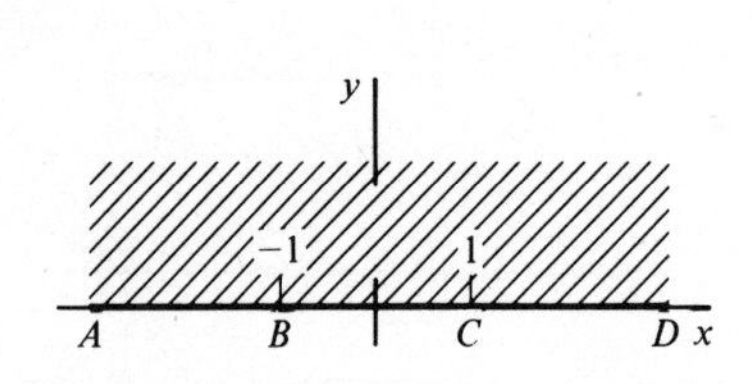

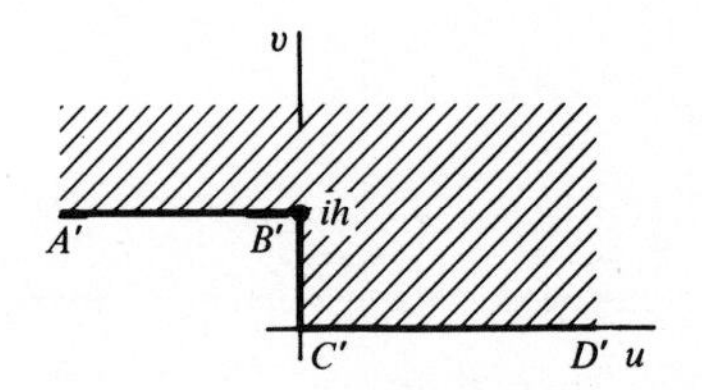

FIGURE 29.

$$w = \frac{h}{\pi} [(z^2 - 1)^{1/2} + \cosh^{-1} z].*$$

* See Exercise 4, Sec. 98.

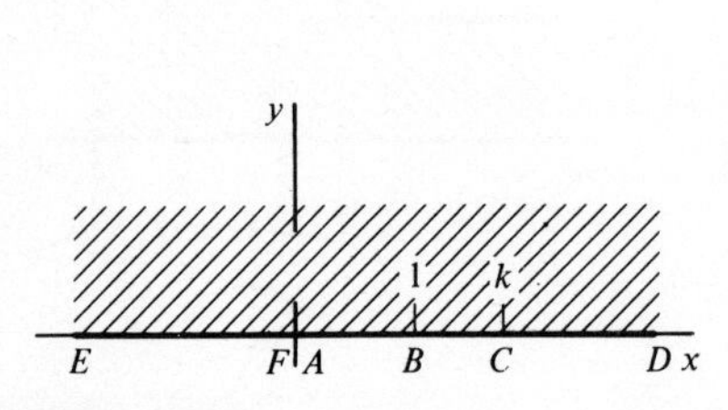

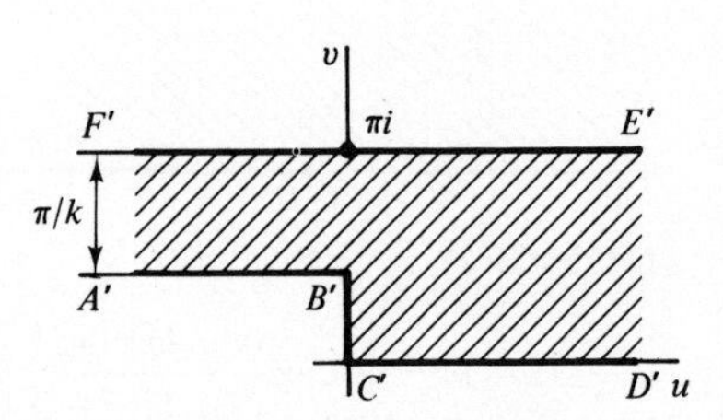

FIGURE 30

$$w = \cosh^{-1} \left(\frac{2z - k - 1}{k - 1}\right) - \frac{1}{k} \cosh^{-1} \left[\frac{(k+1)z - 2k}{(k-1)z}\right].$$

INDEX

INDEX